Date Due

14 APR 1976			
MAY 13 1976			
19 JUL 1977			
JAN 27 1978			
AUG 5 1982			
JUL 23 1982			
RETURNED			
JUN 12 1993			
JUN 11 1993			
RETURNED			
NOV 17 1994			
JAN 05 1994			
RETURNED			
JUL 26 1995			
APR 29 1995			
JAN 04 1996			
RETURNED			
JAN 19 1999			
JAN 13 1999			

DEMCO NO. 38-298

FOURIER TRANSFORMS OF DISTRIBUTIONS AND THEIR INVERSES

A Collection of Tables

Probability and Mathematical Statistics

A Series of Monographs and Textbooks

---

1. Thomas Ferguson. Mathematical Statistics: A Decision Theoretic Approach. 1967
2. Howard Tucker. A Graduate Course in Probability. 1967
3. K. R. Parthasarathy. Probability Measures on Metric Spaces. 1967
4. P. Révész. The Laws of Large Numbers. 1968
5. H. P. McKean, Jr. Stochastic Integrals. 1969
6. B. V. Gnedenko, Yu. K. Belyayev, and A. D. Solovyev. Mathematical Methods of Reliability Theory. 1969
7. Demetrios A. Kappos. Probability Algebras and Stochastic Spaces. 1969
8. Ivan N. Pesin. Classical and Modern Integration Theories. 1970
9. S. Vajda. Probabilistic Programming. 1972
10. Sheldon M. Ross. Introduction to Probability Models. 1972
11. Robert B. Ash. Real Analysis and Probability. 1972
12. V. V. Fedorov. Theory of Optimal Experiments. 1972
13. K. V. Mardia. Statistics of Directional Data. 1972
14. H. Dym and H. P. McKean. Fourier Series and Integrals. 1972
15. Tatsuo Kawata. Fourier Analysis in Probability Theory. 1972
16. Fritz Oberhettinger. Fourier Transforms of Distributions and Their Inverses: A Collection of Tables. 1973
17. Paul Erdös and Joel Spencer. Probabilistic Methods in Combinatorics. 1973
18. K. Sarkadi and I. Vincze. Mathematical Methods of Statistical Quality Control. 1973
19. Michael R. Anderberg. Cluster Analysis for Applications. 1973

In Preparation

L. E. Maistrov. Probability Theory: A Historical Sketch

W. Hengartner and R. Theodorescu. Concentration Functions

William F. Stout. Almost Sure Convergence

L. H. Koopmans. The Spectral Analysis of Time Series

Fourier Transforms of Distributions and Their Inverses

A COLLECTION OF TABLES

Fritz Oberhettinger

Department of Mathematics
Oregon State University
Corvallis, Oregon

ACADEMIC PRESS New York and London 1973

A Subsidiary of Harcourt Brace Jovanovich, Publishers

ACADEMIC PRESS, INC.
111 Fifth Avenue, New York, New York 10003

United Kingdom Edition published by
ACADEMIC PRESS, INC. (LONDON) LTD.
24/28 Oval Road, London NW1

LIBRARY OF CONGRESS CATALOG CARD NUMBER: 73-182604

AMS (MOS) 1970 Subject Classification: 42A68

PRINTED IN THE UNITED STATES OF AMERICA

CONTENTS

Table II. Functions Vanishing Identically for Negative Values of the Argument

Table III. Functions Not Belonging to Either of These Classes

PART II. TABLES OF THE INVERSE TRANSFORMS OF PART I

Table IA. Even Functions

APPENDIX: DISTRIBUTION FUNCTIONS AND THEIR FOURIER TRANSFORMS FOUND IN THE STATISTICAL LITERATURE

PREFACE

The material in this book originated in a report prepared and submitted by the author to the National Bureau of Standards and sponsored by the Office of Naval Research. It was felt that the information gathered there should be made more widely available. The result is this book, a collection of integrals of the Fourier transform type (including their inverses) involving the class of functions which are nonnegative and integrable over the interval $(-\infty, \infty)$.

Most of the results have been extracted from information already available and scattered through the literature. An earlier publication by this author ("Tabellen zur Fourier Transformation," Springer Verlag, 1957) contained many of the Fourier transforms. In contrast, in this volume we concentrate on the probability densities. In addition, a number of new examples have been added.

A sizable amount of effort had to be spent over many years to recognize the functions belonging to the class dealt with in this book. While it is true that a particular function may not fulfill the necessary conditions, it is possible that a suitable combination of a number of them may meet the demanded requirement. It was in the course of these investigations that a number of hitherto unknown results, particularly involving higher functions, were found.

The Author wishes to express his gratitude to the institutions mentioned above and especially to Professor Eugene Lukacs for a series of helpful discussions.

FOURIER TRANSFORMS OF DISTRIBUTIONS AND THEIR INVERSES

A Collection of Tables

INTRODUCTION

Fourier transforms of distribution functions are an important tool in the theory of probability. In this connection they are usually called "characteristic functions of probability distributions." They are very useful in that part of probability theory which can be studied independently of the measure-theoretic foundations. The present book contains tables of distribution functions and of their Fourier transforms. This introduction first lists those properties of characteristic functions which are important in probability theory; it then describes the tables and their use.

Characteristic Functions

A real-valued function $F(x)$ of a real variable x which is

(i) nondecreasing,

(ii) right-continuous,*

and which satisfies the condition

(iii) $\lim_{x \to +\infty} F(x) = 1, \qquad \lim_{x \to -\infty} F(x) = 0$

is called a "distribution function."

* Some authors postulate instead of (ii) that the function $F(x)$ be left-continuous.

Let $F(x)$ be a distribution function; its Fourier transform

$$g(y) = \int_{-\infty}^{\infty} e^{iyx}\, dF(x) \tag{1}$$

is called the "characteristic function of $F(x)$." This is in general a complex-valued function of the real variable y.

Distribution functions are denoted here by the letter F, with or without subscripts, and characteristic functions by the letters g or G with the corresponding subscripts or without subscripts.

The following terminology and notation are used: The integral

$$\alpha_k = \int_{-\infty}^{\infty} x^k\, dF(x) \tag{2}$$

is called the "kth moment of the distribution $F(x)$." Similarly,

$$\beta_k = \int_{-\infty}^{\infty} |\, x\, |^k\, dF(x) \tag{3}$$

is called the "kth absolute moment of $F(x)$," provided that the integrals (2) and (3) are absolutely convergent. If this is the case, then we say that the moments of order k of $F(x)$ exist. It is easily seen that the existence of the kth moment of a distribution function implies the existence of all moments of order not exceeding k.

In the main tables, only absolutely continuous distribution functions are considered. The frequency function* (probability density) of the distribution function $F(x)$ is denoted by $f(x) = F'(x)$. A function $f(x)$ is a frequency function if and only if it is nonnegative and if

$$\int_{-\infty}^{\infty} f(x)\, dx = 1.$$

If $F(x)$ is absolutely continuous, then formulas (1), (2), and (3) reduce to

$$g(y) = \int_{-\infty}^{\infty} e^{iyx} f(x)\, dx, \tag{1a}$$

$$\alpha_k = \int_{-\infty}^{\infty} x^k f(x)\, dx, \tag{2a}$$

$$\beta_k = \int_{-\infty}^{\infty} |\, x\, |^k f(x)\, dx, \tag{3a}$$

respectively.

* Frequency functions are denoted by the letter f; attached to it is the same subscript (if any) which is carried by the corresponding distribution function.

If the function $\varphi(x)$ is nonnegative and if

$$\int_{-\infty}^{\infty} \varphi(x)\, dx = N < \infty,$$

then $f(x) = (1/N)\varphi(x)$ is a frequency function. The corresponding characteristic function is

$$g(y) = N^{-1} \int_{-\infty}^{\infty} e^{iyx}\varphi(x)\, dx.$$

We mention next a few properties of characteristic functions.

Theorem 1 Let $F(x)$ be a distribution function and let $g(y)$ be its characteristic function. Then $g(y)$ is uniformly continuous; moreover $|g(y)| \leq g(0) = 1$ and $g(-y) = \overline{g(y)}$.

Here $\overline{g(y)}$ is the complex conjugate of $g(y)$. There is a close connection between characteristic functions and moments, which is described by the following statements.

Theorem 2 If the characteristic function $G(y)$ of a distribution function $F(x)$ has a derivative of order k at $y = 0$, then all the moments of $F(x)$ up to order k exist if k is even but only up to order $k - 1$ if k is odd.

Theorem 3 If the moment α_k of order k of a distribution exists, then the corresponding characteristic function $g(y)$ can be differentiated s times $(s \leq k)$ and

$$\frac{d^s g(y)}{dy^s} = g^{(s)}(y) = i^s \int_{-\infty}^{\infty} x^s e^{iyx}\, dF(x) \qquad (s = 1, 2, \ldots, k).$$

Moreover $\alpha_s = i^{-s} g^{(s)}(0)$ for $a = 1, 2, \ldots, k$.

Theorem 4 Let $F(x)$ be a distribution function and assume that its nth moment exists. The characteristic function $g(y)$ of $F(x)$ can then be expanded in the form

$$g(y) = 1 + \sum_{j=1}^{n} \frac{\alpha_j}{j!}(iy)^j + O(y^n) \qquad \text{as} \quad y \to 0.$$

The following group of theorems account for the importance of characteristic functions in probability theory.

Theorem 5 (The Inversion Theorem) Let $g(y)$ be the characteristic function of the distribution function $F(x)$. Then

$$F(a+h) - F(a) = \lim_{T\to\infty} (2\pi)^{-1} \int_{-T}^{T} \frac{1 - e^{-iyh}}{iy} e^{-iya} g(y)\, dy,$$

provided that $F(x)$ is continuous at the points a and $a + h$.

As an immediate consequence of Theorem 5 and of Formula (1), the following result is obtained.

Theorem 6 (The Uniqueness Theorem) Two distribution functions are identical if and only if their characteristic functions are identical.

A particular case of the inversion formula is of great interest.

Theorem 7 Let $g(y)$ be a characteristic function and suppose that it is absolutely integrable over $(-\infty, +\infty)$. The corresponding distribution function $F(x)$ is then absolutely continuous and

$$f(x) = F'(x) = (2\pi)^{-1} \int_{-\infty}^{\infty} e^{-iyx} g(y)\, dy \tag{4}$$

is the frequency function (probability density) of the distribution $F(x)$.

It should be noted that the condition of Theorem 7 is only sufficient but not necessary. A distribution may be absolutely continuous and its frequency function may be given by (4) even if the corresponding characteristic function is not absolutely integrable. Naturally, other conditions which insure the validity of the Fourier inversion formula (4) must then be satisfied.

Let $F_1(x)$ and $F_2(x)$ be two distribution functions; it is easily seen that the function

$$F(x) = \int_{-\infty}^{\infty} F_1(x-z)\, dF_2(z) = \int_{-\infty}^{\infty} F_2(x-z)\, dF_1(z) \tag{5}$$

is also a distribution function. Formula (5) defines an operation, called *convolution*, between distribution functions. The resulting distribution function $F(x)$ is called the "convolution of F_1 and F_2." If, in particular, F_1 and F_2 are absolutely continuous, then $F(x)$ is also absolutely continuous and its frequency function is given by

$$f(x) = F'(x) = \int_{-\infty}^{\infty} f_1(x-z) f_2(z)\, dz = \int_{-\infty}^{\infty} f_2(x-z) f_1(z)\, dz. \tag{5a}$$

We say then also that $f(x)$ is the convolution of f_1 and f_2.

Theorem 8 (Convolution Theorem) A distribution function $F(x)$ [respectively, a frequency function $f(x)$] is the convolution of two distribution functions F_1 and F_2 [respectively, two frequency functions f_1 and f_2] if and only if the corresponding characteristic functions satisfy the relation $g(y) = g_1(y)g_2(y)$.

The multiplication of characteristic functions corresponds therefore uniquely to the convolution operation. In probability theory one studies frequently the distribution of the sum of independent random variables. It is well-known that the distribution function of the sum of two independent random variables is the convolution of the distributions of the summands. The direct determination of the distribution of sums will often lead to very complicated integrations while the characteristic function of this distribution can be found easily. It is, according to Theorem 8, the product of the characteristic functions of the summands. In view of the uniqueness theorem it is therefore often advantageous to reformulate a problem concerning distribution functions in terms of characteristic functions.

The next theorem is important in connection with the study of limit distributions. It indicates that the one-to-one correspondence between distribution functions and characteristic functions is continuous.

Theorem 9 (Continuity Theorem) Let $\{F_n(x)\}$ be a sequence of distribution functions and denote the corresponding sequence of characteristic functions by $\{g_n(y)\}$. The sequence $\{F_n(x)\}$ converges to a distribution function $F(x)$ in all points at which $F(x)$ is continuous if and only if the sequence $\{g_n(y)\}$ converges to a function $g(y)$ which is continuous at $y = 0$. The limiting function $g(y)$ is then the characteristic function of $F(x)$.

Only some of the important properties of characteristic functions have been listed here. For details, see [1, 2, 3].

Description and Use of the Tables

The first part of this volume, consisting of Tables I, II, and III, gives Fourier transforms of absolutely continuous distribution functions. The transform pairs are numbered consecutively and are arranged systematically according to the analytical character of the frequency function.

The second part of the volume, consisting of Tables IA and IIIA, gives the inverse transforms of the functions listed in Tables I and III, respectively. The entries in the second column of Tables IA and IIIA are characteristic functions (Fourier transforms) of absolutely continuous distribution functions, arranged systematically. The corresponding frequency functions can be found in the third column. The number in the first column coincides with the number given to the same pair in

Tables I and III, respectively. The inverse of the characteristic functions listed in Table II can be found by means of Tables I and IA.

Next we discuss in some detail the individual tables.

Table I (pp. 15–73) gives the Fourier transforms (characteristic function) of even frequency functions (that is, of frequency functions belonging to symmetric distributions).

Let $f(x)$ be an even frequency function. Then $f(-x) = f(x) = f(|x|)$ has the Fourier transform

$$g(y) = \int_{-\infty}^{\infty} f(|x|) e^{iyx}\, dx = 2 \int_0^{\infty} f(x) \cos xy\, dx. \tag{6}$$

Therefore $g(y) = g(-y)$ so that the transform of an even frequency function is always real and even.

It was found convenient not to list the frequency functions and their transforms directly but to tabulate instead a suitable constant multiple. The second column of Table I contains in each box the function $2Nf(x)$ and also the normalizing factor N. The third column yields the function $Ng(y)$. Here $f(x)$ is an even frequency function while $g(x)$ is its transform given by (6). The first column is used to assign (for reference in the other tables) a number to each pair of transforms. It should be noted that the arguments of the functions in Table I are always positive. Since $f(x)$ as well as $g(y)$ are even functions it is not necessary to list them for negative values of x and y.

Examples of frequency functions, often used in statistics and probability theory, which can be found in this table, are:

Uniform (rectangular) distribution over $(-a, +a)$, No. 1
Cauchy distribution, No. 7
Laplace distribution, No. 60
Normal distribution with mean 0 and standard deviation $(2a)^{-\frac{1}{2}}$, No. 73

Example 1 In item 1 of Table I, we find

$$2Nf(x) = \begin{cases} 1 & \text{if } \ x < a, \\ 0 & \text{if } \ x > a, \end{cases}$$

while $N = a$. Taking into account that $f(x)$ is an even function, we obtain

$$f(x) = \begin{cases} 1/2a & \text{if } \ |x| < a, \\ 0 & \text{if } \ |x| > a. \end{cases}$$

The corresponding characteristic function is found from the third column: Since

$$Ng(y) = ag(y) = y^{-1} \sin(ay),$$

we see that

$$g(y) = (ay)^{-1} \sin(ay).$$

EXAMPLE 2 In item 7 of Table I,

$$2Nf(x) = (a^2 + x^2)^{-1} \qquad \text{and} \qquad N = \pi(2a)^{-1};$$

hence the frequency function is

$$f(x) = a/\pi(a^2 + x^2).$$

Since

$$Ng(y) = (2a)^{-1}\pi e^{-ay},$$

we see that

$$g(y) = e^{-ay} \qquad \text{for} \quad y > 0;$$

hence

$$g(y) = e^{-a|y|} \qquad \text{for} \quad -\infty < y < \infty.$$

Table II (pp. 74–96) permits the determination of the characteristic function of frequency functions $f(x)$ that are zero for negative x. The Fourier transform $G(y)$ of $f(x)$ then becomes

$$G(y) = \int_{-\infty}^{+\infty} f(x) e^{ixy}\, dx = \int_0^\infty f(x) e^{ixy}\, dx = g(y) + ih(y). \tag{7}$$

Here

$$g(y) = \int_0^\infty f(x) \cos xy\, dx, \tag{8}$$

$$h(y) = \int_0^\infty f(x) \sin xy\, dx, \tag{9}$$

and

$$g(0) = G(0) = \int_0^\infty f(x)\, dx = 1.$$

The second column of Table II contains $Nf(x)$; the third, $Nh(y)$. The expressions for N and $g(y)$ can be found in Table I under the number indicated in the first column.

Three examples to illustrate the use of Table II follow.

EXAMPLE 3 Item 1 of Table II yields

$$Nf(x) = \begin{cases} 1 & \text{for} \quad x < a \\ 0 & \text{for} \quad x > a \end{cases} \qquad \text{and} \qquad Nh(y) = 2y^{-1} \sin^2(ya/2).$$

From item 1 in Table I, $N = a$, while $Ng(y) = y^{-1} \sin(ay)$. We are therefore dealing with the frequency function

$$f(x) = \begin{cases} 1/a, & 0 < x < a, \\ 0, & \text{otherwise.} \end{cases}$$

This is the frequency function of a rectangular distribution over $(0, a)$. Its characteristic function is obtained from the tables as

$$G(y) = g(y) + ih(y) = \frac{\sin(ay) + 2i \sin^2(ay/2)}{ay}.$$

From this it follows easily that

$$G(y) = (e^{iay} - 1)/iay.$$

EXAMPLE 4 Item 60 of Table II contains the functions

$$Nf(x) = e^{-ax} \qquad \text{and} \qquad Nh(y) = y(a^2 + y^2)^{-1},$$

while we see from No. 60 in Table I that

$$N = a^{-1} \qquad \text{and} \qquad Ng(y) = a(a^2 + y^2)^{-1}.$$

The frequency function

$$f(x) = \begin{cases} ae^{-ax} & \text{if} \quad x > 0, \\ 0 & \text{if} \quad x < 0, \end{cases}$$

is the density of the exponential distribution. The corresponding Fourier transform is obtained by putting

$$G(y) = g(y) + ih(y) = (a^2 + iay)/(a^2 + y^2) = a/(a - iy).$$

EXAMPLE 5 Item 65 in Table II contains the functions

$$Nf(x) = x^{v-1}e^{-ax} \qquad \text{and} \qquad Nh(y) = \Gamma(v)(a^2 + y^2)^{-\frac{1}{2}v} \sin[v \arctan(y/a)].$$

From 65 in Table I,

$$N = a^{-v}\Gamma(v) \qquad \text{and} \qquad Ng(y) = a^{-v}\Gamma(v)[1 + (y^2/a^2)]^{-\frac{1}{2}v} \cos[v \arctan(y/a)].$$

The frequency function for item 65 is given by

$$f(x) = \begin{cases} [a^v/\Gamma(v)]x^{v-1}e^{-ax}, & x > 0, \\ 0, & x < 0. \end{cases}$$

This is the frequency function of the gamma distribution. The corresponding Fourier transform is obtained from the tables as

$$G(y) = g(y) + ih(y) = [1 + (y^2/a^2)]^{-\frac{1}{2}v} \exp[iv \arctan(y/a)].$$

This is not the form in which this characteristic function is familiar to statisticians. However, it can easily be reduced to the customary form by noting that

$$a + iy = (a^2 + y^2)^{\frac{1}{2}} \exp[i \arctan(y/a)].$$

Using this relation, we obtain

$$G(y) = a^v (a - iy)^{-v}.$$

Table III (p. 97–102) gives Fourier transforms of frequency functions that do not belong to the classes listed in Tables I and II. The second column of Table III gives the function $Nf(x)$ and the normalizing constant N, the third column gives the transform $NG(y)$. Here

$$G(y) = \int_{-\infty}^{+\infty} f(x) e^{ixy} \, dx.$$

The first column gives the serial number of the transform. As an example we mention item 1, the uniform distribution over the interval (a, G).

Table IA (p. 105–144) contains the inverse transforms of the frequency functions listed in Table I. It can be used to determine the frequency functions of even characteristic functions. Column 2 of Table IA contains the function $Ng(y)$ as well as the normalizing factor N; column 3 gives the function $2Nf(x)$; while column 1 indicates the serial number which identifies the same pair of functions in Table I.

Example 6 The function $g(y) = 1/(1 + y^2)$ is an even characteristic function (namely, the characteristic function of the Laplace distribution). To find its frequency function, use Table IA. Item 60 of this table contains

$$Ng(y) = a(a^2 + y^2)^{-1}, \qquad N = a^{-1}$$

$$2Nf(x) = e^{-ax}.$$

Note that, for $a = 1$, $g(y)$ is the desired function. Hence $f(x) = \frac{1}{2}e^{-x}$ for $x > 0$. Since $f(x)$ is necessarily even, we see that the frequency function corresponding to $g(y)$ is $f(x) = \frac{1}{2}e^{-|x|}$.

Tables IA and II can also be used to determine the frequency functions that belong to certain characteristic functions $G(y)$ which are not even. This is possible if the corresponding frequency function vanishes for negative values of the argument. If $G(y)$ is an analytic characteristic function, then it is possible to decide whether the corresponding frequency function has this property. In this connection we mention the following result.

Theorem 10 Let $F(x)$ be a distribution function with analytic characteristic function

$$G(y) = \int_{-\infty}^{+0} e^{iyx} \, dF(x)$$

and suppose that $F(x)$ is bounded to the left. Then

$$\text{lext}[F] = -\limsup_{t\to\infty} t^{-1} \log G(it).$$

Here lext[F] denotes the left extremity of the distribution function $F(x)$; for details the reader is referred to [3] and it is noted here only that Tables IA and II can be used to determine the frequency function $f(x) = F'(x)$ of $G(y)$ if lext[F] $= 0$, that is, if

$$\limsup_{t\to\infty} t^{-1} \log G(it) = 0. \tag{10}$$

The use of the tables is illustrated in the following example.

EXAMPLE 7 Let $G(y) = 4y^{-2} \sin^2 \frac{1}{2}y e^{iy}$. It is easily seen that this function satisfies condition (10) so that the corresponding frequency function vanishes for negative values of its argument. Decompose $G(y)$ into its real and imaginary parts and obtain

$$g(y) = 4y^{-2} \sin^2 \tfrac{1}{2}y \cos y,$$
$$h(y) = 4y^{-2} \sin^2 \tfrac{1}{2}y \sin y.$$

Enter Table IA with $g(y)$ and find under item 2 that $N = 1$ while

$$2Nf(x) = \begin{cases} x & \text{if } 0 < x < 1, \\ 2 - x & \text{if } 1 < x < 2, \\ 0 & \text{if } x > 2. \end{cases}$$

In Table II, item 2, $Nh(y)$ is listed with

$$Nf(x) = \begin{cases} x & \text{if } x < 1, \\ 2 - x & \text{if } 1 < x < 2, \\ 0 & \text{if } x > 2. \end{cases}$$

It can therefore be concluded that

$$f(x) = \begin{cases} 0 & \text{if } x < 0, \\ x & \text{if } 0 < x < 1, \\ 2 - x & \text{if } 1 < x < 2, \\ 0 & \text{if } x > 2, \end{cases}$$

is the frequency function corresponding to $G(y)$. This is the frequency function of the triangular (Simpson's) distribution.

It should be noted that a frequency function $f(x)$ which vanishes for $x < 0$ need not have an analytic characteristic function. If $G(y)$ is not an analytic characteristic function, then no simple criterion is known to determine whether $f(x) = 0$ for $x < 0$. However, it may still be worthwhile to try using Tables IA and II. The next example shows a case where $G(y)$ is not an analytic characteristic function but where the method is nevertheless applicable.

EXAMPLE 8 Let

$$G(y) = \exp\{ -|y|^{\frac{1}{2}} [1 - i(y/|y|)]\};$$

this is the characteristic function of a stable distribution with exponent $\frac{1}{2}$. Decomposing $G(y)$ into real and imaginary parts, we obtain for $y > 0$,

$$g(y) = \exp(-y^{\frac{1}{2}}) \cos y^{\frac{1}{2}} \qquad h(y) = \exp(-y^{\frac{1}{2}}) \sin y^{\frac{1}{2}}.$$

Entering Table IA with $g(y)$, find in item 84

$$Ng(y) = (\pi/a)^{\frac{1}{2}} \exp[-(2ay)^{\frac{1}{2}}] \cos[(2ay)^{\frac{1}{2}}]$$

and

$$N = (\pi/a)^{\frac{1}{2}}, \qquad 2Nf(x) = x^{-\frac{3}{2}} \exp(-a/x).$$

We look up item 84 in Table II and find

$$Nf(x) = x^{-\frac{3}{2}} \exp(-a/x), \qquad Nh(y) = (\pi/a)^{\frac{1}{2}} \exp[-(2ay)^{\frac{1}{2}}] \sin[(2ay)^{\frac{1}{2}}].$$

This agrees with the given functions, and putting $a = \frac{1}{2}$, the desired frequency function is obtained as

$$f(x) = \begin{cases} 0 & \text{for } x < 0, \\ (2\pi)^{-\frac{1}{2}} x^{-\frac{3}{2}} \exp(-1/2x). \end{cases}$$

Table IIIA (p. 146–150) contains the inverse transforms of the functions given in Table III. The second column contains the function $NG(y)$ and the normalizing constant N; the third column gives $Nf(x)$. Here

$$G(y) = \int_{-\infty}^{+\infty} e^{iyx} f(x)\, dx.$$

The notations used in the tables are listed following the Appendix, beginning on p. 162.

Tables of the Appendix

We saw in Example 5 that the tables sometimes give the characteristic function in a form which is unfamiliar to the statistician. It appears therefore desirable to list separately characteristic functions which occur frequently in the statistical literature and to write those in their customary form.

Tables of these functions can be found in the Appendix which was compiled by R. G. Laha. The appendix consists of three tables which contain characteristic functions of univariate density functions (Table A), of discrete distribution functions (Table B), and of multivariate distributions (Table C). The tables of the appendix do not provide a complete coverage of the statistical literature; only the sources listed in the References were consulted. It is certain that many characteristic functions not listed in the tables of the Appendix occur in the literature. Nevertheless, it is hoped that the most important distributions are included.

References

1. H. Cramér (1946). "Mathematical Methods of Statistics." Princeton Univ. Press, Princeton, New Jersey.
2. M. Loève (1955). "Probability Theory." Van Nostrand-Reinhold, Princeton, New Jersey.
3. E. Lukacs (1960). "Characteristic Functions." Griffin, London.

PART I

TABLES OF FOURIER TRANSFORMS

TABLE I

EVEN FUNCTIONS

Definitions

In these tables $f(x)$ is an even function of x, i.e., $f(-x) = f(x)$, such that

$$\int_{-\infty}^{\infty} f(|x|)\, dx = 2\int_{0}^{\infty} f(x)\, dx = 1.$$

Consequently, its Fourier transform is

$$g(y) = \int_{-\infty}^{\infty} f(|x|) e^{ixy}\, dx = 2\int_{0}^{\infty} f(x)\cos(xy)\, dx.$$

Therefore $g(y)$ is an even function of y, i.e., $g(-y) = g(y)$.

A real function $f(x)$ of the real variable x is called a "frequency function" or "probability density" if it is defined in the interval $-\infty < x < \infty$ and if it satisfies the properties:

(1) $f(x)$ is nonnegative in this interval;

(2) $\int_{-\infty}^{\infty} f(x)\, dx = 1.$

Tables of integrals of the form

$$G(y) = \int_{-\infty}^{\infty} f(x) e^{ixy} \, dx$$

are listed here for the cases:

(a) $f(x)$ is an even function of x (Table I, pp. 17–73);
(b) $f(x)$ vanishes identically for negative x (Table II, pp. 75–96);
(c) $f(x)$ does not belong to any of the classes a or b (Table III, pp. 98–102).

A list of notations can be found on pp. 162–167.

1. Algebraic Functions

	$2Nf(x)$	$Ng(y)$
1	$1, \quad x<a$ $0, \quad x>a$ $N=a$	$y^{-1}\sin(ay)$
2	$x, \quad x<1$ $2-x, \quad 1<x<2$ $0, \quad x>2$ $N=1$	$4y^{-2}\cos y\,\sin^2(\frac{1}{2}y)$
3	$x^{-\frac{1}{2}}, \quad x<1$ $0, \quad x>1$ $N=2$	$(2\pi/y)^{\frac{1}{2}}C(y)$
4	$(a+x)^{-1}, \quad x<b$ $0, \quad x>b$ $N=\log(1+b/a)$	$\cos(ay)[\mathrm{Ci}(ay+by)-\mathrm{Ci}(ay)]$ $+\sin(ay)[\mathrm{si}(ay+by)-\mathrm{si}(ay)]$
5	$0, \quad x<b$ $(a+x)^{-n}, \quad x>b$ $n=2, 3, 4, \ldots$ $N=(a+b)^{1-n}(n-1)^{-1}$	$\sum_{m=1}^{n-1}\frac{(m-1)!}{(n-1)!}(a+b)^{-m}(-y)^{n-m-1}$ $\cdot\sin[\frac{1}{2}\pi(n-m)-by]$ $-(-y)^{n-1}[(n-1)!]^{-1}[\sin(ay+\frac{1}{2}\pi n)$ $\cdot\mathrm{Ci}(ay+by)-\cos(ay+\frac{1}{2}\pi n)\,\mathrm{si}(ay+by)]$
6	$0, \quad x<b$ $[x(a+x)]^{-1}, \quad x>b$ $N=a^{-1}\log(1+a/b)$	$a^{-1}[\cos(ay)\,\mathrm{Ci}(ay+by)$ $+\sin(ay)\,\mathrm{si}(ay+by)-\mathrm{Ci}(by)]$
7	$(a^2+x^2)^{-1}$ $N=\pi(2a)^{-1}$	$\pi(2a)^{-1}e^{-ay}$
8	$[b^2+(a-x)^2]^{-1}+[b^2+(a+x)^2]^{-1}$ $N=\pi b^{-1}$	$\pi b^{-1}\cos(ay)e^{-by}$
9	$[(a^2+x^2)(b^2+x^2)]^{-1}$ $N=\pi[2ab(a+b)]^{-1}$	$\frac{1}{2}\pi(a^2-b^2)^{-1}[b^{-1}e^{-by}-a^{-1}e^{-ay}]$
10	$(a^4+x^4)^{-1}$ $N=2^{-\frac{1}{2}}\pi a^{-3}$	$\frac{1}{2}\pi a^{-3}\exp(-ay2^{-\frac{1}{2}})\sin(\frac{1}{4}\pi+2^{-\frac{1}{2}}ay)$

	$2Nf(x)$	$Ng(y)$
11	$[x^4+2a^2x^2\cos(2\vartheta)+a^4]^{-1}-\frac{1}{2}\pi<\vartheta<\frac{1}{2}\pi$ $N=\frac{1}{4}\pi a^{-3}\sec\vartheta$	$\frac{1}{2}\pi a^{-3}\csc(2\vartheta)\exp(-ay\cos\vartheta)$ $\cdot\sin(\vartheta+ay\sin\vartheta)$
12	$x^2[x^4+2a^2x^2\cos(2\vartheta)+a^4]^{-1}$ $-\frac{1}{2}\pi<\vartheta<\frac{1}{2}\pi$ $N=\frac{1}{4}\pi a^{-1}\sec\vartheta$	$\frac{1}{2}\pi a^{-1}\csc(2\vartheta)\exp(-ay\cos\vartheta)$ $\cdot\sin(\vartheta-ay\sin\vartheta)$
13	$x^{-\frac{1}{2}}(a+x)^{-1}$ $N=\pi a^{-\frac{1}{2}}$	$\pi a^{-\frac{1}{2}}\{\cos(ay)[1-C(ay)-S(ay)]$ $+\sin(ay)[C(ay)-S(ay)]\}$
14	$(a+x)^{-\frac{3}{2}}$ $N=2a^{-\frac{1}{2}}$	$2a^{-\frac{1}{2}}-(2\pi y)^{\frac{1}{2}}\{\cos(ay)[1-2S(ay)^{\frac{1}{2}}]$ $-\sin(ay)[1-2C(ay)^{\frac{1}{2}}]\}$
15	$x^{-\frac{1}{2}}(a^2+x^2)^{-\frac{1}{2}}$ $N=\frac{1}{2}(a\pi)^{-\frac{1}{2}}\Gamma^2(\frac{1}{4})$	$(\frac{1}{2}\pi y)^{\frac{1}{2}}I_{-\frac{1}{4}}(\frac{1}{2}ay)K_{\frac{1}{4}}(\frac{1}{2}ay)$
16	$(a-x)^{-\frac{1}{2}},\quad x<a$ $0,\quad x>a$ $N=2a^{\frac{1}{2}}$	$(2\pi)^{\frac{1}{2}}y^{-\frac{1}{2}}[\cos(ay)C(ay)+\sin(ay)S(ay)]$
17	$(a-x)^{-1},\quad x<b$ $0,\quad x>b$ $b<a$ $N=-\log[1-(b/a)]$	$\cos(ay)[\mathrm{Ci}(ay)-\mathrm{Ci}(ay-by)]$ $+\sin(ay)[\mathrm{si}(ay)-\mathrm{si}(ay-by)]$
18	$(a^2-x^2)^{-1},\quad x<b$ $0,\quad x>b$ $b<a$ $N=(2a)^{-1}\log[(a+b)(a-b)^{-1}]$	$(2a)^{-1}\{\cos(ay)[\mathrm{Ci}(ay+by)-\mathrm{Ci}(ay-by)]$ $+\sin(ay)[\mathrm{si}(ay+by)-\mathrm{si}(ay-by)]\}$
19	$0,\quad x<b$ $(x^2-a^2)^{-1},\quad x>b$ $b>a$ $N=(2a)^{-1}\log[(b+a)(b-a)^{-1}]$	$(2a)^{-1}\{\sin(ay)[\mathrm{si}(by-ay)+\mathrm{si}(by+ay)]$ $-\cos(ay)[\mathrm{Ci}(by-ay)-\mathrm{Ci}(by+ay)]\}$
20	$0,\quad x<b$ $x^{-1}(x-b)^{-\frac{1}{2}},\quad x>b$ $N=b^{-\frac{1}{2}}\pi$	$b^{-\frac{1}{2}}\pi[1-C(by)-S(by)]$
21	$x^2(a^2-x^2)^{-\frac{1}{2}},\quad x<a$ $0,\quad x>a$ $N=\frac{1}{4}\pi a^2$	$\frac{1}{2}\pi a^2[J_0(ay)-(ay)^{-1}J_1(ay)]$

	$2Nf(x)$	$Ng(y)$
22	$0, \quad x<b$ $(x-b)^{-\frac{1}{2}}(x+b)^{-1}, \quad x>b$ $N=(2b)^{-\frac{1}{2}}\pi$	$(2b)^{-\frac{1}{2}}\pi\{\cos(by)[1-C(2by)-S(2by)]$ $+\sin(by)[C(2by)-S(2by)]\}$
23	$0, \quad x<b$ $(x-b)^{-\frac{1}{2}}(x+a)^{-1}, \quad x>b$ $N=(a+b)^{-\frac{1}{2}}\pi$	$(a+b)^{-\frac{1}{2}}\pi\{\cos(ay)[1-C(ay+by)$ $-S(ay+by)]+\sin(ay)[C(ay+by)$ $-S(ay+by)]\}$
24	$(a^2-x^2)^{-\frac{1}{2}}, \quad x<a$ $0, \quad x>a$ $N=\frac{1}{2}\pi$	$\frac{1}{2}\pi J_0(ay)$
25	$x(a^2-x^2)^{-\frac{1}{2}}, \quad x<a$ $0, \quad x>a$ $N=a$	$a[1-\frac{1}{2}\pi\mathbf{H}_1(ay)]$
26	$x^{-\frac{1}{2}}(a^2-x^2)^{-\frac{1}{2}}, \quad x<a$ $0, \quad x>a$ $N=2^{-\frac{1}{2}}\pi^{\frac{3}{2}}[\Gamma(\frac{3}{4})]^{-2}$	$(\frac{1}{2}\pi)^{\frac{3}{2}}y^{\frac{1}{2}}[J_{-\frac{1}{4}}(\frac{1}{2}ay)]^2$
27	$0, \quad x<b$ $(a^2-x^2)^{-\frac{1}{2}}, \quad b<x<a$ $0, \quad x>a$ $N=\arccos(b/a)$	$\pi\sum_{n=0}^{\infty} J_{n+\frac{1}{2}}(\frac{1}{2}ay-\frac{1}{2}by)$ $\cdot J_{-n-\frac{1}{2}}(\frac{1}{2}ay+\frac{1}{2}by)$
28	$x^{-\frac{1}{2}}[a+x+(2ax)^{\frac{1}{2}}]^{-1}$ $N=\pi(2a)^{-\frac{1}{2}}$	$\pi(2a)^{-\frac{1}{2}}e^{ay}\operatorname{Erfc}[(ay)^{\frac{1}{2}}]$
29	$x^{-\frac{1}{2}}(a^2+x^2)^{-\frac{1}{2}}[x+(a^2+x^2)^{\frac{1}{2}}]^{-\frac{3}{2}}$ $N=2^{-\frac{1}{2}}a^{-2}$	$2^{-\frac{1}{2}}a^{-2}\sinh(\frac{1}{2}ay)K_1(\frac{1}{2}ay)$
30	$0, \quad x<a$ $x^{-\frac{1}{2}}(x^2-a^2)^{-\frac{1}{2}}, \quad x>a$ $N=2^{-\frac{1}{2}}\pi^{\frac{3}{2}}[\Gamma(\frac{3}{4})]^{-2}$	$-(\frac{1}{2}\pi)^{\frac{3}{2}}y^{\frac{1}{2}}J_{-\frac{1}{4}}(\frac{1}{2}ay)Y_{-\frac{1}{4}}(\frac{1}{2}ay)$
31	$0, \quad x<a$ $x^{-1}(x^2-a^2)^{-\frac{1}{2}}, \quad x>a$ $N=\frac{1}{2}a^{-1}\pi$	$\frac{1}{2}a^{-1}\pi-\frac{1}{4}\pi^2y[J_0(ay)\mathbf{H}_{-1}(ay)+\mathbf{H}_0(ay)J_1(ay)]$
32	$(a^2+x^2)^{-\frac{1}{2}}[a+(a^2+x^2)^{\frac{1}{2}}]^{-\frac{1}{2}}$ $N=(2a)^{-\frac{1}{2}}\pi$	$(2a)^{-\frac{1}{2}}\pi\operatorname{Erfc}[(ay)^{\frac{1}{2}}]$

	$2Nf(x)$	$Ng(y)$
33	$x^{2m}(x^2+z)^{-n-1}$ $n, m=0, 1, 2, \ldots, \quad n \geq m$ $N=(-)^{m+n}\pi \dfrac{(2m)!(2n-2m)!}{2^{2n+1}n!m!(n-m)!}$ $\cdot z^{m-n-\frac{1}{2}}$	$\frac{1}{2}\pi[(-1)^{m+n}/n!](d^n/dz^n)$ $\cdot[z^{m-\frac{1}{2}}\exp(-yz^{\frac{1}{2}})]$
34	$x^{2m}(x^{2n}+a^{2n})^{-1}$ $m, n=1, 2, 3, \ldots, \quad 2m<2n-1$ $N=(\pi/2n)a^{2m-2n-1}\csc[(\pi/2n)(2m+1)]$	$(\pi/2n)a^{2m-2n-1}\sum_{k=1}^{n}\{\exp(-ay\sin[(2k-1)$ $\cdot(\pi/2n)])\sin[(2k-1)(2m+1)(\pi/2n)$ $+ay\cos(2k-1)(\pi/2n)]\}$
35	$x^{-\frac{1}{2}}(a^2+x^2)^{-\frac{1}{2}}[x+(a^2+x^2)^{\frac{1}{2}}]^{-\frac{1}{2}}$ $N=2^{-\frac{1}{2}}a^{-1}\pi$	$2^{-\frac{1}{2}}a^{-1}\pi e^{-\frac{1}{2}ay}I_0(\frac{1}{2}ay)$
36	$(x^2-a^2)^{-\frac{1}{2}}[x+(x^2-a^2)^{\frac{1}{2}}]^{-n}, \quad x>a$ $0, \quad x<a$ $N=(1/n)a^{-n}$	$-\frac{1}{2}a^{-n}\pi[\sin(\frac{1}{2}n\pi)J_n(ay)$ $+\cos(\frac{1}{2}n\pi)Y_n(ay)]$ $-a^{-n}\sum_{k=1}^{n}k!(n+k-1)![(2k)!(n-k)!]^{-1}$ $\cdot(\frac{1}{2}ay)^{-k}\sin(ay+\frac{1}{2}\pi k)$

2. Arbitrary Powers

	$2Nf(x)$	$Ng(y)$
37	$x^{\nu-1}, \quad x<1, \quad \nu>0$ $0, \quad x>1$ $N=\nu^{-1}$	$\frac{1}{2}\nu^{-1}[{}_1F_1(\nu;\nu+1;iy)+{}_1F_1(\nu;\nu+1;-iy)]$
38	$(b-x)^{\nu}, \quad x<b$ $0, \quad x>b, \quad \nu>-1$ $N=b^{\nu+1}(\nu+1)^{-1}$	$-\frac{1}{2}iy^{-\nu-1}\{\exp[-i(\nu\pi/2-by)]\gamma(\nu+1, iby)$ $-\exp[i(\nu\pi/2-by)]\gamma(\nu+1, -iby)\}$
39	$x^{\nu-1}(b-x)^{\mu-1}, \quad x<b$ $0, \quad x>b$ $\nu>0, \quad \mu>0$ $N=b^{\nu+\mu-1}B(\nu, \mu)$	$\frac{1}{2}b^{\nu+\mu-1}B(\nu, \mu)[{}_1F_1(\nu;\mu+\nu;iby)$ $+{}_1F_1(\nu;\mu+\nu;-iby)]$

	$2Nf(x)$	$Ng(y)$
40	$x^{\nu}(a+x)^{-1}, \quad -1<\nu<0$ $N=-\pi\csc(\nu\pi)$	$(2a)^{\nu}(a\pi y)^{\frac{1}{2}}\left\{\frac{\Gamma(\frac{1}{2}+\frac{1}{2}\nu)}{\Gamma(-\frac{1}{2}\nu)}S_{-\nu-\frac{1}{2},\frac{1}{2}}(ay)\right.$ $\left.-2\frac{\Gamma(1+\frac{1}{2}\nu)}{\Gamma(-\frac{1}{2}-\frac{1}{2}\nu)}S_{-\nu-\frac{3}{2},\frac{1}{2}}(ay)\right\}$
41	$x^{\nu}(a^2+x^2)^{-1}, \quad -1<\nu<1$ $N=\frac{1}{2}a^{\nu-1}\pi\sec(\frac{1}{2}\nu\pi)$	$\frac{1}{2}\pi a^{\nu-1}\sec(\frac{1}{2}\nu\pi)\cosh(ay)$ $+\pi^{\frac{1}{2}}2^{\nu-2}y^{1-\nu}\Gamma(\frac{1}{2}\nu-\frac{1}{2})[\Gamma(1-\frac{1}{2}\nu)]^{-1}$ $\cdot{}_1F_2(1;1-\frac{1}{2}\nu,\frac{3}{2}-\frac{1}{2}\nu;\frac{1}{4}a^2y^2)$
42	$(a^2+x^2)^{-\nu-\frac{1}{2}}, \quad \nu>0$ $N=\frac{1}{2}a^{-2\nu}\pi^{\frac{1}{2}}[\Gamma(\nu)/\Gamma(\nu+\frac{1}{2})]$	$[(2a)^{-\nu}\pi^{\frac{1}{2}}/\Gamma(\nu+\frac{1}{2})]y^{\nu}K_{\nu}(ay)$
43	$(x^2+2ax)^{-\nu-\frac{1}{2}}, \quad 0<\nu<\frac{1}{2}$ $N=\frac{1}{2}a^{-2\nu}\pi^{-\frac{1}{2}}\Gamma(\nu)\Gamma(\frac{1}{2}-\nu)$	$2^{-1-\nu}a^{-\nu}\pi^{\frac{1}{2}}\Gamma(\frac{1}{2}-\nu)$ $\cdot y^{\nu}[J_{\nu}(ay)\sin(ay)-Y_{\nu}(ay)\cos(ay)]$
44	$x^{-\nu-\frac{1}{2}}(a^2+x^2)^{-\frac{1}{2}}$ $\cdot[a+(a^2+x^2)^{\frac{1}{2}}]^{\nu}, \quad \nu<\frac{1}{2}$ $N=2^{-\frac{1}{2}}a^{-\frac{1}{2}}\pi^{\frac{1}{2}}[\Gamma(\frac{1}{4}-\frac{1}{2}\nu)/\Gamma(\frac{3}{4}-\frac{1}{2}\nu)]$	$2^{-\frac{1}{2}}a^{-1}\Gamma(\frac{1}{4}-\frac{1}{2}\nu)$ $\cdot y^{-\frac{1}{2}}W_{\frac{1}{2}\nu,\frac{1}{4}}(ay)M_{-\frac{1}{2}\nu,-\frac{1}{4}}(ay)$
45	$x^{-\frac{1}{2}}(a^2+x^2)^{-\frac{1}{2}}[(a^2+x^2)^{\frac{1}{2}}-x]^{\nu}, \quad \nu>-\frac{1}{2}$ $N=a^{\nu-\frac{1}{2}}(\frac{1}{2}\pi)^{\frac{1}{2}}[\Gamma(\frac{1}{4}+\frac{1}{2}\nu)/\Gamma(\frac{3}{4}+\frac{1}{2}\nu)]$	$a^{\nu}(\frac{1}{2}\pi)^{\frac{1}{2}}y^{\frac{1}{2}}I_{-\frac{1}{4}+\frac{1}{2}\nu}(\frac{1}{2}ay)K_{\frac{1}{4}+\frac{1}{2}\nu}(\frac{1}{2}ay)$
46	$x^{-\frac{1}{2}}(a^2+x^2)^{-\frac{1}{2}}[x+(a^2+x^2)^{\frac{1}{2}}]^{\nu}, \quad \nu<\frac{1}{2}$ $N=a^{\nu-\frac{1}{2}}(\frac{1}{2}\pi)^{\frac{1}{2}}[\Gamma(\frac{1}{4}-\frac{1}{2}\nu)/\Gamma(\frac{3}{4}-\frac{1}{2}\nu)]$	$a^{\nu}(\frac{1}{2}\pi)^{\frac{1}{2}}y^{\frac{1}{2}}I_{-\frac{1}{4}-\frac{1}{2}\nu}(\frac{1}{2}ay)K_{\frac{1}{4}-\frac{1}{2}\nu}(\frac{1}{2}ay)$
47	$(a^2+x^2)^{-\frac{1}{2}}[(a^2+x^2)^{\frac{1}{2}}+x]^{-\nu}, \quad \nu>0$ $N=a^{-\nu}\nu^{-1}$	$a^{-\nu}\pi\csc(\nu\pi)[\frac{1}{2}\mathbf{J}_{\nu}(iay)$ $+\frac{1}{2}\mathbf{J}_{\nu}(-iay)-I_{\nu}(ay)\cos(\frac{1}{2}\nu\pi)]$
48	$[(a^2+x^2)^{\frac{1}{2}}+x]^{-\nu}, \quad \nu>1$ $N=a^{1-\nu}\nu(\nu^2-1)^{-1}$	$a^{-\nu}\nu\pi\csc(\nu\pi)y^{-1}\{I_{\nu}(ay)\sin(\frac{1}{2}\nu\pi)$ $+\frac{1}{2}i[\mathbf{J}_{\nu}(iay)-\mathbf{J}_{\nu}(-iay)]\}$
49	$x^{\nu}(a^2+x^2)^{-\mu-1}, \quad -1<\nu<2\mu+1$ $N=\frac{1}{2}a^{\nu-2\mu-1}$ $\cdot B(\frac{1}{2}+\frac{1}{2}\nu,\frac{1}{2}-\frac{1}{2}\nu+\mu)$	$\frac{1}{2}a^{\nu-2\mu-1}B(\frac{1}{2}+\frac{1}{2}\nu,\frac{1}{2}-\frac{1}{2}\nu+\mu)$ $\cdot{}_1F_2(\frac{1}{2}\nu+\frac{1}{2};\frac{1}{2}\nu+\frac{1}{2}-\mu,\frac{1}{2};\frac{1}{4}a^2y^2)$ $+\frac{1}{2}\pi^{\frac{1}{2}}(\frac{1}{2}y)^{2\mu-\nu+1}\Gamma(\frac{1}{2}\nu-\mu-\frac{1}{2})$ $\cdot[\Gamma(\mu-\frac{1}{2}\nu+1)]^{-1}$ $\cdot{}_1F_2(\mu+1;\mu+1-\frac{1}{2}\nu;\mu+\frac{3}{2}-\frac{1}{2}\nu;\frac{1}{4}a^2y^2)$
50	$(a^2-x^2)^{\nu-\frac{1}{2}}, \quad x<a$ $0, \quad x>a, \quad \nu>-\frac{1}{2}$ $N=\frac{1}{2}a^{2\nu}\pi^{\frac{1}{2}}[\Gamma(\nu+\frac{1}{2})/\Gamma(\nu+1)]$	$2^{\nu-1}a^{\nu}\pi^{\frac{1}{2}}\Gamma(\nu+\frac{1}{2})y^{-\nu}J_{\nu}(ay)$

	$2Nf(x)$	$Ng(y)$
51	$x(a^2-x^2)^{\nu-\frac{1}{2}}, \quad x<a$ $0, \quad x>a,$ $\nu>-\frac{1}{2}$ $N=a^{2\nu+1}(2\nu+1)^{-1}$	$a^{2\nu+1}(2\nu+1)^{-1}$ $\cdot[1-(\frac{1}{2}a)^{-\nu}\pi^{\frac{1}{2}}\Gamma(\nu+\frac{3}{2})y^{-\nu}\mathbf{H}_{\nu+1}(ay)]$
52	$(2ax-x^2)^{\nu-\frac{1}{2}}, \quad x<a$ $0, \quad x>a,$ $\nu>-\frac{1}{2}$ $N=a^{2\nu}\pi^{\frac{1}{2}}[\Gamma(\nu+\frac{1}{2})/\Gamma(\nu+1)]$	$(2a)^{\nu}\pi^{\frac{1}{2}}\Gamma(\nu+\frac{1}{2})$ $\cdot y^{-\nu}J_{\nu}(ay)\cos(ay)$
53	$x^{\nu}(a^2-x^2)^{\mu}, \quad x<a$ $0, \quad x>a,$ $\mu>-1,$ $\nu>-1$ $N=\frac{1}{2}a^{2\mu+\nu+1}B(\mu+1,\frac{1}{2}+\frac{1}{2}\nu)$	$\frac{1}{2}a^{2\mu+\nu+1}B(\mu+1,\frac{1}{2}+\frac{1}{2}\nu)$ $\cdot {}_1F_2(\frac{1}{2}+\frac{1}{2}\nu;\frac{1}{2},\frac{1}{2}\nu+\frac{3}{2}+\mu;-\frac{1}{4}a^2y^2)$
54	$x^{-\nu-\frac{1}{2}}(a^2-x^2)^{-\frac{1}{2}}$ $\cdot\{[a+(a^2-x^2)^{\frac{1}{2}}]^{\nu}+[a-(a^2-x^2)^{\frac{1}{2}}]^{\nu}\},$ $x<a$ $0, \quad x>a$ $N=(2a)^{-\frac{1}{2}}B(\frac{1}{4}+\frac{1}{2}\nu,\frac{1}{4}-\frac{1}{2}\nu)$	$(2a)^{-\frac{1}{2}}B(\frac{1}{4}+\frac{1}{2}\nu,\frac{1}{4}-\frac{1}{2}\nu)$ $\cdot {}_1F_1(\frac{1}{4}-\frac{1}{2}\nu;\frac{1}{2};-iay)$ $\cdot {}_1F_1(\frac{1}{4}-\frac{1}{2}\nu;\frac{1}{2};iay)$
55	$x^{-\frac{1}{2}}(b^2-x^2)^{-\frac{1}{2}}$ $\cdot\{[(b+x)^{\frac{1}{2}}+i(b-x)^{\frac{1}{2}}]^{4\nu}$ $+[(b+x)^{\frac{1}{2}}-i(b-x)^{\frac{1}{2}}]^{4\nu}\}, \quad x<b$ $0, \quad x>b$ $N=2^{2\nu}b^{2\nu-\frac{1}{2}}\pi^{\frac{3}{2}}[\Gamma(\frac{3}{4}+\nu)\Gamma(\frac{3}{4}-\nu)]^{-1}$	$2^{2\nu-\frac{1}{2}}b^{2\nu}\pi^{\frac{3}{2}}$ $\cdot y^{\frac{1}{2}}J_{\nu-\frac{1}{4}}(\frac{1}{2}by)J_{-\nu-\frac{1}{4}}(\frac{1}{2}by)$
56	$(x^2-a^2)^{-\nu-\frac{1}{2}}, \quad x>a$ $0, \quad x<a$ $0<\nu<\frac{1}{2}$ $N=\frac{1}{2}a^{-2\nu}\pi^{-\frac{1}{2}}\Gamma(\nu)\Gamma(\frac{1}{2}-\nu)$	$-\frac{1}{2}(2a)^{-\nu}\pi^{\frac{1}{2}}\Gamma(\frac{1}{2}-\nu)$ $\cdot y^{\nu}Y_{\nu}(ay)$
57	$0, \quad x<2a$ $(x^2-2ax)^{-\nu-\frac{1}{2}}, \quad x>2a$ $0<\nu<\frac{1}{2}$ $N=\frac{1}{2}a^{-2\nu}\pi^{\frac{1}{2}}\Gamma(\nu)\Gamma(\frac{1}{2}-\nu)$	$-\frac{1}{2}(2a)^{-\nu}\pi^{\frac{1}{2}}\Gamma(\frac{1}{2}-\nu)$ $\cdot y^{\nu}[J_{\nu}(ay)\sin(ay)+Y_{\nu}(ay)\cos(ay)]$
58	$0, \quad x<a,$ $x^{-1}(x^2-a^2)^{-\nu-\frac{1}{2}}, \quad x>a$ $-\frac{1}{2}<\nu<\frac{1}{2}$ $N=\frac{1}{2}a^{-2\nu-1}\pi\sec(\nu\pi)$	$\frac{1}{2}a^{-2\nu-1}\pi\sec(\nu\pi)$ $\cdot\{1-\frac{1}{2}a\pi y[J_{\nu}(ay)\mathbf{H}_{\nu-1}(ay)$ $-\mathbf{H}_{\nu}(ay)J_{\nu-1}(ay)]\}$

	$2Nf(x)$	$Ng(y)$
59	$0, \quad x<a$ $x^{-\frac{1}{2}}(x^2-a^2)^{-\frac{1}{2}}\{[x+(x^2-a^2)^{\frac{1}{2}}]^{\nu}$ $+[x-(x^2-a^2)^{\frac{1}{2}}]^{\nu}\}, \quad x>a$ $-\frac{1}{2}<\nu<\frac{1}{2}$ $N=(2\pi)^{-\frac{1}{2}}a^{\nu-\frac{1}{2}}\Gamma(\frac{1}{4}-\frac{1}{2}\nu)\Gamma(\frac{1}{4}-\frac{1}{2}\nu)$	$-\frac{1}{2}a^{\nu}\pi(\frac{1}{2}\pi y)^{\frac{1}{2}}$ $\cdot[J_{\frac{1}{2}\nu-\frac{1}{4}}(\frac{1}{2}ay)Y_{-\frac{1}{2}\nu-\frac{1}{4}}(\frac{1}{2}ay)$ $+J_{-\frac{1}{2}\nu-\frac{1}{4}}(\frac{1}{2}ay)Y_{\frac{1}{2}\nu-\frac{1}{4}}(\frac{1}{2}ay)]$

3. Exponential Functions

	$2Nf(x)$	$Ng(y)$
60	e^{-ax} $N=a^{-1}$	$a(a^2+y^2)^{-1}$
61	$x^{-1}(e^{-bx}-e^{-ax}), \quad a>b$ $N=\log(a/b)$	$\frac{1}{2}\log[(a^2+y^2)(b^2+y^2)^{-1}]$
62	$x^{\frac{1}{2}}e^{-ax}$ $N=\frac{1}{2}\pi^{\frac{1}{2}}a^{-\frac{3}{2}}$	$\frac{1}{2}\pi^{\frac{1}{2}}(a^2+y^2)^{-\frac{3}{4}}\cos[\frac{3}{2}\arctan(y/a)]$
63	$x^{-\frac{1}{2}}e^{-ax}$ $N=(\pi/a)^{\frac{1}{2}}$	$(\frac{1}{2}\pi)^{\frac{1}{2}}(a^2+y^2)^{-\frac{1}{2}}[a+(a^2+y^2)^{\frac{1}{2}}]^{\frac{1}{2}}$
64	$0, \quad x<b$ $(x-b)^{\nu}e^{-ax}, \quad x>b$ $\nu>-1$ $N=a^{-\nu-1}\Gamma(1+\nu)e^{-ab}$	$\Gamma(1+\nu)(a^2+y^2)^{-\frac{1}{2}-\frac{1}{2}\nu}e^{-ab}$ $\cdot\cos[by+(\nu+1)\arctan(y/a)]$
65	$x^{\nu-1}e^{-ax}$ $N=a^{-\nu}\Gamma(\nu)$	$a^{-\nu}\Gamma(\nu)[1+(y^2/a^2)]^{-\frac{1}{2}\nu}\cos[\nu\arctan(y/a)]$
66	$(e^{ax}+1)^{-1}$ $N=a^{-1}\log 2$	$\frac{1}{4}a^{-1}[\psi(i\frac{1}{2}y/a)+\psi(-i\frac{1}{2}y/a)$ $-\psi(\frac{1}{2}+i\frac{1}{2}y/a)-\psi(\frac{1}{2}-i\frac{1}{2}y/a)]$
67	$x(e^{ax}-1)^{-1}$ $N=\frac{1}{6}(\pi/a)^2$	$\frac{1}{2}y^{-2}-\frac{1}{2}(\pi/a)^2[\operatorname{csch}(\pi y/a)]^2$
68	$x^{-2}(1-e^{-ax})^2$ $N=2a\log 2$	$a[\log(y^2+4a^2)-\log(y^2+a^2)]$ $-y\operatorname{arcctn}[\frac{1}{2}(y/a)^3+\frac{3}{2}y/a]$
69	$0, \quad x<b$ $(x-b)^{\nu}e^{-ax}, \quad x>b$ $\nu>-1$ $N=a^{-\nu-1}e^{-ab}\Gamma(1+\nu)$	$a^{-\nu-1}e^{-ab}\Gamma(1+\nu)$ $\cdot(1+y^2/a^2)^{-\frac{1}{2}\nu-\frac{1}{2}}\cos[by+(\nu+1)$ $\cdot\arctan(y/a)]$

	$2Nf(x)$	$Ng(y)$
70	$e^{-ax}(1-e^{-bx})^{\nu-1}, \quad \nu>0$ $N=b^{-1}B(\nu, a/b)$	$\frac{1}{2}b^{-1}\{B[\nu, (a-iy)/b]+B[\nu, (a+iy)/b]\}$
71	$x^{\nu-1}(e^{ax}+1)^{-1}, \quad \nu>0$ $N=(1-2^{1-\nu})\Gamma(\nu)\zeta(\nu)$	$\Gamma(\nu)(y^{-\nu}\cos(\frac{1}{2}\nu\pi)+\frac{1}{2}(2a)^{-\nu}\{\zeta[\nu, \frac{1}{2}+(iy/2a)]$ $+\zeta[\nu, \frac{1}{2}-i(y/2a)]-\zeta[\nu, i(y/2a)]$ $-\zeta[\nu, -i(y/2a)]\})$
72	$x^{\nu-1}(e^{ax}-1)^{-1}, \quad \nu>1$ $N=a^{-\nu}\Gamma(\nu)\zeta(\nu)$	$\frac{1}{2}a^{-\nu}\Gamma(\nu)\{\zeta[\nu, 1+(iy/a)]$ $+\zeta[\nu, 1-(iy/a)]\}$
73	$\exp(-ax^2)$ $N=\frac{1}{2}(\pi/a)^{\frac{1}{2}}$	$\frac{1}{2}(\pi/a)^{\frac{1}{2}}\exp(-y^2/4a)$
74	$x^{-\frac{1}{2}}\exp(-ax^2)$ $N=2^{-\frac{1}{2}}a^{-\frac{1}{4}}\pi[\Gamma(\frac{3}{4})]^{-1}$	$2^{-\frac{3}{2}}a^{-\frac{1}{2}}\pi y^{\frac{1}{2}}\exp(-y^2/8a)I_{-\frac{1}{4}}(y^2/8a)$
75	$x^{\frac{1}{2}}\exp(-ax^2)$ $N=2^{-\frac{1}{2}}a^{-\frac{3}{4}}[\Gamma(\frac{1}{4})]^{-1}$	$2^{-\frac{1}{2}}a^{-\frac{3}{2}}\pi y^{\frac{3}{2}}\exp(-y^2/8a)$ $\cdot[I_{-\frac{3}{4}}(y^2/8a)-I_{\frac{1}{4}}(y^2/8a)]$
76	$x^{2n}\exp(-a^2x^2)$ $N=(2a)^{-2n-1}\pi^{\frac{1}{2}}[(2n)!/n!]$	$(-1)^n 2^{-n-1}a^{-2n-1}\pi^{\frac{1}{2}}$ $\cdot\exp(-y^2/4a^2)He_{2n}(2^{-\frac{1}{2}}y/a)$
77	$x^{\nu}\exp(-ax^2), \quad \nu>-1$ $N=\frac{1}{2}a^{-\frac{1}{2}(1+\nu)}\Gamma(\frac{1}{2}+\frac{1}{2}\nu)$	$\frac{1}{2}a^{-\frac{1}{2}(1+\nu)}\Gamma(\frac{1}{2}+\frac{1}{2}\nu)$ $\cdot{}_1F_1[\frac{1}{2}+\frac{1}{2}\nu; \frac{1}{2}; -(y^2/4a)]$
78	$(b^2+x^2)^{-1}\exp(-a^2x^2)$ $N=\frac{1}{2}b^{-1}\pi\exp(a^2b^2)\ \mathrm{Erfc}(ab)$	$\frac{1}{4}b^{-1}\pi\exp(a^2b^2)\{\exp(-by)\ \mathrm{Erfc}[ab-(y/2a)]$ $+\exp(by)\ \mathrm{Erfc}[ab+(y/2a)]\}$
79	$\exp(-ax-b^2x^2)$ $N=\frac{1}{2}b^{-1}\pi^{\frac{1}{2}}\exp(a^2/4b^2)$ $\cdot\mathrm{Erfc}(a/2b)$	$\frac{1}{4}b^{-1}\pi^{\frac{1}{2}}\left\{\exp\left[\left(\frac{a-iy}{2b}\right)^2\right]\mathrm{Erfc}\left(\frac{a-iy}{2b}\right)\right.$ $\left.+\exp\left[\left(\frac{a+iy}{2b}\right)^2\right]\mathrm{Erfc}\left(\frac{a+iy}{2b}\right)\right\}$
80	$x^{\nu-1}\exp(-ax-bx^2), \quad \nu>0$ $N=(2b)^{-\nu/2}\exp(-a^2/8b)$ $\cdot\Gamma(\nu)D_{-\nu}[a(2b)^{-\frac{1}{2}}]$	$\frac{1}{2}(2b)^{-\nu/2}\exp[(a^2-y^2)/8b]\Gamma(\nu)$ $\cdot\{\exp(-iay/4b)D_{-\nu}[(a-iy)(2b)^{-\frac{1}{2}}]$ $+\exp(iay/4b)D_{-\nu}[(a+iy)(2b)^{-\frac{1}{2}}]\}$
81	$\exp[-(ax)^3]$ $N=\Gamma(\frac{1}{3})(3a)^{-1}$	$(3a)^{-\frac{3}{2}}y^{\frac{1}{2}}\{\exp(i\frac{1}{4}\pi)S_{0,\frac{1}{3}}[(y/3a)^{\frac{3}{2}}$ $\cdot 2\exp(i\frac{3}{4}\pi)]+\exp(-i\frac{1}{4}\pi)$ $\cdot S_{0,\frac{1}{3}}[2\exp(-i\frac{3}{4}\pi)(y/3a)^{\frac{3}{2}}]\}$

	$2Nf(x)$	$Ng(y)$
82	$x^{\mu}\exp(-ax^{c}),\quad \mu>-1,$ $0<c\leq 1$ $N=c^{-1}a^{-(\mu+1)/c}\Gamma[(\mu+1)c^{-1}]$	$-\sum_{n=0}^{\infty}\{(-a)^{n}(n!)^{-1}\Gamma(\mu+1+nc)$ $\cdot\sin[\frac{1}{2}\pi(\mu+nc)]y^{-\mu-1-nc}\}$
83	$x^{\mu}\exp(-ax^{c}),\quad \mu>-1,$ $c\geq 1$ $N=c^{-1}a^{-(\mu+1)/c}\Gamma[(\mu+1)c^{-1}]$	$C^{-1}\sum_{n=0}^{\infty}\{(-1)^{n}a^{-(2n+\mu+1)/c}[(2n)!]^{-1}$ $\cdot\Gamma[(2n+1+\mu)c^{-1}]y^{2n}\}$
84	$x^{-\frac{3}{2}}e^{-a/x}$ $N=(\pi/a)^{\frac{1}{2}}$	$(\pi/a)^{\frac{1}{2}}\exp[-(2ay)^{\frac{1}{2}}]\cos[(2ay)^{\frac{1}{2}}]$
85	$x^{-\nu-1}\exp(-a^{2}/4x),\quad \nu>0$ $N=2^{2\nu}a^{-\nu}\Gamma(\nu)$	$2^{\nu}a^{-\nu}y^{\frac{1}{2}\nu}\{\exp[i(\nu\pi/4)]K_{\nu}(ae^{i\frac{1}{4}\pi}y^{\frac{1}{2}})$ $+\exp[-i(\nu\pi/4)]K_{\nu}(ae^{-i\frac{1}{4}\pi}y^{\frac{1}{2}})\}$
86	$x^{-\frac{1}{2}}\exp[-ax-(b/x)]$ $N=a^{-\frac{1}{2}}\pi^{\frac{1}{2}}\exp[-2(ab)^{\frac{1}{2}}]$	$\pi^{\frac{1}{2}}(a^{2}+y^{2})^{-\frac{1}{2}}\exp(-2ub^{\frac{1}{2}})$ $\cdot[u\cos(2b^{\frac{1}{2}}v)-v\sin(2b^{\frac{1}{2}}v)]$ $u=2^{-\frac{1}{2}}[(a^{2}+y^{2})^{\frac{1}{2}}+a]^{\frac{1}{2}}$ $v=2^{-\frac{1}{2}}[(a^{2}+y^{2})^{\frac{1}{2}}-a]^{\frac{1}{2}}$
87	$x^{-\frac{3}{2}}\exp[-ax-(b/x)]$ $N=(\pi/b)^{\frac{1}{2}}\exp[-2(ab)^{\frac{1}{2}}]$	$(\pi/b)^{\frac{1}{2}}\exp(-2b^{\frac{1}{2}}u)\cos(2b^{\frac{1}{2}}v)$ $u=2^{-\frac{1}{2}}[(a^{2}+y^{2})^{\frac{1}{2}}+a]^{\frac{1}{2}}$ $v=2^{-\frac{1}{2}}[(a^{2}+y^{2})^{\frac{1}{2}}-a]^{\frac{1}{2}}$
88	$x^{\nu-1}\exp[-ax-(b^{2}/x)]$ $N=2b^{\nu}a^{-\nu/2}K_{\nu}(2a^{\frac{1}{2}}b)$	$b^{2}\{(a+iy)^{-\nu/2}K_{\nu}[2b(a+iy)^{\frac{1}{2}}]$ $+(a-iy)^{-\nu/2}K_{\nu}[2b(a-iy)^{\frac{1}{2}}]\}$
89	$x^{-2}\exp(-a^{2}x^{-2})$ $N=\frac{1}{2}a^{-1}\pi^{\frac{1}{2}}$	$\frac{1}{2}a^{-1}\pi\sum_{n=0}^{\infty}\frac{(-ay)^{n}}{n!\Gamma(\frac{1}{2}+\frac{1}{2}n)}$
90	$\exp(-ax^{\frac{1}{2}})$ $N=2a^{-2}$	$2^{-\frac{1}{2}}a\pi^{\frac{1}{2}}y^{-\frac{3}{2}}$ $\cdot\{\cos(a^{2}/4y)[\frac{1}{2}-C(a^{2}/4y)]$ $+\sin(a^{2}/4y)[\frac{1}{2}-S(a^{2}/4y)]\}$
91	$x^{-\frac{1}{2}}\exp(-ax^{\frac{1}{2}})$ $N=2a^{-1}$	$(2\pi/y)^{\frac{1}{2}}\{\cos(\frac{1}{4}a^{2}/y)[\frac{1}{2}-S(\frac{1}{4}a^{2}/y)]$ $-\sin(\frac{1}{4}a^{2}/y)[\frac{1}{2}-C(\frac{1}{4}a^{2}/y)]\}$
92	$x^{-\frac{3}{4}}\exp(-ax^{\frac{1}{2}})$ $N=2a^{-\frac{1}{2}}\pi^{\frac{1}{2}}$	$\frac{1}{2}a^{\frac{3}{2}}\pi y^{-\frac{3}{4}}\{J_{\frac{1}{4}}(a^{2}/8y)\sin(a^{2}/8y+\frac{1}{8}\pi)$ $-Y_{\frac{1}{4}}(a^{2}/8y)\cos(a^{2}/8y+\frac{1}{8}\pi)\}$
93	$x^{\nu-1}\exp(-ax^{\frac{1}{2}}),\quad \nu>0$ $N=2a^{-2\nu}\Gamma(2\nu)$	$\Gamma(2\nu)(2y)^{-\nu}(\exp\{-i[\frac{1}{2}\nu\pi+(a^{2}/8y)]\}$ $\cdot D_{-2\nu}[a(-i/2y)^{\frac{1}{2}}]$ $+\exp\{i[\frac{1}{2}\nu\pi+(a^{2}/8y)]\}D_{-2\nu}[a(i/2y)^{\frac{1}{2}}])$

	$2Nf(x)$	$Ng(y)$
94	$\exp[-b(a^2+x^2)^{\frac{1}{2}}]$ $N=aK_1(ab)$	$ab(b^2+y^2)^{-\frac{1}{2}}K_1[a(b^2+y^2)^{\frac{1}{2}}]$
95	$(a^2+x^2)^{-\frac{1}{2}}\exp[-b(a^2+x^2)^{\frac{1}{2}}]$ $N=K_0(ab)$	$K_0[a(b^2+y^2)^{\frac{1}{2}}]$
96	$x^{-\frac{1}{2}}(b^2+x^2)^{-\frac{1}{2}}\exp[-a(b^2+x^2)^{\frac{1}{2}}]$ $N=2^{\frac{3}{4}}a^{\frac{1}{4}}b^{-\frac{1}{4}}\pi^{\frac{1}{2}}K_{\frac{1}{4}}(ab)[\Gamma(\frac{3}{4})]^{-1}$	$(\frac{1}{2}\pi y)^{\frac{1}{2}}I_{-\frac{1}{4}}\{\frac{1}{2}b[(a^2+y^2)^{\frac{1}{2}}-a]\}$ $\cdot K_{\frac{1}{4}}\{\frac{1}{2}b[(a^2+y^2)^{\frac{1}{2}}+a]\}$
97	$(a^2+x^2)^{-\frac{3}{4}}\exp[-b(a^2+x^2)^{\frac{1}{2}}]$ $N=b^{\frac{1}{2}}(2\pi)^{-\frac{1}{2}}[K_{\frac{1}{4}}(\frac{1}{2}ab)]^2$	$b^{\frac{1}{2}}(2\pi)^{-\frac{1}{2}}K_{\frac{1}{4}}\{\frac{1}{2}a[(b^2+y^2)^{\frac{1}{2}}-y]\}$ $\cdot K_{\frac{1}{4}}\{\frac{1}{2}a[(b^2+y^2)^{\frac{1}{2}}+y]\}$
98	$(a^2+x^2)^{-\frac{5}{4}}\exp[-b(a^2+x^2)^{\frac{1}{2}}]$ $N=2^{\frac{1}{2}}b^{\frac{3}{2}}\pi^{-\frac{1}{2}}\{[K_{\frac{3}{4}}(\frac{1}{2}ab)]^2-[K_{\frac{1}{4}}(\frac{1}{2}ab)]^2\}$	$2^{\frac{1}{2}}b^{\frac{3}{2}}\pi^{-\frac{1}{2}}\{K_{\frac{3}{4}}[\frac{1}{2}a((b^2+y^2)^{\frac{1}{2}}-y)]$ $\cdot K_{\frac{3}{4}}[\frac{1}{2}a((b^2+y^2)^{\frac{1}{2}}+y)]$ $-K_{\frac{1}{4}}[\frac{1}{2}a((b^2+y^2)^{\frac{1}{2}}-y)]$ $\cdot K_{\frac{1}{4}}[\frac{1}{2}a((b^2+y^2)^{\frac{1}{2}}+y)]\}$
99	$(b^2+x^2)^{-\frac{1}{2}}[(b^2+x^2)^{\frac{1}{2}}+b]^{\frac{1}{2}}$ $\cdot\exp[-a(b^2+x^2)^{\frac{1}{2}}]$ $N=(\pi/a^{-\frac{1}{2}})\ e^{-ab}$	$(\frac{1}{2}\pi)^{\frac{1}{2}}(a^2+y^2)^{-\frac{1}{2}}$ $\cdot[a+(a^2+y^2)^{\frac{1}{2}}]^{\frac{1}{2}}\exp[-b(a^2+y^2)^{\frac{1}{2}}]$
100	$(a^2+x^2)^{-\frac{1}{2}}\{[(a^2+x^2)^{\frac{1}{2}}+x]^{\nu}$ $+[(a^2+x^2)^{\frac{1}{2}}-x]^{\nu}\}\exp[-b(a^2+x^2)^{\frac{1}{2}}]$ $N=2a^{\nu}K_{\nu}(ab)$	$2a^{\nu}\cos[\nu\arctan(y/b)]$ $\cdot K_{\nu}[a(b^2+y^2)^{\frac{1}{2}}]$
101	$x^{\nu-\frac{1}{2}}(a^2+x^2)^{-\frac{1}{2}}$ $\cdot[(a^2+x^2)^{\frac{1}{2}}+a]^{-\nu}$ $\cdot\exp[-b(a^2+x^2)^{\frac{1}{2}}],\quad \nu>-\frac{1}{2}$ $N=2^{(\nu/2)-\frac{1}{4}}a^{-\frac{1}{4}}\Gamma(\frac{1}{4}+\frac{1}{2}\nu)D_{-\nu-\frac{1}{2}}[2(ab)^{\frac{1}{2}}]$	$2^{(\nu/2)-\frac{1}{4}}a^{-\frac{3}{4}}\Gamma(\frac{1}{4}+\frac{1}{2}\nu)y^{-\frac{1}{2}}$ $\cdot[(b^2+y^2)^{\frac{1}{2}}+b]^{\frac{1}{4}}D_{-\nu-\frac{1}{2}}$ $\cdot\{[2a((b^2+y^2)^{\frac{1}{2}}+b)]^{\frac{1}{2}}\}$ $\cdot M_{(\nu/2),-\frac{1}{4}}\{[a(b^2+y^2)^{\frac{1}{2}}-b]\}$
102	$(a^2+x^2)^{-\frac{1}{2}}[x+(a^2+x^2)^{\frac{1}{2}}]^{\nu}$ $\cdot\exp[-b(a^2+x^2)^{\frac{1}{2}}]$ $N=a^{\nu}\csc(\nu\pi)\{\pi I_{-\nu}(ab)$ $-\int_0^{\pi}\exp(ab\cos t)\cos(\nu t)\,dt\}$	$a^{\nu}\csc(\nu\pi)\{\pi\cos[\nu\arctan(y/b)]$ $\cdot I_{-\nu}[a(b^2+y^2)^{\frac{1}{2}}]$ $-\int_0^{\pi}\exp(ab\cos t)\cosh(ay\sin t)$ $\cdot\cos(\nu t)\,dt\}$
103	$(a^2+x^2)^{-\frac{1}{2}}[(a^2+x^2)^{\frac{1}{2}}+a]^{-\frac{1}{2}}$ $\cdot\exp[-b(a^2+x^2)^{\frac{1}{2}}]$ $N=(2a)^{-\frac{1}{2}}\pi\ e^{ab}\,\mathrm{Erfc}[(2ab)^{\frac{1}{2}}]$	$(2a)^{-\frac{1}{2}}\pi\ e^{ab}$ $\cdot\mathrm{Erfc}\{a^{\frac{1}{2}}[(b^2+y^2)^{\frac{1}{2}}+b]^{\frac{1}{2}}\}$

	$2Nf(x)$	$Ng(y)$
104	$(a^2-x^2)^{-\frac{3}{4}} \exp[-b(a^2-x^2)^{\frac{1}{2}}], \quad x<a$ $0, \quad x>a$ $N=b^{\frac{1}{2}}(\frac{1}{2}\pi)^{\frac{3}{2}}$ $\cdot\{[I_{-\frac{1}{4}}(\frac{1}{2}ab)]^2-[I_{\frac{1}{4}}(\frac{1}{2}ab)]^2\}$	$-\frac{1}{4}b^{\frac{1}{2}}\pi^{\frac{3}{2}}$ $\cdot[J_{-\frac{1}{4}}(z_1)Y_{\frac{1}{4}}(z_2)+Y_{-\frac{1}{4}}(z_1)J_{\frac{1}{4}}(z_2)]$ $z_{\substack{1\\2}}=\frac{1}{2}a[y\pm(y^2-b^2)^{\frac{1}{2}}]$
105	$x^{-2\nu-(\frac{1}{2})}(a^2-x^2)^{-\frac{1}{2}}\{[a-(a^2-x^2)^{\frac{1}{2}}]^{2\nu}$ $\cdot\exp[b(a^2-x^2)^{\frac{1}{2}}]+[a+(a^2-x^2)^{\frac{1}{2}}]^{2\nu}$ $\cdot\exp[-b(a^2-x^2)^{\frac{1}{2}}]\}, \quad \frac{1}{4}\pm\nu>0$ $N=(\frac{1}{2}a)^{\frac{3}{4}}b^{\frac{1}{4}}\pi^{\frac{1}{2}}$ $\cdot\Gamma(\frac{1}{4}+\nu)\Gamma(\frac{1}{4}-\nu)M_{\nu,-\frac{1}{4}}(2ab)$	$a^{-1}\Gamma(\frac{1}{4}+\nu)\Gamma(\frac{1}{4}-\nu)(2\pi y)^{-\frac{1}{2}}$ $M_{\nu,-\frac{1}{4}}\{a[b+(b^2-y^2)^{\frac{1}{2}}]\}$ $\cdot M_{-\nu,-\frac{1}{4}}\{a[b-(b^2-y^2)^{\frac{1}{2}}]\}, \quad b>y$

4. Logarithmic Functions

	$2Nf(x)$	$Ng(y)$
106	$-\log x, \quad x<1$ $0, \quad x>1$ $N=1$	$y^{-1}\,\mathrm{Si}(y)$
107	$\log(a+x), \quad x<b$ $0, \quad x>b$ $a\geq 1$ $N=(a+b)\log(a+b)-a\log a-b$	$y^{-1}\{\sin(by)\log(a+b)$ $-\cos(ay)[\mathrm{si}(ay+by)-\mathrm{si}(ay)]$ $+\sin(ay)[\mathrm{Ci}(ay+by)-\mathrm{Ci}(ay)]\}$
108	$-(1-x^2)^{\frac{1}{2}}\log x, \quad x<1$ $0, \quad x>1$ $N=\frac{1}{2}\pi\log 2$	$\frac{1}{2}\pi\log 2J_0(y)+\frac{1}{2}\pi\sum_{n=1}^{\infty}n^{-1}J_{2n}(y)$
109	$(a^2+x^2)^{-1}\log(a^2+x^2), \quad a\geq 1$ $N=a^{-1}\pi\log(2a)$	$-\frac{1}{2}a^{-1}\pi\{e^{-ay}[\gamma-\log(2a/y)]$ $-e^{ay}\,\mathrm{Ei}(-2ay)\}$
110	$(a^2+x^2)^{-n-\frac{1}{2}}\log(a^2+x^2),$ $a\geq 1, \quad n=1,2,3,\ldots$ $N=a^{-2n}B(\frac{1}{2},n)$ $\cdot[\log(a/2)-\sum_{m=1}^{2n-1}(-)^m m^{-1}]$	$-n![(2n)!]^{-1}(2y/a)^n$ $\cdot\{K_n(ay)[\gamma-2\sum_{m=1}^{n}(2m-1)^{-1}+\log(2y/a)]$ $+\frac{1}{2}n!\sum_{m=0}^{n-1}(\frac{1}{2}ay)^{m-n}K_m(ay)[m!(n-m)]^{-1}\}$

	$2Nf(x)$	$Ng(y)$
111	$\log[(a^2+x^2)(b^2+x^2)^{-1}]$ $N=(a-b)\pi$	$\pi y^{-1}(e^{-by}-e^{-ay})$ $\pm$ as $a \gtrless b$
112	$(a^2+x^2)^{-\frac{1}{2}}\log\{x^{-1}[a+(a^2+x^2)^{\frac{1}{2}}]\}$ $N=\frac{1}{4}\pi^2$	$\frac{1}{4}\pi^2[I_0(ay)-\mathbf{L}_0(ay)]$
113	$\log\{\frac{1}{2}x^{-1}[x+(a^2+x^2)^{\frac{1}{2}}]\}$ $N=a$	$\frac{1}{2}\pi y^{-1}[1+\mathbf{L}_0(ay)-I_0(ay)]$
114	$(x^2-a^2)^{-1}\log(x/a)$ $N=\frac{1}{4}a^{-1}\pi^2$	$\frac{1}{2}a^{-1}\pi[\sin(ay)\ \mathrm{Ci}(ay)$ $-\cos(ay)\ \mathrm{si}(ay)]$
115	$\log(1+a^2x^{-2})$ $N=\pi a$	$\pi y^{-1}(1-e^{-ay})$
116	$-\log(a-x), \quad x<a$ $0, \quad x>a$ $a\leq 1$ $N=a(1-\log a)$	$-y^{-1}\{\sin(ay)[\mathrm{Ci}(ay)-\gamma-\log y]$ $-\cos(ay)\ \mathrm{Si}(ay)\}$
117	$-\log(a-x), \quad x<b$ $0, \quad x>b$ $b<a\leq 1$ $N=-a\log a+b+(a-b)\log(a-b)$	$-y^{-1}\{\sin(by)\log(a-b)$ $+\sin(ay)[\mathrm{Ci}(ay)-\mathrm{Ci}(ay-by)]$ $-\cos(ay)[\mathrm{Si}(ay)-\mathrm{Si}(ay-by)]\}$
118	$-\log(a^2-x^2), \quad x<a$ $0, \quad x>a$ $0<a\leq 1$ $N=2a(1-\log a)$	$y^{-1}\{\cos(ay)\ \mathrm{Si}(2ay)$ $+\sin(ay)[\gamma+\log(y/2a)$ $-\mathrm{Ci}(2ay)]\}$
119	$-(a^2-x^2)^{-\frac{1}{2}}\log(a^2-x^2), \quad x<a$ $0, \quad x>a$ $0<a\leq 1$ $N=\pi\log(2/a)$	$\frac{1}{2}\pi\{[\gamma+\log(2y/a)]J_0(ay)$ $-\frac{1}{2}\pi Y_0(ay)\}$
120	$-\log(a^2-x^2), \quad x<b$ $0, \quad x>b$ $b<a\leq 1$ $N=-(a+b)\log(a+b)$ $+(a-b)\log(a-b)+2b$	$-y^{-1}\{\sin(by)\log(a^2-b^2)$ $-\cos(ay)[\mathrm{si}(ay+by)-\mathrm{si}(ay-by)]$ $+\sin(ay)[\mathrm{Ci}(ay+by)-\mathrm{Ci}(ay-by)]\}$

	$2Nf(x)$	$Ng(y)$
121	$-[x(1-x)]^{-\frac{1}{2}}\log[x(1-x)], \quad x<1$ $0, \quad x>1$ $N=4\pi\log 2$	$-\pi\cos(\frac{1}{2}y)[\frac{1}{2}\pi Y_0(\frac{1}{2}y)$ $-(\gamma+\log 4y)J_0(\frac{1}{2}y)]$
122	$-(a^2-x^2)^{n-\frac{1}{2}}\log(a^2-x^2), \quad x<a$ $0, \quad x>a$ $a\leq 1, \quad n=1,2,3,\ldots$ $N=-a^{2n}B(\frac{1}{2},\frac{1}{2}+n)$ $\cdot[\log(a/2)-\sum_{m=1}^{2n}(-1)^m m^{-1}]$	$-\frac{1}{2}\pi(2n)!(n!)^{-1}(2y/a)^{-n}$ $\cdot\{\frac{1}{2}n!\sum_{m=0}^{n-1}(\frac{1}{2}ay)^{m-n}J_m(ay)[m!(n-m)]^{-1}$ $+J_n(ay)[2\sum_{m=1}^{n}(2m-1)^{-1}-\gamma-\log(2y/a)]$ $+\frac{1}{2}\pi Y_n(ay)\}$
123	$-(2ax-x^2)^{n-\frac{1}{2}}\log(2ax-x^2), \quad x<2a$ $0, \quad x>2a$ $a\leq 1, \quad n=0,1,2,\ldots$ $N=-2a^{2n}B(\frac{1}{2},\frac{1}{2}+n)$ $\cdot[\log(2a)+\sum_{m=1}^{2n}(-1)^m m^{-1}]$ For $n=0, \quad \sum_{m=1}^{n-1}(\ )=0$	$-\pi(2n)!(n!)^{-1}\cos(ay)(2y/a)^{-n}$ $\cdot\{\frac{1}{2}n!\sum_{m=0}^{n-1}(\frac{1}{2}ay)^{m-n}[m!(n-m)]^{-1}J_m(ay)$ $+J_n(ay)[2\sum_{m=0}^{n-1}(2m+1)^{-1}-\gamma$ $-\log(2y/a)]+\frac{1}{2}\pi Y_n(ay)\}$ For $n=0, \quad \sum_{m=0}^{n-1}(\ )=0$
124	$0, \quad x<a$ $-\log\{\frac{1}{2}x^{-\frac{1}{2}}[(x+a)^{\frac{1}{2}}+(x-a)^{\frac{1}{2}}]\}, \quad x>a$ $N=\frac{1}{2}a$	$\frac{1}{2}y^{-1}[\frac{1}{2}\pi J_0(ay)+\mathrm{si}(ay)]$
125	$e^{-ax}(\log x)^2$ $N=a^{-1}[\frac{1}{6}\pi^2+(\gamma+\log a)^2]$	$(a^2+y^2)^{-1}\{\frac{1}{6}a\pi^2+2y\arctan(y/a)$ $+a[\gamma+\frac{1}{2}\log(a^2+y^2)]^2$ $-a[\arctan(y/a)]^2\}$
126	$(1+x^2)^{-1}\log[x+(1+x^2)^{\frac{1}{2}}]$ $\cdot\{[x+(1+x^2)^{\frac{1}{2}}]^n$ $-[x+(1+x^2)^{\frac{1}{2}}]^{-n}\}$ $\cdot\exp[-b(1+x^2)^{\frac{1}{2}}]$ $N=n!\sum_{m=0}^{n-1}\{(\frac{1}{2}b)^{m-n}$ $\cdot K_m(b)[m!(n-m)]^{-1}\},$ $n=1,2,3,\ldots$	$n!\cos[n\arctan(y/b)]$ $\cdot\sum_{m=0}^{n-1}[\frac{1}{2}(b^2+y^2)^{\frac{1}{2}}]^{m-n}K_m[(b^2+y^2)^{\frac{1}{2}}]$ $\cdot[m!(n-m)]^{-1}-2\arctan(y/b)$ $\cdot\sin[n\arctan(y/b)]K_n[(b^2+y^2)^{\frac{1}{2}}]$

	$2Nf(x)$	$Ng(y)$
127	$\log(1+e^{-ax})$ $N=(12a)^{-1}\pi^2$	$\frac{1}{2}[ay^{-2}-\pi y^{-1}\operatorname{csch}(a^{-1}\pi y)]$
128	$-\log(1-e^{-ax})$ $N=(6a)^{-1}\pi^2$	$-\frac{1}{2}[ay^{-2}-\pi y^{-1}\operatorname{ctnh}(a^{-1}\pi y)]$

5. Trigonometric Functions

	$2Nf(x)$	$Ng(y)$
129	$x^{-2}[\sin(ax)]^2$ $N=\frac{1}{2}\pi a$	$\frac{1}{2}\pi(a-\frac{1}{2}y), \quad y<2a$ $0, \quad y>2a$
130	$(\sin ax/x)^{2m}, \quad m=1,2,3,\ldots$ $N=(-1)^m 2^{1-2m} m\pi \cdot \sum_{n=1}^{m}(-1)^n \frac{(2an)^{2m-1}}{(m+n)!(m-n)!}$	$(-1)^m 2^{-2m} m\pi \left\{(m!)^{-2} y^{2m-1} + \sum_{n=1}^{m} \frac{(-1)^n[(2an+y)^{2m-1}+(\lvert 2an-y\rvert)^{2m-1}]}{(m+n)!(m-n)!}\right\},$ $y\leq 2am$ $0, \quad y\geq 2am$
131	$e^{-ax}(\sin x)^{2n}$ $N=i(-1)^n 2^{-2n-2}(2n+1)^{-1} \cdot\left\{\left[\binom{n+i\frac{1}{2}a}{2n+1}\right]^{-1}-\left[\binom{n-i\frac{1}{2}a}{2n+1}\right]^{-1}\right\}$	$i(-1)^n 2^{-2n-2}(2n+1)^{-1} \cdot\left\{\left[\binom{n+\frac{1}{2}y+i\frac{1}{2}a}{2n+1}\right]^{-1} -\left[\binom{n+\frac{1}{2}y-i\frac{1}{2}a}{2n+1}\right]^{-1}\right\}$
132	$(\cosh a-\cos x)^{-1}, \quad x<\pi$ $0, \quad x>\pi$ $N=\pi \operatorname{csch} a$	$-y\frac{\sin(\pi y)}{\sinh a}\sum_{n=0}^{\infty}(-1)^n\epsilon_n(n^2-y^2)^{-1}e^{-an}$
133	$(a^2+x^2)^{-1}(1-2b\cos x+b^2)^{-1}, \quad b<1$ $N=a^{-1}\pi(e^a+b)/2(e^a-b)(1-b^2)$	$\frac{1}{2}a^{-1}\pi(1-b^2)^{-1}(e^a-b)^{-1} \cdot(e^{a-ay}+be^{ay}), \quad y<1$
134	$(a^2+x^2)^{-1}(1-2b\cos x+b^2)^{-1}, \quad b<1$ $N=\pi(e^a+b)/2a(e^a-b)(1-b^2)$	$\frac{1}{2}a^{-1}\pi(1-b^2)^{-1} \cdot[e^{-ay}+(be^{-ay}-b^{n+1}e^{-\delta a})/(e^{-a}-b) +(be^{-an-a\delta}+b^{n+1}e^{a\delta})/(e^a-b)], \quad y=n+\delta$ $0\leq\delta<1, \quad n=1,2,3,\ldots$

	$2Nf(x)$	$Ng(y)$
135	$(\cos x-\cos\delta)^{-\frac{1}{2}}, \quad x<\delta$ $0, \quad x>\delta$ $\delta<\pi$ $N=2^{\frac{1}{2}}K(\sin\frac{1}{2}\delta)$	$2^{-\frac{1}{2}}\pi P_{-\frac{1}{2}+\nu}(\cos\delta)$
136	$(\cos x-\cos\delta)^{\nu-\frac{1}{2}}, \quad x<\delta$ $0, \quad x>\delta$ $0<\delta<\pi, \quad \nu>-\frac{1}{2}$ $N=(\frac{1}{2}\pi)^{\frac{1}{2}}\sin^{\nu}\delta\Gamma(\nu+\frac{1}{2})P^{-\nu}_{-\frac{1}{2}}(\cos\delta)$	$(\frac{1}{2}\pi)^{\frac{1}{2}}\sin^{\nu}\delta\Gamma(\nu+\frac{1}{2})P^{-\nu}_{y-\frac{1}{2}}(\cos\delta)$
137	$(\sin x)^{\alpha}, \quad x<\pi$ $0, \quad x>\pi$ $\alpha>-1$ $N=2^{-\alpha}\pi\Gamma(1+\alpha)[\Gamma(1+\frac{1}{2}\alpha)]^{-2}$	$2^{-\alpha}\pi\cos(\frac{1}{2}y)\Gamma(1+\alpha)$ $\cdot[\Gamma(1+\frac{1}{2}\alpha+\frac{1}{2}y)\Gamma(1+\frac{1}{2}\alpha-\frac{1}{2}y)]^{-1}$
138	$(\cos x)^{\alpha}, \quad x<\frac{1}{2}\pi$ $0, \quad x>\frac{1}{2}\pi$ $\alpha>-1$ $N=2^{-\alpha-1}\pi\Gamma(1+\alpha)[\Gamma(1+\frac{1}{2}\alpha)]^{-2}$	$2^{-\alpha-1}\pi\Gamma(1+\alpha)$ $\cdot[\Gamma(1+\frac{1}{2}\alpha+\frac{1}{2}y)\Gamma(1+\frac{1}{2}\alpha-\frac{1}{2}y)]^{-1}$
139	$\sin[b(a^2-x^2)^{\frac{1}{2}}], \quad x<a$ $0, \quad x>a$ $ab\leq\pi$ $N=\frac{1}{2}a\pi J_1(ab)$	$\frac{1}{2}ab\pi(b^2+y^2)^{-\frac{1}{2}}$ $\cdot J_1[a(b^2+y^2)^{\frac{1}{2}}]$
140	$(a^2-x^2)^{-\frac{3}{4}}\sin[b(a^2-x^2)^{\frac{1}{2}}], \quad x<a$ $0, \quad x>a$ $ab\leq\pi$ $N=(\frac{1}{2}\pi)^{\frac{3}{2}}b^{\frac{1}{2}}[J_{\frac{1}{4}}(\frac{1}{2}ab)]^2$	$b^{\frac{1}{2}}(\frac{1}{2}\pi)^{\frac{3}{2}}J_{\frac{1}{4}}\{\frac{1}{2}a[(b^2+y^2)^{\frac{1}{2}}-y]\}$ $\cdot J_{\frac{1}{4}}\{\frac{1}{2}a[(b^2+y^2)^{\frac{1}{2}}+y]\}$
141	$(a^2-x^2)^{-\frac{1}{2}}\cos[b(a^2-x^2)^{\frac{1}{2}}], \quad x<a$ $0, \quad x>a$ $ab\leq\frac{1}{2}\pi$ $N=\frac{1}{2}\pi J_0(ab)$	$\frac{1}{2}\pi J_0[a(b^2+y^2)^{\frac{1}{2}}]$
142	$(a^2-x^2)^{-\frac{3}{4}}\cos[b(a^2-x^2)^{\frac{1}{2}}], \quad x<a$ $0, \quad x>a$ $ab\leq\frac{1}{2}\pi$ $N=b^{\frac{1}{2}}(\frac{1}{2}\pi)^{\frac{3}{2}}[J_{-\frac{1}{4}}(\frac{1}{2}ab)]^2$	$b^{\frac{1}{2}}(\frac{1}{2}\pi)^{\frac{3}{2}}J_{-\frac{1}{4}}\{\frac{1}{2}a[(b^2+y^2)^{\frac{1}{2}}-y]\}$ $\cdot J_{-\frac{1}{4}}\{\frac{1}{2}a[(b^2+y^2)^{\frac{1}{2}}+y]\}$

	$2Nf(x)$	$Ng(y)$
143	$x^{-\frac{1}{2}}(b-x)^{-\frac{1}{2}}\cos[ax^{\frac{1}{2}}(b-x)^{\frac{1}{2}}],\quad x<b$ $0,\quad x>b$ $ab\le\pi$ $N=\pi J_0(\frac{1}{2}ab)$	$\pi\cos(\frac{1}{2}by)$ $\cdot J_0[\frac{1}{2}b(a^2+y^2)^{\frac{1}{2}}]$
144	$x^{-\frac{1}{2}}(a^2-x^2)^{-\frac{1}{2}}\cos[b(a^2-x^2)^{\frac{1}{2}}],\quad x<a$ $0,\quad x>a$ $ab\le\frac{1}{2}\pi$ $N=2^{-\frac{3}{4}}(b/a)^{\frac{1}{4}}\pi^{\frac{3}{2}}J_{-\frac{1}{4}}(ab)+[\Gamma(\frac{3}{4})]^{-1}$	$(\frac{1}{2}\pi)^{\frac{3}{2}}y^{\frac{1}{2}}$ $\cdot J_{-\frac{1}{4}}\{\frac{1}{2}a[(b^2+y^2)^{\frac{1}{2}}-b]\}$ $\cdot J_{-\frac{1}{4}}\{\frac{1}{2}[(b^2+y^2)^{\frac{1}{2}}+b]\}$
145	$\exp(a\cos x),\quad x<\pi$ $0,\quad x>\pi$ $N=\pi I_0(a)$	$\pi I_y(a)+\sin(\pi y)\int_0^\infty \exp(-a\cosh t-yt)\,dt$
146	$(\cos x)^{-\frac{1}{2}}\exp(-a\cos x),\quad x<\frac{1}{2}\pi$ $0,\quad x>\frac{1}{2}\pi$ $N=a^{\frac{1}{2}}(\frac{1}{2}\pi)^{\frac{3}{2}}\{[I_{-\frac{1}{4}}(\frac{1}{2}a)]^2-[I_{\frac{1}{4}}(\frac{1}{2}a)]^2\}$	$a^{\frac{1}{2}}(\frac{1}{2}\pi)^{\frac{3}{2}}$ $\cdot[I_{-\frac{1}{4}-\frac{1}{2}y}(\frac{1}{2}a)I_{-\frac{1}{4}+\frac{1}{2}y}(\frac{1}{2}a)$ $-I_{\frac{1}{4}+\frac{1}{2}y}(\frac{1}{2}a)I_{\frac{1}{4}-\frac{1}{2}y}(\frac{1}{2}a)]$
147	$(\cos x)^{-\frac{1}{2}}\exp(a\cos x),\quad x<\frac{1}{2}\pi$ $0,\quad x>\frac{1}{2}\pi$ $N=a^{\frac{1}{2}}(\frac{1}{2}\pi)^{\frac{3}{2}}\{[I_{-\frac{1}{4}}(\frac{1}{2}a)]^2+[I_{\frac{1}{4}}(\frac{1}{2}a)]^2\}$	$a^{\frac{1}{2}}(\frac{1}{2}\pi)^{\frac{3}{2}}$ $\cdot[I_{-\frac{1}{4}-\frac{1}{2}y}(\frac{1}{2}a)I_{-\frac{1}{4}+\frac{1}{2}y}(\frac{1}{2}a)$ $+I_{\frac{1}{4}-\frac{1}{2}y}(\frac{1}{2}a)I_{\frac{1}{4}+\frac{1}{2}y}(\frac{1}{2}a)]$
148	$(\sin x)^{-\frac{1}{2}}\exp(-a\sin x),\quad x<\pi$ $0,\quad x>\pi$ $N=(\frac{1}{2}a)^{\frac{1}{2}}\pi^{\frac{3}{2}}\{[I_{-\frac{1}{4}}(\frac{1}{2}a)]^2-[I_{\frac{1}{4}}(\frac{1}{2}a)]^2\}$	$(\frac{1}{2}a)^{\frac{1}{2}}\pi^{\frac{3}{2}}\cos(\frac{1}{2}\pi y)$ $\cdot[I_{-\frac{1}{4}-\frac{1}{2}y}(\frac{1}{2}a)I_{-\frac{1}{4}+\frac{1}{2}y}(\frac{1}{2}a)$ $-I_{\frac{1}{4}-\frac{1}{2}y}(\frac{1}{2}a)I_{\frac{1}{4}+\frac{1}{2}y}(\frac{1}{2}a)]$
149	$(\sin x)^{-\frac{1}{2}}\exp(a\sin x),\quad x<\pi$ $0,\quad x>\pi$ $N=(\frac{1}{2}a)^{\frac{1}{2}}\pi^{\frac{3}{2}}\{[I_{-\frac{1}{4}}(\frac{1}{2}a)]^2+[I_{\frac{1}{4}}(\frac{1}{2}a)]^2\}$	$(\frac{1}{2}a)^{\frac{1}{2}}\pi^{\frac{3}{2}}\cos(\frac{1}{2}\pi y)$ $\cdot[I_{-\frac{1}{4}-\frac{1}{2}y}(\frac{1}{2}a)I_{-\frac{1}{4}+\frac{1}{2}y}(\frac{1}{2}a)$ $+I_{\frac{1}{4}-\frac{1}{2}y}(\frac{1}{2}a)I_{\frac{1}{4}+\frac{1}{2}y}(\frac{1}{2}a)]$
150	$\log[\sec(\frac{1}{2}\pi x)],\quad x<1$ $0,\quad x>1$ $N=\log 2$	$y^{-1}\sin y$ $\cdot\{\gamma+\log 2+\frac{1}{2}\psi[1+(y/\pi)]$ $+\frac{1}{2}\psi[1-(y/\pi)]\}$
151	$\log[\csc(\pi x)],\quad x<1$ $0,\quad x>1$ $N=\log 2$	$y^{-1}\sin y$ $\cdot\{\gamma+\log 2+\frac{1}{2}\psi[1+(2\pi)^{-1}y]$ $+\frac{1}{2}\psi[1-(y/2\pi)]\}$
152	$[\cos(\pi x/2)]^{\nu-1}\log[\sec(\pi x/2)],\quad x<1$ $0,\quad x>1$ $\nu>0$ $N=2^{1-\nu}\Gamma(\nu)[\Gamma(\frac{1}{2}+\frac{1}{2}\nu)]^{-2}$ $\cdot[\log 2-\psi(\nu)+\psi(\frac{1}{2}+\frac{1}{2}\nu)]$	$2^{1-\nu}\Gamma(\nu)\{\Gamma[\frac{1}{2}+\frac{1}{2}\nu+(y/\pi)]\}^{-1}$ $\cdot\{\Gamma[\frac{1}{2}+\frac{1}{2}\nu-(y/\pi)]\}^{-1}$ $\cdot\{\log 2+\frac{1}{2}\psi[\frac{1}{2}+\frac{1}{2}\nu+(y/\pi)]$ $+\frac{1}{2}\psi[\frac{1}{2}+\frac{1}{2}\nu-(y/\pi)]-\psi(\nu)\}$

	$2Nf(x)$	$Ng(y)$
153	$(\sin\pi x)^{\nu-1}\log[\csc(\pi x)],\quad x<1$ $0,\quad x>1$ $\nu>0$ $N=2^{1-\nu}\Gamma(\nu)[\Gamma(\frac{1}{2}+\frac{1}{2}\nu)]^{-2}$ $\cdot[\log 2-\psi(\nu)+\psi(\frac{1}{2}+\frac{1}{2}\nu)]$	$2^{1-\nu}\Gamma(\nu)\cos(\frac{1}{2}y)$ $\cdot\{\Gamma[\frac{1}{2}+\frac{1}{2}\nu+(y/2\pi)]\Gamma[\frac{1}{2}+\frac{1}{2}\nu-(y/2\pi)]\}^{-1}$ $\cdot\{\log 2+\frac{1}{2}\psi[\frac{1}{2}+\frac{1}{2}\nu+(y/2\pi)]$ $+\frac{1}{2}\psi[\frac{1}{2}+\frac{1}{2}\nu-(y/2\pi)]-\psi(\nu)\}$
154	$-(a^2+x^2)^{-1}\log(C^2\,\mathrm{sm}^2 bx),\quad C\leq 1,$ $b>0$ $N=-a^{-1}\pi$ $\cdot\log\{\frac{1}{2}C[1-\exp(-2ab)]\}$	$-a^{-1}\pi\{\cosh(ay)\log(1-e^{-2ab})$ $+\sum_{n=1}^{m} n^{-1}\sinh[a(y-2bn)]$ $+\log(\frac{1}{2}C)e^{-ay}\},$ $m\leq(y/2b)<m+1,\quad m=1,2,3,\ldots$ For $m=0,\quad \sum_{n=1}^{m}(\ )=0$
155	$-(a^2+x^2)^{-1}\log(C^2\,\mathrm{sm}^2 bx),\quad C\leq 1,$ $b>0$ $N=-a^{-1}\pi$ $\cdot\log[\frac{1}{2}C(1+e^{-2ab})]$	$-a^{-1}\pi\{\cosh(ay)\log(1+e^{-2ab})$ $+\log(\frac{1}{2}C)e^{-ay}$ $+\sum_{n=1}^{m}(-1)^n n^{-1}\sinh[a(y-2bn)]\},$ $m\leq(y/2b)<m+1,\quad m=0,1,2,3,\ldots$ For $m=0, \sum_{n=1}^{m}(\ )=0$
156	$-(a^2+x^2)^{-1}$ $\cdot\log[\frac{1}{2}c^2\cosh\delta\pm\frac{1}{2}c^2\cos(bx)],$ $c\cosh(\delta/2)\leq 1$ $N=-a^{-1}\pi\{\frac{1}{2}\delta$ $+\log[\frac{1}{2}c(1\pm e^{-\delta-ab})]\}$	$-a^{-1}\pi\{\cosh(ay)\log(1\pm e^{-\delta-ab})$ $+(\frac{1}{2}\delta+\log\frac{1}{2}c)e^{-ay}$ $+\sum_{n=1}^{m}(\mp 1)^n n^{-1}e^{-n\delta}\sinh[a(y-bn)]\},$ $m\leq(y/b)<m+1,\quad m=0,1,2,3,\ldots$ For $m=0,\quad \sum_{n=1}^{m}(\ )=0$
157	$(a^2+x^2)^{-1}$ $\cdot\log[\frac{1}{2}c^2\cosh\delta\pm\frac{1}{2}c^2\cos(bx)],$ $c\sinh(\frac{1}{2}\delta)\geq 1$ $N=a^{-1}\pi\{\frac{1}{2}\delta$ $+\log[\frac{1}{2}c(1\pm e^{-\delta-ab})]$	$a^{-1}\pi\{\cosh(ay)\log(1\pm e^{-\delta-ab})$ $+(\frac{1}{2}\delta+\log\frac{1}{2}c)e^{-ay}$ $+\sum_{n=1}^{m}(\mp)^n n^{-1}e^{-n\delta}\sinh[a(y-bn)]\},$ $m\leq(y/b)<m+1,\quad m=0,1,2,3,\ldots$ For $m=0,\quad \sum_{n=1}^{m}(\ )=0$

	$2Nf(x)$	$Ng(y)$
158	$-x^{-2}\log[\cos^2(ax)],\quad a>0$ $N=a\pi$	$a\pi\{1-(a^{-1}\log 2)y$ $-a^{-1}\sum_{n=1}^{m}(-1)^n n^{-1}(y-2an)\}$ $m\leq y/2a<m+1,\quad m=0,1,2,3,\ldots$ For $m=0,\quad \sum_{n=1}^{m}(\ )=0$
159	$x^{-2}(a^2+x^2)^{-1}\log[\cos^2(bx)]$ $N=\pi a^{-3}\log[\cosh(ab)]$	$\pi a^{-3}\{ab-\log 2(ay+e^{-ay})$ $+\cosh(ay)\log(1+e^{-2ab})$ $+\sum_{n=1}^{m}(-1)^n n^{-1}\cdot[\sinh(ay-2abn)$ $+\sinh(2abn)-ay\cosh(2abn)]\}$ $m\leq y/2b<m+1,\quad m=0,1,2,\ldots$ For $m=0,\quad \sum_{n=1}^{m}(\ )=0$
160	$(\cos x-\cos\delta)^{-\frac{1}{2}}$ $\cdot\log(\cos x-\cos\delta),\quad x<\delta$ $0,\quad x>\delta$ $0\leq\delta\leq 2\pi$ $N=\pi 2^{-\frac{1}{2}}[2\pi^{-1}K(\sin\delta/2)\log(\sin\delta)$ $-K(\cos\frac{1}{2}\delta)]$	$\pi 2^{-\frac{1}{2}}\{P_{-\frac{1}{2}+y}(\cos\delta)$ $\cdot[\log(\sin\delta)-\gamma-\log 4-\psi(\frac{1}{2}+y)]$ $-Q_{-\frac{1}{2}-y}(\cos\delta)\}$
161	$\sin(a\sin x),\quad x<\pi$ $0,\quad x>\pi$ $a\leq\pi$ $N=\pi\mathbf{H}_0(a)$	$\frac{1}{2}\pi\,\mathrm{ctn}(\frac{1}{2}\pi y)[\mathbf{J}_y(a)-\mathbf{J}_{-y}(a)]$
162	$\sin(a\cos x),\quad x<\pi/2$ $0,\quad x>\pi/2$ $a\leq\pi$ $N=\frac{1}{2}\pi\mathbf{H}_0(a)$	$\frac{1}{4}\pi\,\mathrm{cosec}(\frac{1}{2}\pi y)[\mathbf{J}_y(a)-\mathbf{J}_{-y}(a)]$
163	$\cos(a\cos x),\quad x<\pi/2$ $0,\quad x>\pi/2$ $a\leq\frac{1}{2}\pi$ $N=\frac{1}{2}\pi J_0(a)$	$\frac{1}{4}\pi\sec(\frac{1}{2}\pi y)[\mathbf{J}_y(a)+\mathbf{J}_{-y}(a)]$

	$2Nf(x)$	$Ng(y)$
164	$\cos(a\sin x)$, $x<\pi$ 0, $x>\pi$ $a\leq\frac{1}{2}\pi$ $N=\pi J_0(a)$	$\frac{1}{2}\pi[\mathbf{J}_y(a)+\mathbf{J}_{-y}(a)]$
165	$(\sin x)^{-\frac{1}{2}}\sin(2a\sin x)$, $x<\pi$ 0, $x>\pi$ $a\leq\frac{1}{2}\pi$ $N=a^{\frac{1}{2}}\pi^{\frac{3}{2}}[J_{\frac{1}{4}}(a)]^2$	$a^{\frac{1}{2}}\pi^{\frac{3}{2}}\cos(\frac{1}{2}\pi y)J_{\frac{1}{4}-\frac{1}{2}y}(a)J_{\frac{1}{4}+\frac{1}{2}y}(a)$
166	$(\sin x)^{-\frac{1}{2}}\cos(2a\sin x)$, $x<\pi$ 0, $x>\pi$ $a\leq\pi/4$ $N=a^{\frac{1}{2}}\pi^{\frac{3}{2}}[J_{-\frac{1}{4}}(a)]^2$	$a^{\frac{1}{2}}\pi^{\frac{3}{2}}\cos(\frac{1}{2}\pi y)J_{-\frac{1}{4}+\frac{1}{2}y}(a)$ $\cdot J_{-\frac{1}{4}-\frac{1}{2}y}(a)$
167	$(\cos x)^{-\frac{1}{2}}\sin(2a\cos x)$, $x<\pi/2$ 0, $x>\pi/2$ $a\leq\pi/2$ $N=\frac{1}{2}a^{\frac{1}{2}}\pi^{\frac{3}{2}}[J_{\frac{1}{4}}(a)]^2$	$\frac{1}{2}a^{\frac{1}{2}}\pi^{\frac{3}{2}}J_{\frac{1}{4}+\frac{1}{2}y}(a)J_{\frac{1}{4}-\frac{1}{2}y}(a)$
168	$(\cos x)^{-\frac{1}{2}}\cos(2a\cos x)$, $x<\pi/2$ 0, $x>\pi/2$ $a\leq\pi/4$ $N=\frac{1}{2}a^{\frac{1}{2}}\pi^{\frac{3}{2}}[J_{\frac{1}{4}}(a)]^2$	$\frac{1}{2}a^{\frac{1}{2}}\pi^{\frac{3}{2}}J_{-\frac{1}{4}+\frac{1}{2}y}(a)J_{-\frac{1}{4}-\frac{1}{2}y}(a)$
169	$(\sin x)^{-\frac{3}{2}}\sin(2a\sin x)$, $x<\pi$ 0, $x>\pi$ $a\leq\pi/2$ $N=2(a\pi)^{\frac{3}{2}}\{[J_{-\frac{1}{4}}(a)]^2+[J_{\frac{3}{4}}(a)]^2\}$	$2(a\pi)^{\frac{1}{2}}\cos(\frac{1}{2}\pi y)$ $\cdot[J_{-\frac{1}{4}-\frac{1}{2}y}(a)J_{-\frac{1}{4}+\frac{1}{2}y}(a)+J_{\frac{3}{4}-\frac{1}{2}y}(a)J_{\frac{3}{4}+\frac{1}{2}y}(a)]$
170	$(\cos\frac{1}{2}x)^{-\frac{3}{2}}\sin(2a\cos\frac{1}{2}x)$, $x<\pi$ 0, $x>\pi$ $a\leq\pi/2$ $N=2(a\pi)^{\frac{3}{2}}\{[J_{-\frac{1}{4}}(a)]^2+[J_{\frac{3}{4}}(a)]^2\}$	$2(a\pi)^{\frac{3}{2}}[J_{-\frac{1}{4}-y}(a)J_{-\frac{1}{4}+y}(a)$ $+J_{\frac{3}{4}+y}(a)J_{\frac{3}{4}-y}(a)]$
171	$(\cos x)^{-\frac{1}{2}}\exp(-\frac{1}{2}a^2\sec x)$, $x<\frac{1}{2}\pi$ 0, $x>\frac{1}{2}\pi$ $N=\frac{1}{2}\pi^{-1}a[K_{\frac{1}{4}}(\frac{1}{4}a^2)]^2$	$\pi^{\frac{1}{2}}D_{y-\frac{1}{2}}(a)D_{-y-\frac{1}{2}}(a)$

6. Inverse Trigonometric Functions

	$2Nf(x)$	$Ng(y)$
172	$\arctan(2a^2/x^2)$ $N=a\pi$	$\pi y^{-1}e^{-ay}\sin(ay)$
173	$\arcsin x,\quad x<1$ $0,\quad x>1$ $N=\frac{1}{2}\pi-1$	$\frac{1}{2}\pi y^{-1}[\sin y-\mathbf{H}_0(y)]$
174	$\arccos x,\quad x<1$ $0,\quad x>1$ $N=1$	$\frac{1}{2}\pi y^{-1}\mathbf{H}_0(y)$
175	$x^{-1}\arcsin x,\quad x<1$ $0,\quad x>1$ $N=\frac{1}{2}\pi\log 2$	$\frac{1}{2}\pi[\mathrm{Ci}(y)-\mathrm{Ji}_0(y)]$
176	$\arctan(a^nx^{-n}),\quad n=2, 4, 6, \ldots$ $N=\frac{1}{2}a\pi$ $\cdot\sum_{m=1}^{n}(-1)^{m+1}\cos[(m+\frac{1}{2})(\pi/n)]$	$-\frac{1}{2}\pi y^{-1}$ $\cdot\sum_{m=1}^{n}(-1)^m\exp\{-ay\sin[(m-\frac{1}{2})(\pi/n)]\}$ $\cdot\sin\{ay\cos[(m-\frac{1}{2})(\pi/n)]\}$

7. Hyperbolic Functions

	$2Nf(x)$	$Ng(y)$
177	$\mathrm{sech}(ax)$ $N=\frac{1}{2}a^{-1}\pi$	$\frac{1}{2}a^{-1}\pi\,\mathrm{sech}(\frac{1}{2}a^{-1}\pi y)$
178	$[\mathrm{sech}(ax)]^2$ $N=a^{-1}$	$\frac{1}{2}a^{-2}\pi y\,\mathrm{csch}(\frac{1}{2}a^{-1}\pi y)$
179	$[\mathrm{sech}(ax)]^3$ $N=\frac{1}{4}a^{-1}\pi$	$\frac{1}{4}a^{-3}\pi(a^2+y^2)\,\mathrm{sech}(\pi y/2a)$
180	$[\mathrm{sech}(ax)]^{2n},\quad n=2, 3, 4, \ldots$ $N=2^{2n-2}[a(2n-1)!]^{-1}$ $\cdot[(n-1)!]^2$	$2^{2n-1}a^{-2}\pi[(2n-1)!]^{-1}y\,\mathrm{csch}(\pi y/2a)$ $\prod_{m=1}^{n-1}[m^2+(v^2/4a^2)]$

	$2Nf(x)$	$Ng(y)$
181	$[\operatorname{sech}(ax)]^{2n+1}, \quad n=1,2,3,\ldots$ $N=2^{-2n-1}a^{-1}\pi(2n)!(n!)^{-2}$	$2^{2n-1}a^{-1}\pi[(2n)!]^{-1}\operatorname{sech}(\pi y/2a)$ $\cdot\prod_{m=1}^{n}[(m-\frac{1}{2})^2+(y^2/4a^2)]$
182	$x\operatorname{csch}(ax)$ $N=(\frac{1}{2}\pi/a)^2$	$\frac{1}{4}(\pi/a)^2[\operatorname{sech}(\frac{1}{2}\pi y/a)]^2$
183	$\cosh(ax)\operatorname{sech}(bx), \quad a<b$ $N=(\pi/2b)\sec(a\pi/2b)$	$(\pi/b)\cos(a\pi/2b)\cosh(\pi y/2b)$ $\cdot[\cos(a\pi/b)+\cosh(\pi y/b)]^{-1}$
184	$\sinh(ax)\operatorname{sech}(bx), \quad a<b$ $N=\frac{1}{2}b^{-1}\left[\pi\tan\left(\frac{a\pi}{2b}\right)\right.$ $\left.+\psi\left(\frac{3}{4}-\frac{1}{4}ab^{-1}\right)-\psi\left(\frac{3}{4}+\frac{a}{4b}\right)\right]$	$\frac{1}{4}b^{-1}\left\{\psi\left(\frac{3b-a+iy}{4b}\right)+\psi\left(\frac{3b-a-iy}{4b}\right)\right.$ $-\psi\left(\frac{3b+a+iy}{4b}\right)-\psi\left(\frac{3b+a-iy}{4b}\right)$ $+2\pi\sin\left(\frac{a\pi}{b}\right)$ $\left.\cdot\left[\cos\left(\frac{a\pi}{b}\right)+\cosh\left(\frac{\pi y}{b}\right)\right]^{-1}\right\}$
185	$\sinh(ax)\operatorname{csch}(bx), \quad a<b$ $N=\frac{1}{2}b^{-1}\pi\tan(a\pi/2b)$	$\frac{1}{2}b^{-1}\pi\sin(a\pi/b)$ $+[\cos(a\pi/b)+\cosh(\pi y/b)]^{-1}$
186	$1-\tanh(ax)$ $N=a^{-1}\log 2$	$\frac{1}{4}a^{-1}[\psi(iy/4a)+\psi(-iy/4a)$ $-\psi(\frac{1}{2}+i\frac{1}{4}a^{-1}y)-\psi(\frac{1}{2}-i\frac{1}{4}a^{-1}y)]$
187	$x^{s-1}(x^{-1}-\operatorname{csch}x), \quad -1<s<1$ $N=2\Gamma(s)(2^{-s}-1)$ $\cdot\zeta(s)$	$-(s-1)^{-1}\Gamma(s)$ $\cdot\{2^{-s}(s-1)[\zeta(s,\frac{1}{2}+i\frac{1}{2}y)+\zeta(s,\frac{1}{2}-i\frac{1}{2}y)]$ $-y^{1-s}\sin(\frac{1}{2}\pi s)\}$
188	$x^{-1}\sinh(ax)\operatorname{sech}(bx), \quad a<b$ $N=\frac{1}{2}\log\left[\frac{1+\sin(a\pi/2b)}{1-\sin(a\pi/2b)}\right]$	$\frac{1}{2}\log\left[\frac{\cosh(\pi y/2b)+\sin(a\pi/2b)}{\cosh(\pi y/2b)-\sin(a\pi/2b)}\right]$
189	$x^{-1}[\sinh(ax)]^2\operatorname{csch}(bx), \quad a<\frac{1}{2}b$ $N=-\frac{1}{2}\log[\cos(a\pi/b)]$	$\frac{1}{4}\log\left[\frac{1+\cosh(\pi y/b)}{\cosh(\pi y/b)+\cos(2a\pi/b)}\right]$

	$2Nf(x)$	$Ng(y)$
190	$x^{-1}(x^{-1} - \operatorname{csch} x)$ $N = \log 2$	$\log(1+e^{-\pi y})$
191	$(a^{i}+x^2)\operatorname{sech}(\pi x/2a)$ $N = 2a^3$	$2a^3[\operatorname{sech}(ay)]^3$
192	$x(a^2+x^2)\operatorname{csch}(\pi x/a)$ $N = \frac{3}{8}a^4$	$\frac{3}{8}a^4[\operatorname{sech}(\frac{1}{2}ay)]^4$
193	$(1+x^2)^{-1}\operatorname{sech}(\pi x)$ $N = 2-\frac{1}{2}\pi$	$2\cosh(\frac{1}{2}y) - e^{y}\arctan(e^{-\frac{1}{2}y})$ $- e^{-y}\arctan(e^{\frac{1}{2}y})$
194	$(1+x^2)^{-1}\operatorname{sech}(\frac{1}{4}\pi x)$ $N = 2^{-\frac{1}{2}}[\pi - 2\log(2^{\frac{1}{2}}+1)]$	$2^{-\frac{1}{2}}\{\pi e^{-y} + 2\sinh y \arctan(2^{-\frac{1}{2}}\operatorname{csch} y)$ $-(\cosh y)\log[(\cosh y + 2^{-\frac{1}{2}})$ $\cdot(\cosh y - 2^{-\frac{1}{2}})^{-1}]\}$
195	$(1+x^2)^{-1}\operatorname{sech}(\frac{1}{2}\pi x)$ $N = \log 2$	$ye^{-y} + \cosh y \log(1+e^{-2y})$
196	$x(1+x^2)^{-1}\operatorname{csch}(\pi x)$ $N = \frac{1}{2}(2\log 2 - 1)$	$\frac{1}{2}ye^{-y} - \frac{1}{2} + \cosh y \log(1+e^{-y})$
197	$(x^2+1)^{-1}\sinh(ax)\operatorname{csch}(\pi x), \quad a \leq \pi$ $N = \sin a \log(2\cos\frac{1}{2}a)$ $-\frac{1}{2}a\cos a$	$\frac{1}{2}e^{-y}(y\sin y - a\cos a)$ $+\frac{1}{2}\sin a \cosh y \log(1+2e^{-y}\cos a + e^{-2y})$ $-\cos a \sinh y \arctan[\sin a(e^{y}+\cos a)^{-1}]$
198	$(x^2+1)^{-1}\cosh(ax)\operatorname{sech}(\frac{1}{2}\pi x), \quad a \leq \frac{1}{2}\pi$ $N = a\sin a$ $+\cos a \log(2\cos a)$	$ye^{-y}\cos a + ae^{-y}\sin a$ $+\sin a \sinh y \arctan\left[\dfrac{e^{-2y}\sin(2a)}{1+e^{-2y}\cos(2a)}\right]$ $+\frac{1}{2}\cos a \cosh y$ $\cdot\log[1+2\cos(2a)e^{-2y}+e^{-4y}]$
199	$(1+x^2)^{-1}\sinh(ax)\operatorname{csch}(\frac{1}{2}\pi x), \quad a \leq \frac{1}{2}\pi$ $N = \frac{1}{2}\pi \sin a$ $-\cos a \log[\operatorname{ctn}(\frac{1}{4}\pi - a)]$	$\frac{1}{2}\pi \sin a e^{-y}$ $-\dfrac{1}{2}\cos a \cosh y \log\left[\dfrac{\cosh y + \sin a}{\cosh y - \sin a}\right]$ $+\sin a \sinh y \arctan(\cos a \operatorname{csch} y)$

	$2Nf(x)$	$Ng(y)$
200	$x^{-1}\operatorname{csch}(cx)[\cosh(ax)$ $-\cosh(bx)],\quad c>a>b$ $N=\log\left[\dfrac{\cos(\frac{1}{2}\pi b/c)}{\cos(\frac{1}{2}\pi a/c)}\right]$	$\dfrac{1}{2}\log\left[\dfrac{\cos(\pi b/c)+\cosh(\pi y/c)}{\cos(\pi a/c)+\cosh(\pi y/c)}\right]$
201	$[\cosh(ax)+\cos b]^{-1},\quad 0<b<\pi$ $N=a^{-1}b\csc b$	$a^{-1}\pi\operatorname{csch}\sinh(by/a)\operatorname{csch}(\pi y/a)$
202	$[\cosh(ax)+\cosh b]^{-1}$ $N=a^{-1}b\operatorname{csch} b$	$a^{-1}\pi\operatorname{csch} b\sin(by/a)\operatorname{csch}(\pi y/a)$
203	$\cosh(\frac{1}{2}ax)[\cosh(ax)+\cosh b]^{-1}$ $N=\frac{1}{2}a^{-1}\pi\operatorname{sech}(\frac{1}{2}b)$	$\frac{1}{2}a^{-1}\pi\operatorname{sech}(\frac{1}{2}b)\cosh(by/a)$ $\cdot\operatorname{sech}(\pi y/a)$
204	$\cosh(\frac{1}{2}ax)$ $\cdot[\cosh(ax)+\cos b]^{-1},\quad 0<b<\pi$ $N=\frac{1}{2}a^{-1}\pi\sec(\frac{1}{2}b)$	$\frac{1}{2}a^{-1}\pi\sec(\frac{1}{2}b)\cosh(a^{-1}by)$ $\cdot\operatorname{sech}(a^{-1}\pi y)$
205	$\{1+2\cosh[(\frac{2}{3}\pi)^{\frac{1}{2}}x]\}^{-1}$ $N=\frac{1}{3}(\frac{1}{2}\pi)^{\frac{1}{2}}$	$(\frac{1}{2}\pi)^{\frac{1}{2}}\{1+2\cosh[(\frac{2}{3}\pi)^{\frac{1}{2}}y]\}^{-1}$
206	$(\cosh x+\cos b)^{-\frac{1}{2}},\quad 0<b<\pi$ $N=2^{\frac{1}{2}}K(\sin\frac{1}{2}b)$	$2^{-\frac{1}{2}}\pi P_{-\frac{1}{2}+iy}(\cos b)\operatorname{sech}(\pi y)$
207	$(\cosh x+\cosh b)^{-\frac{1}{2}}$ $N=2^{\frac{1}{2}}\operatorname{sech}(\frac{1}{2}b)K(\tanh\frac{1}{2}b)$	$2^{-\frac{1}{2}}\pi\mathfrak{P}_{-\frac{1}{2}+iy}(\cosh b)\operatorname{sech}(\pi y)$
208	$(\cosh x-\cos b)^{-\frac{1}{2}},\quad 0<b<\pi$ $N=2^{\frac{1}{2}}K(\cos\frac{1}{2}b)$	$2^{-\frac{1}{2}}[Q_{-\frac{1}{2}+iy}(\cos b)+Q_{-\frac{1}{2}-iy}(\cos b)]$
209	$(\cosh a-\cosh x)^{-\frac{1}{2}},\quad x<a$ $0,\quad x>a$ $N=2^{\frac{1}{2}}\pi\operatorname{sech}(\frac{1}{2}a)K[\tanh(\frac{1}{2}a)]$	$2^{-\frac{1}{2}}\pi\mathfrak{P}_{-\frac{1}{2}+iy}(\cosh a)$
210	$0,\quad x<a$ $(\cosh x-\cosh a)^{-\frac{1}{2}},\quad x>a$ $N=2^{\frac{1}{2}}\operatorname{sech}(\frac{1}{2}a)K[\operatorname{sech}(\frac{1}{2}a)]$	$2^{-\frac{1}{2}}[\mathfrak{Q}_{-\frac{1}{2}+iy}(\cosh a)+\mathfrak{Q}_{-\frac{1}{2}-iy}(\cosh a)]$

	$2Nf(x)$	$Ng(y)$
211	$[\operatorname{sech}(ax)]^{\nu}, \quad \nu>0$ $N=\frac{1}{2}a^{-1}\pi^{\frac{1}{2}}\Gamma(\frac{1}{2}\nu)[\Gamma(\frac{1}{2}+\frac{1}{2}\nu)]^{-1}$	$\frac{1}{2}a^{-1}\pi^{\frac{1}{2}}[\Gamma(\frac{1}{2}\nu)\Gamma(\frac{1}{2}+\frac{1}{2}\nu)]^{-1}$ $\cdot\Gamma(\frac{1}{2}\nu+i\frac{1}{2}a^{-1}y)\Gamma[\frac{1}{2}\nu-i(y/2a)]$
212	$(\operatorname{csch} ax)^{\nu}, \quad 0<\nu<1$ $N=\frac{1}{2}a^{-1}\Gamma(\frac{1}{2}\nu)\Gamma(\frac{1}{2}-\frac{1}{2}\nu)$	$2^{\nu}a^{-1}\pi\sin(\frac{1}{2}\nu\pi)\Gamma(1-\nu)$ $\cdot[\Gamma(1-\frac{1}{2}\nu+\frac{1}{2}iy/a)\Gamma(1-\frac{1}{2}\nu-\frac{1}{2}iy/a)]^{-1}$ $\cdot\cosh(\pi y/2a)[\cosh(\pi y/a)-\cos(\nu\pi)]^{-1}$
213	$(\cosh a+\cosh x)^{-\nu}, \quad \nu>0$ $N=(\sinh a)^{-\nu}\mathfrak{Q}_{\nu-1}(\operatorname{ctnh} a)$	$(\pi/2)^{\frac{1}{2}}[\Gamma(\nu)]^{-1}(\sinh a)^{\frac{1}{2}-\nu}$ $\cdot\Gamma(\nu+iy)\Gamma(\nu-iy)\mathfrak{P}^{\frac{1}{2}-\nu}_{-\frac{1}{2}+iy}(\cosh a)$
214	$[\cos b+\cosh x]^{-\nu}, \quad \nu>0,$ $0<b<\pi$ $N=(\frac{1}{2}\pi)^{\frac{1}{2}}\Gamma(\nu)(\sin b)^{\frac{1}{2}-\nu}P^{\frac{1}{2}-\nu}_{-\frac{1}{2}}(\cos b)$	$(\frac{1}{2}\pi)^{\frac{1}{2}}[\Gamma(\nu)]^{-1}(\sin b)^{\frac{1}{2}-\nu}$ $\cdot\Gamma(\nu+iy)\Gamma(\nu-iy)\cdot P^{\frac{1}{2}-\nu}_{-\frac{1}{2}+iy}(\cos b)$
215	$(\cosh x-\cosh a)^{-\nu}, \quad x>a$ $0, \quad x<a$ $0<\nu<1$ $N=i(2\pi)^{\frac{1}{2}}e^{-i\pi\nu}(\sinh a)^{\frac{1}{2}-\nu}\Gamma(1-\nu)$ $\cdot Q^{\nu-\frac{1}{2}}_{-\frac{1}{2}}(\cosh a)$	$i(2\pi)^{-\frac{1}{2}}e^{-i\nu\pi}\Gamma(1-\nu)(\sinh a)^{\frac{1}{2}-\nu}$ $\cdot[\mathfrak{Q}^{\nu-\frac{1}{2}}_{-\frac{1}{2}-iy}(\cosh a)+\mathfrak{Q}^{\nu-\frac{1}{2}}_{-\frac{1}{2}+iy}(\cosh a)]$
216	$(\cosh a-\cosh x)^{-\nu}, \quad x<a$ $0, \quad x>a$ $\nu<1$ $N=(\sinh a)^{-\nu}\mathfrak{Q}_{-\nu}(\operatorname{ctnh} a)$	$(\frac{1}{2}\pi)^{\frac{1}{2}}\Gamma(1-\nu)(\sinh a)^{\frac{1}{2}-\nu}$ $\cdot\mathfrak{P}^{\nu-\frac{1}{2}}_{-\frac{1}{2}+iy}(\cosh a)$
217	$\sinh[b(a^2-x^2)^{\frac{1}{2}}], \quad x<a$ $0, \quad x>a$ $N=\frac{1}{2}a\pi I_1(ab)$	$\frac{1}{2}ab\pi(y^2-b^2)^{-\frac{1}{2}}J_1[a(y^2-b^2)^{\frac{1}{2}}], \quad y>b$ $\frac{1}{2}\pi ab(b^2-y^2)^{-\frac{1}{2}}I_1[a(b^2-y^2)^{\frac{1}{2}}], \quad y<b$
218	$(a^2-x^2)^{-\frac{1}{2}}\cosh[b(a^2-x^2)^{\frac{1}{2}}], \quad x<a$ $0, \quad x>a$ $N=\frac{1}{2}\pi I_0(ab)$	$\frac{1}{2}\pi J_0[a(y^2-b^2)^{\frac{1}{2}}], \quad y>b$ $\frac{1}{2}\pi I_0[a(b^2-y^2)^{\frac{1}{2}}], \quad y<b$
219	$(a^2-x^2)^{-\frac{3}{4}}\sinh[b(a^2-x^2)^{\frac{1}{2}}], \quad x<a$ $0, \quad x>a$ $N=b^{\frac{1}{2}}(\frac{1}{2}\pi)^{\frac{3}{2}}[I_{\frac{1}{4}}(\frac{1}{2}ab)]^2$	$2^{-\frac{3}{2}}\pi^{\frac{3}{2}}b^{\frac{1}{2}}J_{\frac{1}{4}}\{\frac{1}{2}a[y-(y^2-b^2)^{\frac{1}{2}}]\}$ $\cdot J_{\frac{1}{4}}\{\frac{1}{2}a[y+(y^2-b^2)^{\frac{1}{2}}]\}$
220	$(a^2-x^2)^{-\frac{3}{4}}\cosh[b(a^2-x^2)^{\frac{1}{2}}], \quad x<a$ $0, \quad x>a$ $N=b^{\frac{1}{2}}(\frac{1}{2}\pi)^{\frac{3}{2}}[I_{-\frac{1}{4}}(\frac{1}{2}ab)]^2$	$b^{\frac{1}{2}}(\frac{1}{2}\pi)^{\frac{3}{2}}J_{-\frac{1}{4}}\{\frac{1}{2}a[y-(y^2-b^2)^{\frac{1}{2}}]\}$ $\cdot J_{-\frac{1}{4}}\{\frac{1}{2}a[y+(y^2-b^2)^{\frac{1}{2}}]\}$

	$2Nf(x)$	$Ng(y)$
221	$x^{-\frac{1}{2}}(a^2-x^2)^{-\frac{1}{2}}\cosh[b(a^2-x^2)^{\frac{1}{2}}],\quad x<a$ $0,\quad x>a$ $N=2^{-\frac{3}{4}}(b/a)^{\frac{1}{4}}\pi^{\frac{3}{2}}[\Gamma(\frac{3}{4})]^{-1}I_{-\frac{1}{4}}(ab)$	$(\frac{1}{2}\pi)^{\frac{3}{2}}y^{\frac{1}{2}}I_{-\frac{1}{4}}\{\frac{1}{2}a[b-(b^2-y^2)^{\frac{1}{2}}]\}$ $\cdot I_{-\frac{1}{4}}\{\frac{1}{2}a[b+(b^2-y^2)^{\frac{1}{2}}]\}$
222	$\exp(-a\cosh x)\cosh(b\sinh x),\quad a>b$ $N=K_0[(a^2-b^2)^{\frac{1}{2}}]$	$K_{iy}[(a^2-b^2)^{\frac{1}{2}}]\cos[y\tanh^{-1}(b/a)]$
223	$x^{-1}e^{-ax}\sinh(bx)$ $N=\frac{1}{2}\log[(a+b)/(a-b)]$	$\dfrac{1}{4}\log\left[\dfrac{y^2+(a+b)^2}{y^2+(a-b)^2}\right]$
224	$(e^{bx}+1)^{-1}\sinh(ax),\quad a<b$ $N=\frac{1}{2}b^{-1}\pi\csc(ab^{-1}\pi)$ $-\frac{1}{2}a^{-1}$	$b^{-1}\pi\sin(ab^{-1}\pi)\cosh(b^{-1}\pi y)$ $\cdot[\cos(2b^{-1}\pi y)-\cos(2ab^{-1}\pi)]^{-1}$ $-\frac{1}{2}a(a^2+y^2)^{-1}$
225	$(e^{bx}-1)^{-1}\sinh(ax),\quad a<b$ $N=\frac{1}{2}[b^{-1}\pi\operatorname{ctn}(a\pi/b)+a^{-1}]$	$\frac{1}{2}b^{-1}\pi\sin(2ab^{-1}\pi)[\cosh(2b^{-1}\pi y)$ $-\cos(2ab^{-1}\pi)]^{-1}+\frac{1}{2}a(a^2+y^2)^{-1}$
226	$e^{-ax}[\sinh(bx)]^{\nu},\quad \nu>-1,\quad b\nu<a$ $N=2^{-\nu-1}b^{-1}\Gamma(\nu+1)$ $\cdot\Gamma(\frac{1}{2}ab^{-1}-\frac{1}{2}\nu)$ $\cdot[\Gamma(1+\frac{1}{2}\nu+\frac{1}{2}ab^{-1})]^{-1}$	$2^{-\nu-2}b^{-1}\Gamma(\nu+1)\left\{\dfrac{\Gamma[\frac{1}{2}b^{-1}(a-b\nu-iy)]}{\Gamma[\frac{1}{2}b^{-1}(a+b\nu-iy)+1]}\right.$ $\left.+\dfrac{\Gamma[\frac{1}{2}b^{-1}(a-b\nu+iy)]}{\Gamma[\frac{1}{2}b^{-1}(a+b\nu+iy)+1]}\right\}$
227	$\exp(-bx^2)\cosh(ax)$ $N=\frac{1}{2}(\pi/b)^{\frac{1}{2}}\exp(-\frac{1}{4}a^2/b)$	$\frac{1}{2}(\pi/b)^{\frac{1}{2}}\cos(\frac{1}{2}ay/b)\exp[\frac{1}{4}(a^2-y^2)/b]$
228	$x^{-2}\exp(-x^2)\sinh(x^2)$ $N=(\frac{1}{2}\pi)^{\frac{1}{2}}$	$(\frac{1}{2}\pi)^{\frac{1}{2}}\{\exp(-\frac{1}{8}y^2)-2^{-\frac{3}{2}}\pi^{\frac{1}{2}}y\operatorname{Erfc}(2^{-\frac{3}{2}}y)\}$
229	$x^{-9/4}\exp(-a^2x^2)\sinh(a^2x^2)$ $N=2^{\frac{1}{2}}a\pi^{-\frac{1}{2}}$	$2^{\frac{1}{2}}a\pi^{-\frac{1}{2}}\{\exp(-\frac{1}{8}a^{-2}y^2)$ $-2^{-\frac{1}{2}}a^{-1}y\exp(-y^2/16a^2)\operatorname{Erfc}(2^{-\frac{3}{2}}a^{-1}y)\}$
230	$\exp(-bx)\sinh(ax^{\frac{1}{2}})$ $N=\frac{1}{2}ab^{\frac{3}{2}}\pi^{\frac{1}{2}}\exp(-a^2/4b)$	$\frac{1}{2}a\pi^{\frac{1}{2}}(b^2+y^2)^{\frac{3}{4}}\exp\{a^2b[4(b^2+y^2)]^{-1}\}$ $\cdot\cos[\frac{3}{2}\arctan(b^{-1}y)-\frac{1}{4}a^2y(b^2+y^2)^{-1}]$
231	$e^{-ax}\begin{cases}\operatorname{ctnh}(bx^{\frac{1}{2}})\\ \tanh(bx^{\frac{1}{2}})\end{cases}$	*See* Mordell, L. J. (1920). *Messenger Math.* **49**, 65–72.
232	$x^{-\frac{1}{2}}\operatorname{sech}(ax^{\frac{1}{2}})$ $N=a^{-1}\pi$	*See* Mordell, L. J. (1939). *Acta Math.* **61**, 323–360.

	$2Nf(x)$	$Ng(y)$
233	$(a^2+x^2)^{-\frac{1}{2}}\sinh[b(a^2+x^2)^{\frac{1}{2}}]$ $\cdot\operatorname{sech}[c(a^2+x^2)^{\frac{1}{2}}],\quad b<c$ $N=\pi c^{-1}\sum_{n=0}^{\infty}\{(-1)^n\sin[(n+\frac{1}{2})b\pi/c]$ $\cdot[(n+\frac{1}{2})^2(\pi^2/c^2)+a^2]^{-\frac{1}{2}}\}$	$\pi c^{-1}\sum_{n=0}^{\infty}\{(-1)^n\sin[(n+\frac{1}{2})(b/c)\pi]$ $\cdot[(n+\frac{1}{2})^2(\pi^2/c^2)+a^2]^{-\frac{1}{2}}$ $\cdot\exp(-y[(n+\frac{1}{2})^2(\pi^2/c^2)+a^2]^{\frac{1}{2}})\}$
234	$(a^2+x^2)^{-\frac{1}{2}}\cos[b(a^2+x^2)^{\frac{1}{2}}]$ $\operatorname{csch}[c(a^2+x^2)^{\frac{1}{2}}],\quad b<c$ $N=\frac{1}{2}\pi c^{-1}\sum_{n=0}^{\infty}\{(-1)^n\epsilon_n$ $\cdot[a^2+(n^2\pi^2/c^2)]^{-\frac{1}{2}}\cos[n\pi(b/c)]\}$	$\frac{1}{2}\pi c^{-1}\sum_{n=0}^{\infty}\{(-1)^n\epsilon_n(a^2+c^{-2}n^2\pi^2)^{-\frac{1}{2}}$ $\cdot\cos[(b/c)n\pi]\exp(-y[a^2+c^{-2}n^2\pi^2]^{\frac{1}{2}})\}$
235	$\cosh[b(a^2+x^2)^{\frac{1}{2}}]$ $\cdot\operatorname{sech}[c(a^2+x^2)^{\frac{1}{2}}],\quad b<c$ $N=c^{-2}\pi^2\sum_{n=0}^{\infty}\{(-1)^n(n+\frac{1}{2})$ $\cdot[a^2+(n+\frac{1}{2})^2(\pi^2/c^2)]^{-\frac{1}{2}}$ $\cdot\cos[(b/c)(n+\frac{1}{2})\pi]\}$	$c^{-2}\pi^2\sum_{n=0}^{\infty}\{(-1)^n(n+\frac{1}{2})[a^2+(n+\frac{1}{2})^2$ $\cdot(\pi^2/c^2)]^{-\frac{1}{2}}\cos[(n+\frac{1}{2})\pi(b/c)]$ $\cdot\exp(-y[a^2+(n+\frac{1}{2})^2(\pi^2/c^2)]^{\frac{1}{2}})\}$
236	$-\log(1-e^{-2x})\cosh x$ $N=1$	$(1-y^2)(1+y^2)^{-2}$ $+\frac{1}{2}\pi y(1+y^2)^{-1}\tanh(\frac{1}{2}\pi y)$
237	$[\cosh(ax)]^{-\nu}\log[\cosh(ax)],\quad \nu>0$ $N=2^{\nu-3}a^{-1}B(\frac{1}{2}\nu,\frac{1}{2}\nu)[\psi(\frac{1}{2}+\frac{1}{2}\nu)-\psi(\frac{1}{2}\nu)]$	$2^{\nu-2}[a\Gamma(\nu)]^{-1}\,\lvert\Gamma[\frac{1}{2}\nu+i\frac{1}{2}(y/a)]\rvert^2$ $\cdot(\psi(\nu)-\log 2-\operatorname{Re}\{\psi[\frac{1}{2}\nu+i\frac{1}{2}(y/a)]\})$
238	$\log(1+\cos a\operatorname{sech}x),\quad a<\pi$ $N=\frac{1}{8}\pi^2-\frac{1}{2}a^2$	$\pi y^{-1}\operatorname{csch}(\pi y)[\cosh(\frac{1}{2}\pi y)-\cosh(ay)]$
239	$\cosh x\log(2\cosh x)-x\sinh x$ $N=\frac{1}{2}\pi$	$\frac{1}{2}\pi(1+y^2)^{-1}\operatorname{sech}(\frac{1}{2}\pi y)$
240	$\log[\operatorname{ctnh}(ax)]$ $N=\frac{1}{8}\pi^2a^{-1}$	$\frac{1}{2}\pi y^{-1}\tanh(\frac{1}{4}a^{-1}\pi y)$
241	$\log(1+a^{-2}\operatorname{sech}^2x)$ $N=\{\log[(1+a^{-2})^{\frac{1}{2}}+a^{-1}]\}^2$	$2\pi y^{-1}\operatorname{csch}(\frac{1}{2}\pi y)$ $\cdot\sin^2\{\frac{1}{2}y\log[(1+a^{-2})^{\frac{1}{2}}+a^{-1}]\}$

	$2Nf(x)$	$Ng(y)$
242	$-\log(1-a^2\,\mathrm{sech}^2 x), \quad a\le 1$ $N=(\arcsin a)^2$	$2\pi y^{-1}\,\mathrm{csch}(\frac{1}{2}\pi y)$ $\cdot\sinh^2(\frac{1}{2}y\arcsin a)$
243	$\log[(b+a\,\mathrm{sech}x)(b-a\,\mathrm{sech}x)^{-1}]$ $N=\pi\arcsin(a/b)$	$\pi y^{-1}\,\mathrm{sech}(\frac{1}{2}\pi y)\sinh[y\arcsin(a/b)]$
244	$(\cosh x+\cos\delta)^{-\frac{1}{2}}$ $\cdot\log(\cosh x+\cos\delta), \quad 0<\delta<\pi$ $N=2^{-\frac{1}{2}}\pi[\log(\sin\delta)$ $\cdot 2\pi^{-1}K(\sin\frac{1}{2}\delta)+K(\cos\frac{1}{2}\delta)]$	$2^{-\frac{1}{2}}\pi\,\mathrm{sech}(\pi y)\{P_{-\frac{1}{2}+iy}(\cos\delta)$ $\cdot[-\gamma+\log(\sin\delta)-\log 4$ $-\frac{1}{2}\psi(\frac{1}{2}+iy)-\frac{1}{2}\psi(\frac{1}{2}-iy)]$ $+\frac{1}{2}Q_{-\frac{1}{2}+iy}(\cos\delta)+\frac{1}{2}Q_{-\frac{1}{2}-iy}(\cos\delta)\}$
245	$(\cosh x+\cosh a)^{-\frac{1}{2}}\log(\cosh x+\cosh a)$ $N=2^{-\frac{1}{2}}\pi[\mathrm{sech}(\frac{1}{2}a)K(\mathrm{sech}\frac{1}{2}a)$ $+\log(\sinh a)$ $\cdot(2/\pi)\,\mathrm{sech}(\frac{1}{2}a)K(\tanh\frac{1}{2}a)]$	$2^{-\frac{1}{2}}\pi\,\mathrm{sech}(\pi y)\{\mathfrak{P}_{-\frac{1}{2}+iy}(\cosh a)$ $\cdot[-\gamma-\log 4+\log\sinh a$ $-\frac{1}{2}\psi(\frac{1}{2}+iy)-\frac{1}{2}\psi(\frac{1}{2}-iy)]$ $+\frac{1}{2}\mathfrak{Q}_{-\frac{1}{2}+iy}(\cosh a)+\frac{1}{2}\mathfrak{Q}_{-\frac{1}{2}-iy}(\cosh a)\}$
246	$-(\cosh a-\cosh x)^{-\frac{1}{2}}$ $\cdot\log(\cosh a-\cosh x), \quad x<a$ $0, \quad x>a$ $\cosh a\le 2$ $N=2^{-\frac{1}{2}}\pi\{2\pi^{-1}\log(\sinh a)$ $\cdot\mathrm{sech}(\frac{1}{2}a)K[\tanh(a\frac{1}{2})]$ $-\mathrm{sech}(\frac{1}{2}a)K[\mathrm{sech}(\frac{1}{2}a)]\}$	$-2^{-\frac{1}{2}}\pi\,\mathrm{sech}(\pi y)\{\mathfrak{P}_{-\frac{1}{2}+iy}(\cosh a)$ $\cdot[\log(\mathrm{sech}a)-\gamma-\log 4$ $-\frac{1}{2}\psi(\frac{1}{2}+iy)-\frac{1}{2}\psi(\frac{1}{2}-iy)]$ $-\frac{1}{2}\mathfrak{Q}_{-\frac{1}{2}+iy}(\cosh a)-\frac{1}{2}\mathfrak{Q}_{-\frac{1}{2}-iy}(\cosh a)\}$
247	$\log\left[\dfrac{\cosh(ax)+\sin b}{\cosh(ax)-\sin b}\right], \quad b\le\frac{1}{2}\pi$ $N=\pi a^{-1}b$	$\pi y^{-1}\sinh(by/a)\,\mathrm{sech}(\frac{1}{2}\pi y/a)$
248	$\log\left[\dfrac{\cosh(ax)+\cos b}{\cosh(ax)+\cos c}\right]$ $N=\frac{1}{2}(c^2-b^2)/a, \quad c>b, \quad c,b\le\pi$	$\pi y^{-1}\,\mathrm{csch}(\pi y/a)$ $\cdot[\cosh(cy/a)-\cosh(by/a)]$
249	$(\cosh x-a)^{-\frac{1}{2}}$ $\cdot\log\left[\dfrac{(\cosh x+1)^{\frac{1}{2}}+(\cosh x-a)^{\frac{1}{2}}}{(\cosh x+1)^{\frac{1}{2}}-(\cosh x-a)^{\frac{1}{2}}}\right],$ $a<1$ $N=2^{\frac{1}{2}}\pi K[(\frac{1}{2}-\frac{1}{2}a)^{\frac{1}{2}}]$	$2^{-\frac{1}{2}}\pi^2[\mathrm{sech}(\pi y)]^2P_{-\frac{1}{2}+iy}(a)$

	$2Nf(x)$	$Ng(y)$
250	$\exp(-a \sinh x)$ $N=\frac{1}{2}\pi[\mathbf{H}_0(a)-Y_0(a)]$	$S_{0,iy}(a)=-i\frac{1}{2}\pi \operatorname{csch}(\pi y)$ $\cdot[J_{iy}(a)-J_{-iy}(a)-\mathbf{J}_{iy}(a)+\mathbf{J}_{-iy}(a)]$
251	$\exp(-a \cosh x)$ $N=K_0(a)$	$K_{iy}(a)$
252	$(\sinh x)^{-\frac{1}{2}} \exp(-2a \sinh x)$ $N=\frac{1}{2}\pi(a\pi)^{\frac{1}{2}}$ $\cdot[J_{\frac{1}{4}}(a)\, Y_{-\frac{1}{4}}(a)-J_{-\frac{1}{4}}(a)\, Y_{\frac{1}{4}}(a)]$	$\frac{1}{4}\pi(a\pi)^{\frac{1}{2}}[J_{\frac{1}{4}-i\frac{1}{2}y}(a)\, Y_{-\frac{1}{4}-i\frac{1}{2}y}(a)$ $-J_{-\frac{1}{4}+i\frac{1}{2}y}(a)\, Y_{\frac{1}{4}+\frac{1}{2}y}(a)$ $+J_{\frac{1}{4}+i\frac{1}{2}y}(a)\, Y_{-\frac{1}{4}+i\frac{1}{2}y}(a)-J_{-\frac{1}{4}-i\frac{1}{2}y}(a)$ $\cdot Y_{\frac{1}{4}-i\frac{1}{2}y}(a)]$
253	$(\cosh x)^{-\frac{1}{2}} \exp(-2a \cosh x)$ $N=(a/\pi)^{\frac{1}{2}}[K_{\frac{1}{4}}(a)]^2$	$(a/\pi)^{\frac{1}{2}}K_{\frac{1}{4}+i\frac{1}{2}y}(a)\, K_{\frac{1}{4}-i\frac{1}{2}y}(a)$
254	$(\cosh x)^{-\frac{3}{2}} \exp(-2a \cosh x)$ $N=2\pi^{-\frac{1}{2}}a^{\frac{3}{2}}\{[K_{\frac{3}{4}}(a)]^2-[K_{\frac{1}{4}}(a)]^2\}$	$2\pi^{-\frac{1}{2}}a^{\frac{3}{2}}[K_{\frac{3}{4}+i\frac{1}{2}y}(a)\, K_{\frac{3}{4}-i\frac{1}{2}y}(a)$ $-K_{\frac{1}{4}+i\frac{1}{2}y}(a)\, K_{\frac{1}{4}-i\frac{1}{2}y}(a)]$
255	$[\operatorname{sech}(ax)]^\nu \exp\{-b[\operatorname{sech}(ax)]^2\}$ $N=2^{\nu-2}a^{-1}B(\frac{1}{2}\nu, \frac{1}{2}\nu)$ $\cdot{}_1F_1(\frac{1}{2}\nu; \frac{1}{2}+\frac{1}{2}\nu; -b)$	$2^{\nu-2}[a\Gamma(\nu)]^{-1}\,\lvert\,\Gamma[\frac{1}{2}\nu+i\frac{1}{2}(y/a)]\,\rvert^2$ $\cdot{}_2F_2[\frac{1}{2}\nu+i\frac{1}{2}(y/a), \frac{1}{2}\nu-i\frac{1}{2}(y/a); \frac{1}{2}\nu,$ $\frac{1}{2}+\frac{1}{2}\nu; -b]$
256	$(\sinh x)^{-\frac{1}{2}} \exp(-a \operatorname{csch} x)$ $N=a(\frac{1}{2}\pi)^{\frac{3}{2}}\{[J_{\frac{1}{4}}(\frac{1}{2}a)]^2+[Y_{\frac{1}{4}}(\frac{1}{2}a)]^2\}$	$2^{\frac{1}{2}} \operatorname{Re}\{\Gamma(\frac{1}{2}+iy)\, D_{-\frac{1}{2}-iy}[(2ia)^{\frac{1}{2}}]$ $\cdot D_{-\frac{1}{2}-iy}[(-2ia)^{\frac{1}{2}}]\}$
257	$\sinh(a \cos x), \quad x<\frac{1}{2}\pi$ $0, \quad x>\frac{1}{2}\pi$ $N=\frac{1}{2}\pi\mathbf{L}_0(a)$	$-\frac{1}{4}i\pi \csc(\frac{1}{2}\pi y)[\mathbf{J}_y(ia)-\mathbf{J}_{-y}(ia)]$
258	$\sinh(a \sin x), \quad x<\pi$ $0, \quad x>\pi$ $N=\pi\mathbf{L}_0(a)$	$-\frac{1}{2}i\pi \operatorname{ctn}(\frac{1}{2}\pi y)[\mathbf{J}_y(ia)-\mathbf{J}_{-y}(ia)]$
259	$\cosh(a \sin x), \quad x<\pi$ $0, \quad x>\pi$ $N=\pi I_0(a)$	$\frac{1}{2}\pi[\mathbf{J}_y(ia)+\mathbf{J}_{-y}(ia)]$
260	$\cosh(a \cos x), \quad x<\frac{1}{2}\pi$ $0, \quad x>\frac{1}{2}\pi$ $N=\frac{1}{2}\pi I_0(a)$	$\frac{1}{4}\pi \operatorname{sech}(\frac{1}{2}\pi y)[\mathbf{J}_y(ia)+\mathbf{J}_{-y}(ia)]$

	$2Nf(x)$	$Ng(y)$
261	$(\sin x)^{-\frac{1}{2}}\sinh(a\sin x),\quad x<\pi$ $0,\quad x>\pi$ $N=(\frac{1}{2}a)^{\frac{1}{2}}\pi^{-\frac{3}{2}}[I_{\frac{1}{4}}(\frac{1}{2}a)]^2$	$(\frac{1}{2}a)^{\frac{1}{2}}\pi^{\frac{3}{2}}\cos(\frac{1}{2}\pi y)I_{\frac{1}{4}+\frac{1}{2}y}(\frac{1}{2}a)$ $\cdot I_{\frac{1}{4}-\frac{1}{2}y}(\frac{1}{2}a)$
262	$(\sin x)^{-\frac{1}{2}}\cosh(a\sin x),\quad x<\pi$ $0,\quad x>\pi$ $N=(\frac{1}{2}a)^{\frac{1}{2}}\pi^{\frac{3}{2}}[I_{-\frac{1}{4}}(\frac{1}{2}a)]^2$	$(\frac{1}{2}a)^{\frac{1}{2}}\pi^{\frac{3}{2}}\cos(\frac{1}{2}\pi y)$ $\cdot I_{-\frac{1}{4}-\frac{1}{2}y}(\frac{1}{2}a)I_{-\frac{1}{4}+\frac{1}{2}y}(\frac{1}{2}a)$
263	$(\sec x)^{\frac{1}{2}}\sinh(a\cos x),\quad x<\frac{1}{2}\pi$ $0,\quad x>\frac{1}{2}\pi$ $N=a^{\frac{1}{2}}(\frac{1}{2}\pi)^{\frac{3}{2}}[I_{\frac{1}{4}}(\frac{1}{2}a)]^2$	$a^{\frac{1}{2}}(\frac{1}{2}\pi)^{\frac{3}{2}}I_{\frac{1}{4}+\frac{1}{2}y}(\frac{1}{2}a)I_{\frac{1}{4}-\frac{1}{2}y}(\frac{1}{2}a)$
264	$(\sec x)^{\frac{1}{2}}\cosh(a\cos x),\quad x<\frac{1}{2}\pi$ $0,\quad x>\frac{1}{2}\pi$ $N=a^{\frac{1}{2}}(\frac{1}{2}\pi)^{\frac{3}{2}}[I_{-\frac{1}{4}}(\frac{1}{2}a)]^2$	$a^{\frac{1}{2}}(\frac{1}{2}\pi)^{\frac{3}{2}}I_{-\frac{1}{4}+\frac{1}{2}y}(\frac{1}{2}a)$ $\cdot I_{-\frac{1}{4}-\frac{1}{2}y}(\frac{1}{2}a)$
265	$(\csc x)^{\frac{3}{2}}\sinh(2a\sin x),\quad x<\pi$ $0,\quad x>\pi$ $N=2(a\pi)^{\frac{3}{2}}\{[I_{-\frac{1}{4}}(a)]^2-[I_{\frac{3}{4}}(a)]^2\}$	$2(a\pi)^{\frac{3}{2}}\cos(\frac{1}{2}\pi y)$ $\cdot[I_{-\frac{1}{4}-\frac{1}{2}y}(a)I_{-\frac{1}{4}+\frac{1}{2}y}(a)$ $-I_{\frac{3}{4}+\frac{1}{2}y}(a)I_{\frac{3}{4}-\frac{1}{2}y}(a)]$
266	$(\sec x)^{\frac{3}{2}}\sinh(2a\cos\frac{1}{2}x),\quad x<\pi$ $0,\quad x>\pi$ $N=2(a\pi)^{\frac{3}{2}}\{[I_{-\frac{1}{4}}(a)]^2-[I_{\frac{3}{4}}(a)]^2\}$	$2(a\pi)^{\frac{3}{2}}\{I_{-\frac{1}{4}-y}(a)I_{-\frac{1}{4}+y}(a)$ $-I_{\frac{3}{4}-y}(a)I_{\frac{3}{4}+y}(a)\}$
267	$[\operatorname{sech}(ax)]^{\nu}$ $\cdot\cosh[b\operatorname{sech}(ax)],\quad \nu>0$ $N=2^{\nu-2}a^{-1}B(\frac{1}{2}\nu,\frac{1}{2}\nu)$ $\cdot{}_1F_2(\frac{1}{2}\nu;\frac{1}{2},\frac{1}{2}+\frac{1}{2}\nu;\frac{1}{4}b^2)$	$2^{\nu-2}[a\Gamma(\nu)]^{-1}\lvert\Gamma(\frac{1}{2}\nu+i\frac{1}{2}y/a)\rvert^2$ $\cdot{}_2F_3(\frac{1}{2}\nu+i\frac{1}{2}y/a,\frac{1}{2}\nu-i\frac{1}{2}y/a;$ $\frac{1}{2},\frac{1}{2}\nu,\frac{1}{2}+\frac{1}{2}\nu;\frac{1}{4}b^2)$
268	$[\operatorname{sech}(ax)]^{\nu}$ $\cdot\sinh[b\operatorname{sech}(ax)],\quad \nu>0$ $N=2^{\nu-1}a^{-1}bB(\frac{1}{2}+\frac{1}{2}\nu,\frac{1}{2}+\frac{1}{2}\nu)$ $\cdot{}_1F_2(\frac{1}{2}+\frac{1}{2}\nu;\frac{3}{2},1+\frac{1}{2}\nu;\frac{1}{4}b^2)$	$2^{\nu-1}b[a\Gamma(1+\nu)]^{-1}\lvert\Gamma(\frac{1}{2}+\frac{1}{2}\nu+i\frac{1}{2}y/a)\rvert^2$ $\cdot{}_2F_3(\frac{1}{2}+\frac{1}{2}\nu+i\frac{1}{2}y/a,\frac{1}{2}+\frac{1}{2}\nu-i\frac{1}{2}y/a;$ $\frac{3}{2},\frac{1}{2}+\frac{1}{2}\nu,1+\frac{1}{2}\nu;\frac{1}{4}b^2)$
269	$\arctan[\sinh a\operatorname{sech}(bx)]$ $N=\frac{1}{2}\pi a/b$	$\frac{1}{2}\pi y^{-1}\sin(ay/b)\operatorname{sech}(\frac{1}{2}\pi y/b)$
270	$(1+c^2\sinh^2x)^{-\frac{1}{2}\nu}$ $\cdot\cosh[\nu\arctan(c\sinh x)],\quad 0<\nu\leq 1$ $N=(\frac{1}{2}\pi/c)^{\frac{1}{2}}\Gamma(\nu)$ $\cdot(1-c^2)^{\frac{1}{4}-\frac{1}{2}\nu}P^{\frac{1}{2}-\nu}_{-\frac{1}{2}}(c^{-1})$	$(\frac{1}{2}\pi/c)^{\frac{1}{2}}[\Gamma(\nu)]^{-1}\cosh(\frac{1}{2}\pi y)$ $\cdot\lvert\Gamma(\nu+iy)\rvert^2(1-c^2)^{\frac{1}{4}-\frac{1}{2}\nu}P^{\frac{1}{2}-\nu}_{-\frac{1}{2}+iy}(c^{-1})$

8. Gamma Functions (Including Incomplete Gamma Functions) and Related Functions

	$2Nf(x)$	$Ng(y)$
271	$\vert\Gamma(a+ix)\vert^2, \quad a>0$ $N=2^{-2a}\pi\Gamma(2a)$	$2^{-2a}\pi\Gamma(2a)[\operatorname{sech}(\frac{1}{2}y)]^{2a}$
272	$\vert\Gamma(a+ibx)\Gamma(\frac{1}{2}-a+ibx)\vert^2, \quad 0<a<\frac{1}{2}$ $N=[b\pi\sin(2a\pi)]^{-1}$	$[b\pi\sin(2a\pi)]^{-1}\mathfrak{P}_{2a-1}[\cosh(\frac{1}{2}y/b)]$
273	$\vert\Gamma(\frac{1}{4}+ibx)\vert^4$ $N=(b\pi)^{-1}$	$2b^{-1}\pi^{-2}\operatorname{sech}(\frac{1}{4}y/b)K[\tanh(\frac{1}{4}y/b)]$
274	$\vert\Gamma(b+iax)\Gamma(c+iax)\vert^2$ $N=2^{1-2b-2c}\pi^{\frac{3}{2}}a^{-1}\Gamma(2b)$ $\cdot\Gamma(2c)\Gamma(b+c)[\Gamma(\frac{1}{2}+b+c)]^{-1}$	$2^{\frac{1}{2}-b-c}a^{-1}\pi^{\frac{3}{2}}\Gamma(2b)\Gamma(2c)$ $\cdot\Gamma(b+c)[\sinh(\frac{1}{2}y/a)]^{\frac{1}{2}-b-c}$ $\cdot\mathfrak{P}_{b-c-\frac{1}{2}}^{\frac{1}{2}-b-c}[\cosh(\frac{1}{2}y/a)]$
275	$\vert\Gamma(a+icx)\vert^2\vert\Gamma(b+icx)\vert^{-2}, \quad b-a>\frac{1}{2}$ $N=2^{2b-2a-2}c^{-1}\Gamma(2a)$ $\cdot B(b-a-\frac{1}{2},\frac{1}{2})[\Gamma(2b-1)]^{-1}$	$\frac{1}{2}(2\pi)^{\frac{1}{2}}2^{b-a}c^{-1}[\Gamma(b-a)]^{-1}$ $\cdot[\sinh(\frac{1}{2}y/c)]^{b-a-\frac{1}{2}}$ $\cdot e^{-i\pi(a-b+\frac{1}{2})}\mathfrak{Q}_{a+b-\frac{3}{2}}^{a-b+\frac{1}{2}}[\cosh(\frac{1}{2}y/c)]$
276	$[\vert\Gamma(\frac{3}{4}+ix)\vert\cosh(\pi x)]^{-2}$ $N=2^{\frac{1}{2}}\pi^{-\frac{3}{2}}\log(1+2^{\frac{1}{2}})$	$2^{\frac{1}{2}}\pi^{-\frac{3}{2}}(\cosh y)^{-\frac{1}{2}}$ $\cdot\log[(1+\cosh y)^{\frac{1}{2}}+(\cosh y)^{\frac{1}{2}}]$
277	$x\operatorname{Erfc}(ax)$ $N=\frac{1}{4}a^{-2}$	$(\frac{1}{2}a^{-2}+y^{-2})\exp(-\frac{1}{4}y^2/a^2)-y^{-2}$
278	$\operatorname{Erfc}(ax)$ $N=a^{-1}\pi^{-\frac{1}{2}}$	$a^{-1}\pi^{-\frac{1}{2}}{}_1F_1(1;\frac{3}{2};-\frac{1}{4}a^{-2}y^2)$ or $-iy^{-1}\exp(-y^2/4a^2)\operatorname{Erf}(\frac{1}{2}ia^{-1}y)$
279	$x^{-1}[\operatorname{Erfc}(ax)-\operatorname{Erfc}(bx)], \quad a<b$ $N=\log(b/a)$	$\frac{1}{2}[\operatorname{Ei}(-y^2/4a^2)-\operatorname{Ei}(-y^2/4b^2)]$
280	$x^{\nu-1}\operatorname{Erfc}(ax), \quad \nu>0$ $N=a^{-\nu}\nu^{-1}\pi^{-\frac{1}{2}}\Gamma(\frac{1}{2}+\frac{1}{2}\nu)$	$a^{-\nu}\nu^{-1}\pi^{-\frac{1}{2}}\Gamma(\frac{1}{2}+\frac{1}{2}\nu)$ $\cdot{}_2F_2[\frac{1}{2}\nu,\frac{1}{2}\nu+\frac{1}{2};\frac{1}{2},1+\frac{1}{2}\nu;-(y^2/4a^2)]$
281	$-ix^{-1}\exp(-a^2x^2)\operatorname{Erf}(iax)$ $N=\frac{1}{2}\pi$	$\frac{1}{2}\pi\operatorname{Erfc}(\frac{1}{2}a^{-1}y)$
282	$\operatorname{Erfc}[(ax)^{\frac{1}{2}}]$ $N=\frac{1}{2}a^{-1}$	$(\frac{1}{2}a)^{\frac{1}{2}}[(a^2+y^2)^{\frac{1}{2}}+a]^{-\frac{1}{2}}(a^2+y^2)^{-\frac{1}{2}}$

	$2Nf(x)$	$Ng(y)$
283	$\operatorname{Erfc}\{a[b+(b^2+x^2)^{\frac{1}{2}}]^{\frac{1}{2}}\}$ $N=\frac{1}{2}a^{-2}\exp(-2a^2b)$	$2^{-\frac{1}{2}}a\exp(-a^2b)(a^4+y^2)^{-\frac{1}{2}}$ $\cdot[a^2+(a^4+y^2)^{\frac{1}{2}}]^{-\frac{1}{2}}\exp[-b(a^4+y^2)^{\frac{1}{2}}]$
284	$\operatorname{Erfc}\left[a\left(\frac{1+\cos x}{\cos x}\right)^{\frac{1}{2}}\right],\quad x<\pi/2$ $0,\quad x>\pi/2$ $N=\frac{1}{2}\pi[\operatorname{Erfc}(a)]$	$\exp(-a^2)D_{y-1}(2^{\frac{1}{2}}a)D_{-y-1}(2^{\frac{1}{2}}a)$
285	$\operatorname{Erfc}(a\cosh x)$ $N=\frac{1}{2}\operatorname{Ei}(-a^2)$	$\frac{1}{2}a^{-1}\exp(-\frac{1}{2}a^2)W_{-\frac{1}{2},\frac{1}{2}iy}(a^2)$
286	$\exp[(a\cosh x)^2]\operatorname{Erfc}(a\cosh x)$ $N=\frac{1}{2}\exp(\frac{1}{2}a^2)K_0(\frac{1}{2}a^2)$	$\frac{1}{2}\exp(\frac{1}{2}a^2)\operatorname{sech}(\frac{1}{2}\pi y)K_{\frac{1}{2}iy}(\frac{1}{2}a^2)$
287	$-i\exp(-a^2\cosh^2 x)\operatorname{Erf}(ia\cosh x)$ $N=\frac{1}{2}\pi\exp(-\frac{1}{2}a^2)I_0(\frac{1}{2}a^2)$	$2^{-2}\pi\exp(-\frac{1}{2}a^2)\operatorname{sech}(\frac{1}{2}\pi y)$ $\cdot[I_{iy}(\frac{1}{2}a^2)+I_{-iy}(\frac{1}{2}a^2)]$
288	$\operatorname{Erf}[b\operatorname{sech}(ax)]$ $N=a^{-1}b\pi^{\frac{1}{2}}{}_2F_2(\frac{1}{2},\frac{1}{2};\frac{3}{2},1;-b^2)$	$a^{-1}b\pi^{\frac{1}{2}}\operatorname{sech}[\frac{1}{2}\pi(y/a)]$ $\cdot{}_2F_2[\frac{1}{2}+\frac{1}{2}i(y/a),\frac{1}{2}-\frac{1}{2}i(y/a);\frac{3}{2},1;-b^2]$
289	$(\operatorname{sech}x)^{\frac{1}{2}}\exp(a^2\operatorname{sech}x)$ $\cdot\operatorname{Erfc}[a(1+\operatorname{sech}x)^{\frac{1}{2}}]$ $N=(2\pi)^{-\frac{1}{2}}a[K_{\frac{1}{4}}(\frac{1}{2}a^2)]^2$	$\pi^{\frac{1}{2}}\operatorname{sech}(\pi y)D_{-\frac{1}{2}+iy}(2^{\frac{1}{2}}a)D_{-\frac{1}{2}-iy}(2^{\frac{1}{2}}a)$
290	$\exp(a^2x^2)\operatorname{Erfc}(ax+b)$ $N=-\frac{1}{2}\pi^{-\frac{1}{2}}a^{-1}\operatorname{Ei}(-b^2)$	$-\frac{1}{2}\pi^{-\frac{1}{2}}a^{-1}\exp(\frac{1}{4}y^2/a^2)\operatorname{Ei}[-(b^2+\frac{1}{4}y^2/a^2)]$
291	$-\operatorname{Ei}(-ax)$ $N=a^{-1}$	$y^{-1}\arctan(y/a)$
292	$-\operatorname{Ei}(-bx),\quad x<a$ $0,\quad x>a$ $N=b^{-1}-a\operatorname{Ei}(-ab)$	$-y^{-1}[\sin(ay)\operatorname{Ei}(-ab)$ $-\arctan(y/b)-\frac{1}{2}i\operatorname{Ei}(-ab-iay)$ $+\frac{1}{2}i\operatorname{Ei}(-ab+iay)]$
293	$-e^{-ax}\operatorname{Ei}(-bx)$ $N=a^{-1}\log(1+a/b),\quad a\geq -b$	$(a^2+y^2)^{-1}\{y\arctan[y(a+b)^{-1}]$ $+\frac{1}{2}a\log[(1+a/b)^2+y^2/b^2]\}$
294	$-\operatorname{Ei}(-ax^2)$ $N=(\pi/a)^{\frac{1}{2}}$	$\pi y^{-1}\operatorname{Erf}(\frac{1}{2}ya^{-\frac{1}{2}})$

	$2Nf(x)$	$Ng(y)$
295	$-\exp(ax^2)\ \mathrm{Ei}[-(ax^2+b)]$ $N=\frac{1}{2}\pi^{\frac{3}{2}}a^{-\frac{1}{2}}\ \mathrm{Erfc}(b^{\frac{1}{2}})$	$\frac{1}{2}\pi^{\frac{3}{2}}a^{-\frac{1}{2}}\exp(\frac{1}{4}y^2/a)\ \mathrm{Erfc}(\frac{1}{2}ya^{-\frac{1}{2}}+b^{\frac{1}{2}})$
296	$x^{-1}[e^{-ax}\ \overline{\mathrm{Ei}}(ax)-e^{ax}\ \mathrm{Ei}(-ax)]$ $N=\frac{1}{2}\pi^2$	$\pi\arctan(a/y)$
297	$\exp(-ax^2)\ \overline{\mathrm{Ei}}(ax^2)$ $-\exp(ax^2)\ \mathrm{Ei}(-ax^2)$ $N=\frac{1}{2}\pi^{\frac{3}{2}}a^{-\frac{1}{2}}$	$\frac{1}{2}\pi^{\frac{3}{2}}a^{-\frac{1}{2}}[\exp(\frac{1}{4}y^2/a)\ \mathrm{Erfc}(\frac{1}{2}ya^{-\frac{1}{2}})$ $+i\exp(-\frac{1}{4}y^2/a)\ \mathrm{Erf}(i\frac{1}{2}ya^{-\frac{1}{2}})]$
298	$-e^{ax}\ \mathrm{Ei}\{-a[(b^2+x^2)^{\frac{1}{2}}+x]\}$ $-e^{-ax}\ \mathrm{Ei}\{-a[(b^2+x^2)^{\frac{1}{2}}-x]\}$ $N=2a^{-1}K_0(ab)$	$2a(a^2+y^2)^{-1}K_0[b(a^2+y^2)^{\frac{1}{2}}]$
299	$-\exp(a\cosh x)\ \mathrm{Ei}(-a\cosh x)$ $N=\frac{1}{4}\pi^2[I_0(a)-\mathbf{L}_0(a)]$ $+\frac{1}{2}[S_{-1,0}(ia)+S_{-1,0}(-ia)]$	$-\frac{1}{2}\pi^2[\mathrm{csch}(\pi y)]^2[I_{iy}(a)+I_{-iy}(a)$ $-\exp(\frac{1}{2}\pi y)\mathbf{J}_{iy}(ia)$ $-\exp(-\frac{1}{2}\pi y)\mathbf{J}_{-iy}(ia)]$
300	$\exp(-a\cosh x)\ \overline{\mathrm{Ei}}(a\cosh x)$ $N=\frac{1}{4}\pi^2[I_0(a)-\mathbf{L}_0(a)]$ $-\frac{1}{2}[S_{-1,0}(ia)+S_{-1,0}(-ia)]$	$\frac{1}{2}\pi^2[\mathrm{csch}(\pi y)]^2\{\cosh(\pi y)[I_{iy}(a)+I_{-iy}(a)]$ $-\exp(\frac{1}{2}\pi y)\mathbf{J}_{iy}(-ia)$ $-\exp(-\frac{1}{2}\pi y)\mathbf{J}_{-iy}(-ia)\}$
301	$\exp(-a\cosh x)\ \overline{\mathrm{Ei}}(a\cosh x)$ $-\exp(a\cosh x)\ \mathrm{Ei}(-a\cosh x)$ $N=\frac{1}{2}\pi^2[I_0(a)-\mathbf{L}_0(a)]$	$\pi^2[\mathrm{csch}(\pi y)]^2\sinh(\frac{1}{2}\pi y)$ $\cdot\{\sinh(\frac{1}{2}\pi y)[I_{iy}(a)-I_{-iy}(a)]$ $+\mathbf{J}_{iy}(ia)-\mathbf{J}_{-iy}(ia)\}$
302	$\mathrm{Si}(bx),\quad x<a$ $0,\quad x>a$ $N=a\ \mathrm{Si}(ab)-2b^{-1}\sin^2(\frac{1}{2}ab)$	$\frac{1}{2}y^{-1}\{2\sin(ay)\ \mathrm{Si}(ab)$ $+\mathrm{Ci}(ay+ab)-\mathrm{Ci}(\|ay-ab\|)$ $+\log(\|(y-b)/(y+b)\|)\},\quad y\neq b$ $\frac{1}{2}b^{-1}[2\sin(ab)\ \mathrm{Si}(ab)+\mathrm{Ci}(2ab)$ $-\gamma-\log(2ab)],\quad y=b$
303	$e^{-bx}\ \mathrm{Si}(ax)$ $N=b^{-1}\arctan(a/b)$	$\frac{1}{2}(b^2+y^2)^{-1}\{b\arctan[(a+y)/b]$ $-b\arctan[(y-a)/b]-\frac{1}{2}y$ $\cdot\log[b^2+(y+a)^2]$ $+\frac{1}{2}y\log[b^2+(a-y)^2]\}$
304	$[\mathrm{si}(ax)]^2$ $N=\frac{1}{2}\pi a^{-1}$	$\frac{1}{2}\pi y^{-1}\log(1+y)a^{-1},\quad y<2a$ $\frac{1}{2}\pi y^{-1}\log[(y+a)(y-a)^{-1}],\quad y>2a$

	$2Nf(x)$	$Ng(y)$
305	$[\mathrm{Ci}(ax)]^2$ $N=\frac{1}{2}\pi a^{-1}$	$\frac{1}{2}\pi y^{-1}\log(1+ya^{-1}),\quad y<2a$ $\frac{1}{2}\pi y^{-1}\log(y^2a^{-2}-1),\quad y>2a$
306	$-[\sin(ax)\ \mathrm{si}(ax)-\cos(ax)\ \mathrm{Ci}(ax)]$ $N=\frac{1}{2}\pi a^{-1}$	$\frac{1}{2}\pi(a+y)^{-1}$
307	$\sin(ax^2)\ \mathrm{Ci}(ax^2)-\cos(ax^2)\ \mathrm{si}(ax^2)$ $N=(\frac{1}{2}\pi)^{\frac{3}{2}}a^{-\frac{1}{2}}$	$(2a)^{-\frac{1}{2}}\pi^{\frac{3}{2}}\{\sin(\frac{1}{4}y^2/a)[\frac{1}{2}-S(\frac{1}{4}y^2/a)]$ $+\cos(\frac{1}{4}y^2/a)[\frac{1}{2}-C(\frac{1}{4}y^2/a)]\}$
308	$-[\sin(ax^2)\ \mathrm{si}(ax^2)$ $+\cos(ax^2)\ \mathrm{Ci}(ax^2)]$ $N=a^{-\frac{1}{2}}(\frac{1}{2}\pi)^{\frac{1}{2}}$	$(2a)^{-\frac{1}{2}}\pi^{\frac{1}{2}}\{\cos(y^2/4a)[\frac{1}{2}-S(y^2/4a)]$ $-\sin(y^2/4a)[\frac{1}{2}-C(y^2/4a)]\}$
309	$\sin(a\cosh x)\ \mathrm{Ci}(a\cosh x)$ $-\cos(a\cosh x)\ \mathrm{si}(a\cosh x)$ $N=\frac{1}{4}\pi^2[\mathbf{H}_0(a)-Y_0(a)]$	$\frac{1}{2}\pi\ \mathrm{sech}(\frac{1}{2}\pi y)\,S_{0,iy}(a)$
310	$-\cos(a\cosh x)\ \mathrm{Ci}(a\cosh x)$ $-\sin(a\cosh x)\ \mathrm{si}(a\cosh x)$ $N=S_{-1,0}(a)$	$\frac{1}{2}\pi y\ \mathrm{csch}(\frac{1}{2}\pi y)\,S_{-1,iy}(a)$
311	$[\frac{1}{2}-C(ax^2)]\cos(ax^2)$ $+[\frac{1}{2}-S(ax^2)]\sin(ax^2)$ $N=\frac{1}{4}(\pi/2a)^{\frac{1}{2}}$	$\frac{1}{2}(2a\pi)^{-\frac{1}{2}}[\sin(y^2/4a)\ \mathrm{Ci}(y^2/4a)$ $-\cos(y^2/4a)\ \mathrm{si}(y^2/4a)]$
312	$S(ax^{-1})$ $N=a$	$\frac{1}{4}y^{-1}\{\sin[2(ay)^{\frac{1}{2}}]-\cos[2(ay)^{\frac{1}{2}}]$ $+\exp[-2(ay)^{\frac{1}{2}}]\}$
313	$x^{-1}\{\cos(ax^2)[C(ax^2)-S(ax^2)]$ $+\sin(ax^2)[C(ax^2)+S(ax^2)-1]\}$ $N=\frac{1}{4}\pi$	$\frac{1}{2}\pi\{\frac{1}{2}+C^2(\frac{1}{2}y^2/a)+S^2(\frac{1}{4}y^2/a)$ $-C(\frac{1}{4}y^2/a)-S(\frac{1}{4}y^2/a)\}$
314	$\cos(a\cosh x)[\frac{1}{2}-S(a\cosh x)]$ $-\sin(a\cosh x)[\frac{1}{2}-c(a\cosh x)]$ $N=\frac{1}{4}\pi^{-1}\Gamma^2(\frac{1}{4})\,S_{\frac{1}{2},0}(a)$	$\frac{1}{4}\pi^{-1}\Gamma(\frac{1}{4}-i\frac{1}{2}y)\,\Gamma(\frac{1}{4}+i\frac{1}{2}y)\,S_{\frac{1}{2},iy}(a)$
315	$\cos(a\cosh x)[\frac{1}{2}-C(a\cosh x)]$ $+\sin(a\cosh x)[\frac{1}{2}-S(a\cosh x)]$ $N=\frac{1}{2}\pi^{-1}\Gamma^2(\frac{3}{4})\,S_{-\frac{1}{2},0}(a)$	$\frac{1}{2}\pi^{-1}\Gamma(\frac{3}{4}-i\frac{1}{2}y)\,\Gamma(\frac{3}{4}+i\frac{1}{2}y)\,S_{-\frac{1}{2},iy}(a)$

	$2Nf(x)$	$Ng(y)$
316	$x^{-2\nu}\gamma(\nu, ax^2)$ $N=\pi^{\frac{1}{2}}a^{\nu-\frac{1}{2}}(2\nu-1)^{-1}, \quad \nu>\frac{1}{2}$	$\frac{1}{2}\pi^{\frac{1}{2}}(\frac{1}{4}y^2)^{\nu-\frac{1}{2}}\Gamma(\frac{1}{2}-\nu, \frac{1}{4}y^2/a)$
317	$x^{-2\nu}\Gamma(\nu, ax^2)$ $N=\pi^{\frac{1}{2}}a^{\nu-\frac{1}{2}}(1-2\nu)^{-1}, \quad \nu<\frac{1}{2}$	$\frac{1}{2}\pi^{\frac{1}{2}}(\frac{1}{4}y^2)^{\nu-\frac{1}{2}}\gamma(\frac{1}{2}-\nu, \frac{1}{4}y^2/a)$
318	$\exp(ax^2)\ \Gamma(\nu, ax^2), \quad -1<\nu<0$ $N=\frac{1}{2}\pi(\pi/a)^{\frac{1}{2}}[\Gamma(1-\nu)\cos(\pi\nu)]^{-1}$	$\frac{1}{2}(\pi/a)^{-\frac{1}{2}}\Gamma(\frac{1}{2}+\nu)[\Gamma(1-\nu)]^{-1}$ $\cdot\exp(\frac{1}{4}y^2/a)\Gamma(\frac{1}{2}-\nu, \frac{1}{4}y^2/a)$

9. Elliptic Integrals and Legendre Functions

	$2Nf(x)$	$Ng(y)$
319	$K[(1-a^{-2}x^2)^{\frac{1}{2}}], \quad x<a$ $0, \quad x>a$ $N=\frac{1}{4}a\pi^2$	$\frac{1}{4}a\pi^2[J_0(\frac{1}{2}ay)]^2$
320	$K[(\frac{1}{2}-\frac{1}{2}x)^{\frac{1}{2}}], \quad x<1$ $0, \quad x>1$ $N=\frac{1}{4}\pi^{\frac{3}{2}}[\Gamma(\frac{5}{4})]^{-2}$	$\frac{1}{16}\pi^{\frac{3}{2}}[\Gamma(\frac{5}{4})]^{-2}y^{-\frac{1}{2}}s_{-\frac{1}{2}i,0}(y)$
321	$(a^2-x^2)^{-\frac{1}{2}}K\{[(a^2-b^2)/(a^2-x^2)]^{\frac{1}{2}}\},$ $x<b$ $(a^2-b^2)^{-\frac{1}{2}}K\{[(a^2-x^2)/(a^2-b^2)]^{\frac{1}{2}}\},$ $b<x<a$ $0, \quad x>a$ $N=\frac{1}{2}\pi^2$	$\frac{1}{2}\pi^2J_0[\frac{1}{2}y(a+b)]J_0[\frac{1}{2}y(a-b)]$
322	$K[\cos(\frac{1}{2}x)], \quad x<\pi$ $0, \quad x>\pi$ $N=\frac{1}{4}\pi[\Gamma(\frac{1}{4})]^2[\Gamma(\frac{3}{4})]^{-2}$	$\frac{1}{4}\pi\cos(\pi y)\Gamma(\frac{1}{4}+\frac{1}{2}y)\Gamma(\frac{1}{4}-\frac{1}{2}y)$ $\cdot[\Gamma(\frac{3}{4}+\frac{1}{2}y)\Gamma(\frac{3}{4}-\frac{1}{2}y)]^{-1}$
323	$(1+a^2\cos^2x)^{-\frac{1}{2}}$ $\cdot K[a\cos x(1+a^2\cos^2x)^{-\frac{1}{2}}], \quad x<\frac{1}{2}\pi$ $0, \quad x>\frac{1}{2}\pi$ $N=2^{-\frac{1}{2}}a^{-1}(p+1)^{-\frac{1}{2}}$ $\cdot K[2^{\frac{1}{2}}(p+1)^{-\frac{1}{2}}], \quad p=(1+a^{-2})^{\frac{1}{2}}$	$\frac{1}{2}a^{-1}\cos(\frac{1}{2}\pi y)\mathfrak{Q}_{-\frac{1}{2}+\frac{1}{2}y}(p)$ $\cdot\mathfrak{Q}_{-\frac{1}{2}-\frac{1}{2}y}(p)$
324	$\operatorname{sech}(ax)K[\tanh(ax)]$ $N=\frac{1}{16}a^{-1}\pi^{-1}[\Gamma(\frac{1}{4})]^4$	$\frac{1}{16}(a\pi)^{-1}\mid\Gamma(\frac{1}{4}+\frac{1}{4}iy/a)\mid^4$

	$2Nf(x)$	$Ng(y)$
325	$\operatorname{sech} x\, K(\operatorname{sech} x)$ $N=\frac{1}{8}\pi[\Gamma(\frac{1}{4})]^2[\Gamma(\frac{3}{4})]^{-2}$	$\dfrac{\pi}{8}\dfrac{\Gamma(\frac{1}{4}+\frac{1}{4}iy)\Gamma(\frac{1}{4}-\frac{1}{4}iy)}{\Gamma(\frac{3}{4}+\frac{1}{4}iy)\Gamma(\frac{3}{4}-\frac{1}{4}iy)}$
326	$\operatorname{sech} x\, K(a \operatorname{sech} x), \quad a \leq 1$ $N=\{K[2^{-\frac{1}{2}}(1-r)^{\frac{1}{2}}]\}^2$	$\frac{1}{4}\pi^2 \operatorname{sech}(\frac{1}{2}\pi y)\{P_{-\frac{1}{2}+\frac{1}{2}iy}(r)\}^2, \quad r=(1-a^2)^{\frac{1}{2}}$
327	$\operatorname{sech} x\, K[(1-a^2 \operatorname{sech}^2 x)^{\frac{1}{2}}], \quad a \leq 1$ $N=K[2^{-\frac{1}{2}}(1-r)^{\frac{1}{2}}]K[2^{-\frac{1}{2}}(1+r)^{\frac{1}{2}}]$	$\frac{1}{4}\pi^2(\operatorname{sech}\frac{1}{2}\pi y)^2 P_{-\frac{1}{2}+\frac{1}{2}iy}(r)$ $\cdot P_{-\frac{1}{2}+\frac{1}{2}iy}(r), \quad r=(1-a^2)^{\frac{1}{2}}$
328	$0, \qquad \sinh x < a^{-1}$ $\operatorname{csch} x\, K[(1-a^{-2}\operatorname{csch}^2 x)^{\frac{1}{2}}],$ $\sinh x > a^{-1}$ $N=(1+z)^{-1}\{K[(\frac{1}{2}+\frac{1}{2}z)^{-\frac{1}{2}}]\}^2$	$\frac{1}{4}\{[\mathfrak{Q}_{-\frac{1}{2}+\frac{1}{2}iy}(z)]^2$ $+[\mathfrak{Q}_{-\frac{1}{2}-\frac{1}{2}iy}(z)]^2\}, \quad z=(1+a^{-2})^{\frac{1}{2}}$
329	$(1+a\cosh x)^{-\frac{1}{2}}K\{2^{\frac{1}{2}}[1+a\cosh x]^{-\frac{1}{2}}\}$ $N=\frac{1}{2}(2\pi a)^{-\frac{1}{2}}[\Gamma(\frac{1}{4})]^2$ $\cdot K\{[\frac{1}{2}-\frac{1}{2}(1-a^{-2})^{\frac{1}{2}}]^{\frac{1}{2}}\}$	$2^{-\frac{5}{2}}(\pi/a)^{\frac{1}{2}}\Gamma(\frac{1}{4}+\frac{1}{2}iy)\Gamma(\frac{1}{4}-\frac{1}{2}iy)$ $\cdot P_{-\frac{1}{2}+iy}[(1-a^{-2})^{\frac{1}{2}}]$
330	$(1+a^2\sinh^2 x)^{-\frac{1}{2}}$ $\cdot K[(1+a^2\sinh^2 x)^{-\frac{1}{2}}], \quad a>1$ $N=a^{-1}K\{[\frac{1}{2}-\frac{1}{2}(1-a^{-2})^{\frac{1}{2}}]^{\frac{1}{2}}\}$ $\cdot K\{[\frac{1}{2}+\frac{1}{2}(1-a^{-2})^{\frac{1}{2}}]^{\frac{1}{2}}\}$	$\frac{1}{4}\pi^2 a^{-1}\operatorname{sech}(\frac{1}{2}\pi y)$ $\cdot P_{-\frac{1}{2}+\frac{1}{2}iy}[(1-a^{-2})^{\frac{1}{2}}]P_{-\frac{1}{2}+\frac{1}{2}iy}[-(1-a^{-2})^{\frac{1}{2}}]$
331	$(1+a^2\cosh^2 x)^{-\frac{1}{2}}$ $\cdot K[a\cosh x(1+a^2\cosh^2 x)^{-\frac{1}{2}}]$ $N=2a^{-1}(p+1)^{-1}K[(p-1)^{\frac{1}{2}}(p+1)^{-\frac{1}{2}}]$ $\cdot K[2^{\frac{1}{2}}(p+1)^{-\frac{1}{2}}], \quad p=(1+a^{-2})^{\frac{1}{2}}$	$\frac{1}{4}a^{-1}\pi\operatorname{sech}(\frac{1}{2}\pi y)$ $\cdot\mathfrak{P}_{-\frac{1}{2}+\frac{1}{2}iy}(p)[\mathfrak{Q}_{-\frac{1}{2}+\frac{1}{2}iy}(p)$ $+\mathfrak{Q}_{-\frac{1}{2}-\frac{1}{2}iy}(p)]$
332	$(a\cosh x+1)^{-\frac{1}{2}}K\left[\left(\dfrac{a\cosh x-1}{a\cosh x+1}\right)^{\frac{1}{2}}\right],$ $a>1$ $N=2^{-\frac{5}{2}}(a\pi)^{-\frac{1}{2}}[\Gamma(\frac{1}{4})]^2$ $\cdot\{K[2^{-\frac{1}{2}}(1-s)]+K[2^{-\frac{1}{2}}(1+s)]\},$ $s=(1-a^{-2})^{\frac{1}{2}}$	$2^{-7/2}(\pi/a)^{\frac{1}{2}}\operatorname{sech}(\pi y)\Gamma(\frac{1}{4}+\frac{1}{2}iy)$ $\cdot\Gamma(\frac{1}{4}-\frac{1}{2}iy)$ $\cdot[P_{-\frac{1}{2}+iy}(s)+P_{-\frac{1}{2}+iy}(-s)]$
333	$(1-x^2)^{-\frac{1}{2}\mu}P_\nu^\mu(x), \quad x<1$ $0, \qquad x>1$ $-\frac{1}{2}<\nu<\frac{1}{2}, \quad \mu<\frac{1}{2}$ $N=2^{\mu-1}\pi^{\frac{1}{2}}$ $\cdot\{\Gamma[\frac{1}{2}(3-\mu+\nu)]\Gamma[\frac{1}{2}(2-\mu-\nu)]\}^{-1}$	$2^{\mu-1}\pi^{\frac{1}{2}}\{\Gamma[\frac{1}{2}(3-\mu-\nu)]\Gamma[\frac{1}{2}(2-\mu-\nu)]\}^{-1}$ $\cdot(\mu+\nu)(\mu-\nu-1)y^{\mu-\frac{1}{2}}$ $\cdot s_{-\mu-\frac{1}{2},\nu+\frac{1}{2}}(y)$

	$2Nf(x)$	$Ng(y)$
334	$x^{\lambda-1}(1-x^2)^{-\frac{1}{2}\mu}P_\nu^\mu(x), \quad x<1$ $0, \quad x>1$ $\lambda>0, \quad \mu<\frac{1}{2}, \quad -\frac{1}{2}<\nu<\frac{1}{2}$ $N=2^{\mu-\lambda}\pi^{\frac{1}{2}}\Gamma(\lambda)$ $\cdot\{\Gamma[\frac{1}{2}(1+\lambda-\mu+\nu)]$ $\cdot\Gamma[\frac{1}{2}(1+\lambda-\mu-\nu)]\}^{-1}$	$2^{\mu-\lambda}\pi^{\frac{1}{2}}\Gamma(\lambda)$ $\cdot\{\Gamma[1+\frac{1}{2}(\lambda-\mu+\nu)]\Gamma[\frac{1}{2}(1+\lambda-\mu-\nu)]\}^{-1}$ $\cdot{}_2F_3[\frac{1}{2}\lambda, \frac{1}{2}(1+\lambda); \frac{1}{2}, \frac{1}{2}(1+\lambda-\mu-\nu),$ $1+\frac{1}{2}(\lambda-\mu+\nu); -y^2/4]$
335	$P_\nu[(2x^2/a^2)-1], \quad x<a$ $0, \quad x>a$ $-\frac{1}{2}<\nu<\frac{1}{2}$ $N=a(2\nu+1)^{-1}\cos(\nu\pi)$	$\frac{1}{2}a\pi J_{\nu+\frac{1}{2}}(ay/2)J_{-\nu-\frac{1}{2}}(ay/2)$
336	$\mathfrak{P}_\nu[(a^2+b^2-x^2)/2ab], \quad x<a-b$ $0, \quad x>a-b$ $a>b$ $N=(ab)^{\frac{1}{2}}(2\nu+1)^{-1}$ $\cdot[(a/b)^{\nu+\frac{1}{2}}-(b/a)^{\nu+\frac{1}{2}}]$	$\frac{1}{2}(ab)^{\frac{1}{2}}\pi[J_{\nu+\frac{1}{2}}(by)Y_{\nu+\frac{1}{2}}(ay)$ $-J_{\nu+\frac{1}{2}}(ay)Y_{\nu+\frac{1}{2}}(by)]$
337	$x^{-1}\mathfrak{P}_\nu(2x^{-2}-1), \quad x<1$ $0, \quad x>1$ $-1<\nu<0$ $N=-\frac{1}{2}\pi\csc(\nu\pi)$	$-\frac{1}{2}\pi\csc(\nu\pi){}_1F_1(1+\nu; 1; iy)$ $\cdot{}_1F_1(1+\nu; 1; -iy)$
338	$\mathfrak{Q}_\nu(1+x^2), \quad \nu>-\frac{1}{2}$ $N=2^{-\frac{1}{2}}\pi(2\nu+1)^{-1}$	$2^{-\frac{1}{2}}\pi I_{\nu+\frac{1}{2}}(2^{-\frac{1}{2}}y)K_{\nu+\frac{1}{2}}(2^{-\frac{1}{2}}y)$
339	$\mathfrak{Q}_\nu[(x^2+a^2+b^2)/2ab], \quad \nu>-\frac{1}{2}$ $N=a^{-\nu}b^{\nu+1}\pi(2\nu+1)^{-1}, \quad a>b$	$(ab)^{\frac{1}{2}}\pi I_{\nu+\frac{1}{2}}(by)K_{\nu+\frac{1}{2}}(ay)$
340	$\mathfrak{P}_\nu(1+2a^{-2}\cos^2 x), \quad x<\frac{1}{2}\pi$ $0, \quad x>\frac{1}{2}\pi$ $N=\frac{1}{2}\pi\{\mathfrak{P}_\nu[(1+a^{-2})^{\frac{1}{2}}]\}^2$	$\frac{1}{2}\pi\mathfrak{P}_\nu^{\frac{1}{2}y}[(1+a^{-2})^{\frac{1}{2}}]$ $\cdot\mathfrak{P}_\nu^{-\frac{1}{2}y}[(1+a^{-2})^{\frac{1}{2}}]$
341	$\mathfrak{P}_\nu(\cosh x), \quad -1<\nu<0$ $N=-\frac{1}{4}\pi^{-2}\sin(\nu\pi)$ $\cdot[\Gamma(-\frac{1}{2}\nu)\Gamma(\frac{1}{2}+\frac{1}{2}\nu)]^2$	$-\frac{1}{4}\pi^{-2}\sin(\nu\pi)\Gamma(-\frac{1}{2}\nu+\frac{1}{2}iy)$ $\cdot\Gamma(-\frac{1}{2}\nu-\frac{1}{2}iy)\Gamma(\frac{1}{2}+\frac{1}{2}\nu+\frac{1}{2}iy)$ $\cdot\Gamma(\frac{1}{2}+\frac{1}{2}\nu-\frac{1}{2}iy)$
342	$\mathfrak{P}_\nu(a\cosh x), \quad a\geq 1,$ $-1<\nu<0$ $N=-2^{\nu-\frac{3}{2}}(a\pi)^{-\frac{1}{2}}\cot(\nu\pi/2)$ $\cdot[\Gamma(\frac{1}{2}+\frac{1}{2}\nu)]^2\{P_{-\frac{1}{2}}^{-\nu-\frac{1}{2}}[(1-a^{-2})^{\frac{1}{2}}]$ $+P_{-\frac{1}{2}}^{-\nu-\frac{1}{2}}[-(1-a^{-2})^{\frac{1}{2}}]\}$	$-2^{\nu-\frac{3}{2}}(a\pi)^{-\frac{1}{2}}\sin(\nu\pi)$ $\cdot[\cosh(\pi y)-\cos(\nu\pi)]^{-1}\Gamma(\frac{1}{2}+\frac{1}{2}\nu+\frac{1}{2}iy)$ $\cdot\Gamma(\frac{1}{2}+\frac{1}{2}\nu-\frac{1}{2}iy)\{P_{-\frac{1}{2}+iy}^{-\nu-\frac{1}{2}}[(1-a^{-2})^{\frac{1}{2}}]$ $+P_{-\frac{1}{2}+iy}^{-\nu-\frac{1}{2}}[-(1-a^{-2})^{\frac{1}{2}}]\}$

	$2Nf(x)$	$Ng(y)$
343	$(\sinh x)^{\mu}\mathfrak{P}_{\nu}^{\mu}(\cosh x),\quad \nu+\mu<0,$ $\mu-\nu<1$ $N=2^{-\mu-2}\pi^{-\frac{1}{2}}\{\Gamma[\frac{1}{2}(1+\nu-\mu)]\}^{2}$ $\cdot[\Gamma(-\nu-\mu)\Gamma(1+\nu-\mu)]^{-1}$ $\cdot[\Gamma(\frac{1}{2}-\mu)]^{-1}\{\Gamma[\frac{1}{2}(-\nu-\mu)]\}^{2}$	$2^{-\mu-2}\pi^{-\frac{1}{2}}\Gamma[\frac{1}{2}(1+\nu-\mu+iy)]$ $\cdot\Gamma[\frac{1}{2}(1+\nu-\mu-iy)]\Gamma[\frac{1}{2}(-\nu-\mu+iy)]$ $\cdot\Gamma[\frac{1}{2}(-\nu-\mu-iy)]$ $\cdot[\Gamma(-\nu-\mu)\Gamma(1+\nu-\mu)\Gamma(\frac{1}{2}-\mu)]^{-1}$
344	$\mathfrak{P}_{\nu}(1+2a^{2}\sinh^{2}x),\quad 0<a<1,$ $-1<\nu<0$ $N=\frac{1}{2}P_{\nu}(r)[Q_{\nu}(r)$ $+Q_{-\nu-1}(r)],\quad r=(1-a^{2})^{\frac{1}{2}}$	$\frac{1}{2}\operatorname{sech}(\frac{1}{2}\pi y)$ $\cdot\mathrm{Re}\{P_{\nu}^{-i\frac{1}{2}y}(r)[Q_{\nu}^{i\frac{1}{2}y}(r)+Q_{-\nu-1}^{i\frac{1}{2}y}(r)]\},$
345	$\mathfrak{P}_{\nu}(1+2a^{2}\sinh^{2}x),\quad a>1,\quad -1<\nu<0$ $N=\frac{1}{2}[a\cos(\nu\pi)]^{-1}$ $\cdot\{P_{-\frac{1}{2}}^{\nu+\frac{1}{2}}(s)Q_{-\frac{1}{2}}^{-\nu-\frac{1}{2}}(s)$ $-P_{-\frac{1}{2}}^{-\nu-\frac{1}{2}}(s)Q_{-\frac{1}{2}}^{\nu+\frac{1}{2}}(s)\}$	$\frac{1}{2}[a\cos(\nu\pi)]^{-1}$ $\cdot\{P_{-\frac{1}{2}+i\frac{1}{2}y}^{\nu+\frac{1}{2}}(s)\ \mathrm{Re}[Q_{-\frac{1}{2}+i\frac{1}{2}y}^{-\nu-\frac{1}{2}}(s)]$ $-P_{-\frac{1}{2}+i\frac{1}{2}y}^{-\nu-\frac{1}{2}}(s)\ \mathrm{Re}[Q_{-\frac{1}{2}+i\frac{1}{2}y}^{\nu+\frac{1}{2}}(s)]\},$ $s=(1-a^{-2})^{\frac{1}{2}}$
346	$\mathfrak{P}_{\nu}(1+2a^{2}\cosh^{2}x)$ $N=\frac{1}{2}\pi^{-1}$ $\cdot\{[\mathfrak{Q}_{\nu}(ap)]^{2}-[\mathfrak{Q}_{-\nu-1}(ap)]^{2}\},$ $-1<\nu<0$	$\frac{1}{4}a^{-1}\tan(\pi\nu)$ $\cdot\{[\vert\Gamma(1+\nu+\frac{1}{2}iy)\vert\mathfrak{P}_{-\frac{1}{2}+\frac{1}{2}iy}^{-\nu-\frac{1}{2}}(p)]^{2}$ $-[\vert\Gamma(-\nu+\frac{1}{2}iy)\vert\mathfrak{P}_{-\frac{1}{2}+\frac{1}{2}iy}^{\nu+\frac{1}{2}}(p)]^{2}\},$ $p=(1+a^{-2})^{\frac{1}{2}}$
347	$\mathfrak{P}_{\nu}(2a^{2}\cosh^{2}x-1),\quad a>1,\quad -1<\nu<0,$ $N=\frac{1}{2}a^{-1}\{P_{-\frac{1}{2}}^{-\nu-\frac{1}{2}}(s)Q_{-\frac{1}{2}}^{\nu+\frac{1}{2}}(s)$ $+P_{-\frac{1}{2}}^{\nu+\frac{1}{2}}(s)Q_{-\frac{1}{2}}^{-\nu-\frac{1}{2}}(s)\}$	$\frac{1}{2}[a\cosh(\frac{1}{2}\pi y)]^{-1}$ $\cdot\{P_{-\frac{1}{2}+i\frac{1}{2}y}^{-\nu-\frac{1}{2}}(s)\ \mathrm{Re}[Q_{-\frac{1}{2}+i\frac{1}{2}y}^{\nu+\frac{1}{2}}(s)]$ $+P_{-\frac{1}{2}+i\frac{1}{2}y}^{\nu+\frac{1}{2}}(s)\ \mathrm{Re}[Q_{-\frac{1}{2}+i\frac{1}{2}y}^{-\nu-\frac{1}{2}}(s)]\},$ $s=(1-a^{-2})^{\frac{1}{2}}$
348	$0,\quad \sinh x<1/a$ $\mathfrak{P}_{\nu}(2a^{2}\sinh^{2}x-1),\quad \sinh x>1/a$ $-1<\nu<0$ $N=-\frac{1}{2}\pi\csc(\nu\pi)[\mathfrak{P}_{\nu}(ap)]^{2}$	$(a\pi)^{-1}\,\mathrm{Re}[\mathfrak{Q}_{-\frac{1}{2}+i\frac{1}{2}y}^{-\nu-\frac{1}{2}}(p)\mathfrak{Q}_{-\frac{1}{2}+i\frac{1}{2}y}^{\nu+\frac{1}{2}}(p)],$ $p=(1+a^{-2})^{\frac{1}{2}}$
349	$\mathfrak{Q}_{\nu}(a\cosh x),\quad a\geq 1,$ $\nu>-1$ $N=2^{\nu-\frac{3}{2}}a^{-\frac{1}{2}}\pi^{\frac{1}{2}}[\Gamma(\frac{1}{2}+\frac{1}{2}\nu)]^{2}$ $\cdot P_{-\frac{1}{2}}^{-\nu-\frac{1}{2}}[(1-a^{-2})^{\frac{1}{2}}]$	$2^{\nu-\frac{3}{2}}(\pi/a)^{\frac{1}{2}}\Gamma(\frac{1}{2}+\frac{1}{2}\nu+\frac{1}{2}iy)\Gamma(\frac{1}{2}+\frac{1}{2}\nu-\frac{1}{2}iy)$ $\cdot P_{-\frac{1}{2}+iy}^{-\nu+\frac{1}{2}}[(1-a^{-2})^{\frac{1}{2}}]$
350	$e^{-i\mu\pi}(\sinh x)^{-\mu}$ $\cdot\mathfrak{Q}_{\nu}^{\mu}(\cosh x),\quad \mu+\nu+1>0,\quad \mu<\frac{1}{2}$ $N=2^{\mu-2}\pi^{\frac{1}{2}}\Gamma(\frac{1}{2}-\mu)$ $\cdot\{\Gamma[\frac{1}{2}(1+\nu+\mu)]\}^{\frac{1}{2}}$ $\cdot\{\Gamma[1+\frac{1}{2}(\nu-\mu)]\}^{-2}$	$2^{\mu-2}\pi^{\frac{1}{2}}\Gamma(\frac{1}{2}-\mu)\Gamma[\frac{1}{2}(1+\nu+\mu+iy)]$ $\cdot\Gamma[\frac{1}{2}(1+\nu+\mu-iy)]$ $\cdot\{\Gamma[1+\frac{1}{2}(\nu-\mu+iy)]$ $\cdot\Gamma[1+\frac{1}{2}(\nu-\mu-iy)]\}^{-1}$

	$2Nf(x)$	$Ng(y)$
351	$e^{-i\pi\mu}(\sinh x)^{-\nu-1}$ $\cdot\mathfrak{Q}_\nu^\mu(\operatorname{ctnh} x),\quad \mu<\frac{1}{2},\quad (\nu\pm\mu)>-1$ $N=2^{\nu-2}[\Gamma(1+\nu)\Gamma(1+\nu-\mu)]^{-1}$ $\cdot\Gamma^2[\frac{1}{2}(1+\nu-\mu)]\Gamma^2[\frac{1}{2}(1+\nu+\mu)]$	$2^{\nu-2}[\Gamma(1+\nu)\Gamma(1+\nu-\mu)]^{-1}$ $\cdot\Gamma[\frac{1}{2}(1+\nu-\mu+iy)]\Gamma[\frac{1}{2}(1+\nu-\mu-iy)]$ $\cdot\Gamma[\frac{1}{2}(1+\nu+\mu+iy)]\Gamma[\frac{1}{2}(1+\nu+\mu-iy)]$
352	$(a^2\cosh^2 x-1)^{\frac{1}{2}\mu}e^{i\mu\pi}$ $\cdot\mathfrak{Q}_\nu^{-\mu}(a\cosh x),\quad \mu-\nu-1\leq 0,\quad a\geq 1$ $N=2^{-\mu-2}\pi^{\frac{1}{2}}a^{\mu-\nu-1}$ $\cdot[\Gamma(\frac{1}{2}+\frac{1}{2}\nu-\frac{1}{2}\mu)]^2[\Gamma(\frac{3}{2}+\nu)]^{-1}$ $\cdot{}_2F_1(\frac{1}{2}+\frac{1}{2}\nu-\frac{1}{2}\mu, \frac{1}{2}+\frac{1}{2}\nu+\frac{1}{2}\mu;$ $\frac{3}{2}+\nu; a^{-2})$	$2^{-\mu-2}a^{\mu-\nu-1}\pi^{\frac{1}{2}}[\Gamma(\frac{3}{2}+\nu)]^{-1}$ $\cdot\lvert\Gamma(\frac{1}{2}+\frac{1}{2}\nu-\frac{1}{2}\mu+i\frac{1}{2}y)\rvert^2$ $\cdot{}_2F_1(\frac{1}{2}+\frac{1}{2}\nu-\frac{1}{2}\mu+i\frac{1}{2}y, \frac{1}{2}+\frac{1}{2}\nu-\frac{1}{2}\mu-i\frac{1}{2}y;$ $\frac{3}{2}+\nu; a^{-2})$
353	$\mathfrak{Q}_\nu(1+2a^2\sinh^2 x),\quad 0<a<1,\quad \nu>-1$ $N=(\pi^2/8)[P_\nu(r)]^2+\frac{1}{2}[Q_\nu(r)]^2$	$\frac{1}{2}\pi\operatorname{csch}(\frac{1}{2}\pi y)[P_\nu^{i\frac{1}{2}y}(r)Q_\nu^{-i\frac{1}{2}y}(r)$ $-P_\nu^{-i\frac{1}{2}y}(r)Q_\nu^{i\frac{1}{2}y}(r)],\quad r=(1-a^2)^{\frac{1}{2}}$
354	$\mathfrak{Q}_\nu(1+2a^2\sinh^2 x),\quad a>1,\quad \nu>-1$ $N=-\frac{1}{2}\pi[a\sin(\nu\pi)]^{-1}P_{-\frac{1}{2}}^{-\nu-\frac{1}{2}}(s)$ $\cdot Q_{-\nu}^{\nu+\frac{1}{2}}(s),\quad s=(1-a^{-2})^{\frac{1}{2}}$	$-\frac{1}{2}\pi[a\sin(\nu\pi)]^{-1}P_{-\frac{1}{2}+i\frac{1}{2}y}^{-\nu-\frac{1}{2}}(s)$ $\cdot\operatorname{Re}[Q_{-\frac{1}{2}+i\frac{1}{2}y}^{\nu+\frac{1}{2}}(s)]$
355	$\mathfrak{Q}_\nu(1+2a^2\cosh^2 x),\quad \nu>-1$ $N=\frac{1}{2}(\mathfrak{Q}_\nu(ap)]^2$	$\frac{1}{4}a^{-1}\pi\,\lvert\Gamma(1+\nu+i\frac{1}{2}y)\rvert^2[\mathfrak{P}_{-\frac{1}{2}+i\frac{1}{2}y}^{-\nu-\frac{1}{2}}(p)]^2,$ $p=(1+a^{-2})^{\frac{1}{2}}$
356	$\mathfrak{Q}_\nu(2a^2\cosh^2 x-1),\quad a>1,\quad \nu>-1$ $N=\frac{1}{4}a^{-1}\pi[\Gamma(1+\nu)]^2[P_{-\frac{1}{2}}^{-\nu-\frac{1}{2}}(s)]^2$	$\frac{1}{4}a^{-1}\pi\,\lvert\Gamma(1+\nu+i\frac{1}{2}y)\rvert^2$ $\cdot[P_{-\frac{1}{2}+i\frac{1}{2}y}^{-\nu-\frac{1}{2}}(s)]^2,\quad s=(1-a^{-2})^{\frac{1}{2}}$
357	$\operatorname{sech}(\pi x)P_{-\frac{1}{2}+ix}(a),\quad -1<a<1$ $N=2^{-\frac{1}{2}}(1+a)^{-\frac{1}{2}}$	$2^{-\frac{1}{2}}(a+\cosh y)^{-\frac{1}{2}}$
358	$\dfrac{\tanh(\pi x)}{\sinh(\alpha x)}P_{-\frac{1}{2}+ix}(a),\quad -1<a<1,\quad \alpha\geq\pi$ $N=\alpha^{-1}\sum_{n=0}^{\infty}(-1)^n\epsilon_n Q_{-\frac{1}{2}+(n\pi/\alpha)}(a)$	$\alpha^{-1}\sum_{n=0}^{\infty}(-1)^n\epsilon_n\cos[(n\pi/\alpha)y]Q_{(n\pi/\alpha)-\frac{1}{2}}(a),$ $-\alpha\leq y\leq\alpha$
359	$(\operatorname{sech}\pi x)^2\dot{P}_{-\frac{1}{2}+ix}(a),\quad -1<a<1$ $N=2^{-\frac{1}{2}}\pi^{-1}(1-a)^{-\frac{1}{2}}$ $\cdot\log\{[2^{\frac{1}{2}}+(1-a)^{\frac{1}{2}}]^2/(1+a)\}$	$2^{-\frac{1}{2}}\pi^{-1}(\cosh y-a)^{-\frac{1}{2}}$ $\cdot\log\left\{\dfrac{[(\cosh y+1)^{\frac{1}{2}}+(\cosh y-a)^{\frac{1}{2}}]^2}{1+a}\right\}$

	$2Nf(x)$	$Ng(y)$
360	$\lvert\Gamma(\frac{1}{4}+i\frac{1}{2}x)\rvert^2 P_{-\frac{1}{2}+ix}(a),\quad 0<a<1$ $N=2^{\frac{3}{2}}\pi^{\frac{1}{2}}(1+r)^{-\frac{1}{2}}K[2^{\frac{1}{2}}(1+r^{-1})^{-\frac{1}{2}}]$	$2^{\frac{3}{2}}\pi^{\frac{1}{2}}(r+\cosh y)^{-\frac{1}{2}}$ $\cdot K[2^{\frac{1}{2}}(1+r^{-1}\cosh y)^{-\frac{1}{2}}],\quad r=(1-a^2)^{\frac{1}{2}}$
361	$\operatorname{sech}(\pi x)P^{\mu}_{-\frac{1}{2}+ix}(a),\quad -1<a<1,\quad \mu<1$ $N=2^{-\mu-\frac{1}{2}}(1+a)^{-\frac{1}{2}}\mathfrak{P}^{\mu}_{\mu}[(\frac{1}{2}+\frac{1}{2}a)^{\frac{1}{2}}]$	$2^{-\mu-\frac{1}{2}}(a+1)^{\frac{1}{2}\mu}(a+\cosh y)^{-\frac{1}{2}-\frac{1}{2}\mu}$ $\cdot\mathfrak{P}^{\mu}_{\mu}[(1+\cosh y)^{\frac{1}{2}}(a+\cosh y)^{-\frac{1}{2}}]$
362	$\lvert\Gamma(\mu+ix)\rvert^2 P^{\frac{1}{2}-\mu}_{-\frac{1}{2}+ix}(a),\quad -1<a<1,$ $\mu>0$ $N=(\frac{1}{2}\pi)^{\frac{1}{2}}\Gamma(\mu)(1-a)^{\frac{1}{2}\mu-\frac{1}{4}}(1+a)^{-\frac{1}{2}\mu-\frac{1}{4}}$	$(\frac{1}{2}\pi)^{\frac{1}{2}}\Gamma(\mu)(1-a^2)^{\frac{1}{2}\mu-\frac{1}{4}}$ $\cdot(a+\cosh y)^{-\mu}$
363	$\lvert\Gamma(\frac{1}{4}-\frac{1}{2}\mu+i\frac{1}{2}x)\rvert^2$ $\cdot P^{\mu}_{-\frac{1}{2}+ix}(a),\quad 0<a<1,\quad \mu<\frac{1}{2}$ $N=\pi^{\frac{1}{2}}2^{\mu+1}r^{-\frac{1}{2}}\mathfrak{Q}_{-\mu-\frac{1}{2}}(r^{-1})$	$\pi^{\frac{1}{2}}2^{\mu+1}r^{-\frac{1}{2}}\mathfrak{Q}_{-\mu-\frac{1}{2}}(r^{-1}\cosh y),\quad r=(1-a^2)^{\frac{1}{2}}$
364	$\cosh(\frac{1}{2}\pi x)\lvert\Gamma(\frac{1}{2}-\mu+ix)\rvert^2$ $\cdot P^{\mu}_{-\frac{1}{2}+ix}(a),\quad \mu<\frac{1}{2},\quad 0<a<1$ $N=(\frac{1}{2}\pi)^{\frac{1}{2}}(a^2-1)^{-\frac{1}{2}\mu}a^{\mu-\frac{1}{2}}\Gamma(\frac{1}{2}-\mu)$	$(\frac{1}{2}\pi)^{\frac{1}{2}}(a^2-1)^{-\frac{1}{2}\mu}\Gamma(\frac{1}{2}-\mu)$ $\cdot(a^2+\sinh^2 y)^{\frac{1}{2}\mu-\frac{1}{4}}$ $\cdot\cos[(\frac{1}{2}-\mu)\arctan(a^{-1}\sinh y)]$
365	$\operatorname{sech}(\pi x)\lvert\Gamma(\mu+ix)\rvert^2$ $\cdot P^{\frac{1}{2}-\mu}_{-\frac{1}{2}+ix}(a),\quad -1<a<1,\quad \mu>0$ $N=2^{-\mu+1}(1+a)^{\frac{1}{2}\mu-\frac{1}{4}}(1-a)^{-\frac{1}{2}}\Gamma(2\mu)$ $\cdot\exp[-i\pi(\frac{1}{2}-\mu)]\mathfrak{Q}^{\frac{1}{2}-\mu}_{\mu-\frac{1}{2}}[2^{\frac{1}{2}}(1-a)^{-\frac{1}{2}}]$	$2^{-\mu+1}(1-a^2)^{\frac{1}{2}\mu-\frac{1}{4}}\Gamma(2\mu)$ $\cdot e^{-i\pi(\frac{1}{2}-\mu)}(\cosh y-a)^{-\frac{1}{2}\mu-\frac{1}{4}}$ $\cdot\mathfrak{Q}^{\frac{1}{2}-\mu}_{\mu-\frac{1}{2}}[2^{\frac{1}{2}}\cosh(\frac{1}{2}y)(\cosh y-a)^{-\frac{1}{2}}]$
366	$\operatorname{sech}(\pi x)[P_{-\frac{1}{2}+ix}(a)]^2,\quad 0<a\le 1$ $N=\pi^{-1}K[(1-a^2)^{\frac{1}{2}}]$	$\pi^{-1}\operatorname{sech}(\frac{1}{2}y)$ $\cdot K[(1-a^2)^{\frac{1}{2}}\operatorname{sech}(\frac{1}{2}y)]$
367	$\lvert\Gamma(\frac{1}{2}-\mu+ix)\rvert^2[P^{\mu}_{-\frac{1}{2}+ix}(a)]^2,\quad \mu<\frac{1}{2},$ $0<a\le 1$ $N=(1-a^2)^{-\frac{1}{2}}\mathfrak{Q}_{-\mu-\frac{1}{2}}[(1+a^2)/(1-a^2)]$	$(1-a^2)^{-\frac{1}{2}}\mathfrak{Q}_{-\mu-\frac{1}{2}}[2(1-a^2)^{-1}\cosh^2(\frac{1}{2}y)-1]$
368	$[\operatorname{sech}(\pi x)]^2P_{-\frac{1}{2}+ix}(a)P_{-\frac{1}{2}+ix}(-a),\quad a<1$ $N=\pi^{-1}K(a)$	$\pi^{-1}\operatorname{sech}(\frac{1}{2}y)K\{[1-(1-a^2)\operatorname{sech}^2(\frac{1}{2}y)]^{\frac{1}{2}}\}$
369	$\operatorname{sech}(\pi x)[\mathfrak{P}_{-\frac{1}{2}+ix}(a)]^2,\quad a>1$ $N=a^{-1}\pi^{-1}K[(1-a^{-2})^{\frac{1}{2}}]$	$\pi^{-1}[a^2+\sinh^2(\frac{1}{2}y)]^{-\frac{1}{2}}$ $\cdot K\{a(1-a^{-2})^{\frac{1}{2}}[a^2+\sinh^2(\frac{1}{2}y)]^{-\frac{1}{2}}\}$
370	$\lvert\Gamma(\frac{1}{2}-\mu+ix)\rvert^2[\mathfrak{P}^{\mu}_{-\frac{1}{2}+ix}(a)]^2,\quad a>1,$ $\mu<\frac{1}{2}$ $N=(a^2-1)^{-\frac{1}{2}}\mathfrak{Q}_{-\mu-\frac{1}{2}}[1+2(a^2-1)^{-1}]$	$(a^2-1)^{-\frac{1}{2}}\mathfrak{Q}_{-\mu-\frac{1}{2}}[1+2(a^2-1)^{-1}\cosh^2(\frac{1}{2}y)]$

10. Bessel Functions

	$2Nf(x)$	$Ng(y)$
371	$x^{-\frac{1}{2}}[J_0(ax)]^2$ $N=(2a/\pi)^{-\frac{1}{2}}\Gamma(\frac{1}{4})[\Gamma(\frac{3}{4})]^{-3}$	$(\frac{1}{2}\pi/y)^{\frac{1}{2}}\{P_{-\frac{1}{4}}[(1-4a^2y^{-2})^{\frac{1}{2}}]\}^2, \quad y>2a$
372	$x^{-\frac{1}{2}}[J_\nu(ax)]^2, \quad \nu>-\frac{1}{4}$ $N=(2a/\pi)^{-\frac{1}{2}}\Gamma(\frac{1}{4}+\nu)$ $\cdot[\Gamma^2(\frac{3}{4})\Gamma(\frac{3}{4}+\nu)]^{-1}$	$(\frac{1}{2}\pi/y)^{\frac{1}{2}}\Gamma(\frac{1}{4}+\nu)[\Gamma(\frac{1}{4}-\nu)]^{-1}$ $\cdot\{P_{-\frac{1}{4}}^{-\nu}[(1-4a^2y^{-2})^{\frac{1}{2}}]\}^2, \quad y>2a$
373	$[x^\nu J_\nu(ax)]^2, \quad -\frac{1}{4}<\nu<0$ $N=\frac{1}{2}[\pi\Gamma(\frac{1}{4}-\nu)]^{-1}$ $\cdot\Gamma(-\nu)\Gamma(\frac{1}{2}+2\nu)$	$2^{3\nu-\frac{1}{2}}\pi^{-\frac{1}{2}}a^{\nu-\frac{1}{2}}y^{-\nu-\frac{1}{2}}[\Gamma(\frac{1}{2}-\nu)]^{-1}$ $\cdot(4a^2-y^2)^{-\nu}\{\pi\mathfrak{P}_{\nu-\frac{1}{2}}^{\nu}[(a/y)+(y/4a)]$ $-2e^{-i\pi\nu}\mathfrak{Q}_{\nu-\frac{1}{2}}^{\nu}[(a/y)+(y/4a)]\}, \quad y<2a$ $-2^{3\nu+\frac{1}{2}}\pi^{-\frac{1}{2}}\sin(\pi\nu)a^{\nu-\frac{1}{2}}[\Gamma(\frac{1}{2}-\nu)]^{-1}$ $\cdot y^{-\nu-\frac{1}{2}}(y^2-4a^2)^{-\nu}\exp(-i\pi\nu)$ $\cdot\mathfrak{Q}_{\nu-\frac{1}{2}}^{\nu}[(a/y)+(y/4a)], \quad y>2a$
374	$x^{-2\nu}[J_\nu(ax)]^2, \quad \nu>0$ $N=\frac{1}{2}a^{2\nu-1}\Gamma(\nu)$ $\cdot[\Gamma(\frac{1}{2}+\nu)\Gamma(\frac{1}{2}+2\nu)]^{-1}$	$2\pi^{-\frac{1}{2}}a^{1-2\nu}[\Gamma(2\nu)]^{-1}\Gamma(\frac{1}{2}+2\nu)$ $\cdot\int_{\arcsin(y/2b)}^{\frac{1}{2}\pi}\cos^{2\nu}t(4a^2\sin^2t-y^2)^{\nu-\frac{1}{2}}\,dt, \quad y<2a$ $0, \quad y>2a$
375	$x^{\frac{1}{2}}[J_{-\frac{1}{8}}(ax^2)]^2$ $N=2^{-\frac{5}{4}}\pi^{-\frac{1}{2}}\Gamma(\frac{1}{8})/\Gamma(\frac{7}{8})$	$-\frac{1}{4}a^{-1}(\frac{1}{2}\pi y)^{\frac{1}{2}}J_{-\frac{1}{8}}(y^2/16a)Y_{\frac{1}{8}}(y^2/16a)$
376	$J_0[b(a^2-x^2)^{\frac{1}{2}}], \quad x<a$ $0, \quad x>a$ $ab\leq\tau_{0,1}$ $N=b^{-1}\sin(ab)$	$z^{-1}\sin(az), \quad z=(b^2+y^2)^{\frac{1}{2}}$
377	$(a^2-x^2)^{-\frac{1}{2}}J_\nu[b(a^2-x^2)^{\frac{1}{2}}], \quad x<a$ $0, \quad x>a$ $\nu>-1,$ $ab\leq\tau_{\nu,1}$ $N=\frac{1}{2}\pi[J_{\frac{1}{2}\nu}(\frac{1}{2}ab)]^2$	$\frac{1}{2}\pi J_{\frac{1}{2}\nu}[\frac{1}{2}a(z+y)]J_{\frac{1}{2}\nu}[\frac{1}{2}a(z-y)],$ $z=(b^2+y^2)^{\frac{1}{2}}$
378	$(a^2-x^2)^{\frac{1}{2}\nu}J_\nu[b(a^2-x^2)^{\frac{1}{2}}], \quad x<a$ $0, \quad x>a$ $ab\leq\tau_{\nu,1}, \quad \nu>-1,$ $N=(\frac{1}{2}\pi a/b)^{\frac{1}{2}}J_{\nu+\frac{1}{2}}(ab)a^\nu$	$(\frac{1}{2}\pi a)^{\frac{1}{2}}(ab)^\nu(b^2+y^2)^{-\frac{1}{2}\nu-\frac{1}{4}}$ $\cdot J_{\nu+\frac{1}{2}}[a(b^2+y^2)^{\frac{1}{2}}]$

	$2Nf(x)$	$Ng(y)$
379	$-\log(a^2-x^2)J_0[b(a^2-x^2)^{\frac{1}{2}}], \quad x<a$ $0, \quad x>a$ $ab\le\tau_{0,1}, \quad a\le 1$ $N=-2b^{-1}\{\sin(ab)[\mathrm{Ci}(2ab)$ $-\mathrm{Ci}(ab)+\log a]$ $-\cos(ab)[\mathrm{Si}(2ab)-\mathrm{Si}(ab)]\}$	$-2z^{-1}\{\sin(az)[\mathrm{Ci}(2az)-\frac{1}{2}\,\mathrm{Ci}(az+ay)$ $-\frac{1}{2}\,\mathrm{Ci}(az-ay)+\log(ab)-\log z]$ $-\cos(az)[\mathrm{Si}(2az)-\frac{1}{2}\,\mathrm{Si}(az+ay)$ $-\frac{1}{2}\,\mathrm{Si}(az-ay)]\}, \quad z=(b^2+y^2)^{\frac{1}{2}}$
380	$J_0[b(ax-x^2)^{\frac{1}{2}}], \quad x<a$ $0, \quad x>a$ $ab\le 2\tau_{0,1}$ $N=2b^{-1}\sin(\frac{1}{2}ab)$	$2\cos(\frac{1}{2}ay)z^{-1}\sin(\frac{1}{2}az), \quad z=(b^2+y^2)^{\frac{1}{2}}$
381	$(ax-x^2)^{-\frac{1}{2}}J_\nu[b(ax-x^2)^{\frac{1}{2}}], \quad x<a$ $0, \quad x>a$ $ab\le 2\tau_{\nu,1}, \quad \nu>-1$ $N=\pi[J_{\frac{1}{2}\nu}(\frac{1}{4}ab)]^2$	$\pi\cos(\frac{1}{2}ay)J_{\frac{1}{2}\nu}[\frac{1}{4}a(z+y)]J_{\frac{1}{2}\nu}[\frac{1}{4}a(z-y)],$ $z=(b^2+y^2)^{\frac{1}{2}}$
382	$(ax-x^2)^{\frac{1}{2}\nu}J_\nu[b(ax-x^2)^{\frac{1}{2}}], \quad x<a$ $0, \quad x>a$ $ab\le 2\tau_{\nu,1}, \quad \nu>-1$ $N=(\frac{1}{2}a)^\nu(\pi a/b)^{\frac{1}{2}}J_{\nu+\frac{1}{2}}(\frac{1}{2}ab)$	$(\pi a)^{\frac{1}{2}}(\frac{1}{2}ab)^\nu\cos(\frac{1}{2}ay)$ $\cdot(b^2+y^2)^{-\frac{1}{2}\nu-\frac{1}{4}}J_{\nu+\frac{1}{2}}[\frac{1}{2}a(b^2+y^2)^{\frac{1}{2}}]$
383	$-Y_0[b(ax-x^2)^{\frac{1}{2}}], \quad x<a$ $0, \quad x>a$ $ab\le 2\rho_{0,1}$ $N=-4\pi^{-1}b^{-1}[\sin(\frac{1}{2}ab)\,\mathrm{Ci}(\frac{1}{2}ab)$ $-\cos(\frac{1}{2}ab)\,\mathrm{Si}(\frac{1}{2}ab)]$	$-2\pi^{-1}\cos(\frac{1}{2}ay)z^{-1}$ $\cdot\{\sin(\frac{1}{2}az)[\mathrm{Ci}(\frac{1}{2}az+\frac{1}{2}ay)$ $+\mathrm{Ci}(\frac{1}{2}az-\frac{1}{2}ay)]-\cos(\frac{1}{2}az)$ $\cdot[\mathrm{Si}(\frac{1}{2}az+\frac{1}{2}ay)+\mathrm{Si}(\frac{1}{2}az-\frac{1}{2}ay)]\},$ $z=(b^2+y^2)^{\frac{1}{2}}$
384	$-(ax-x^2)^{-\frac{1}{2}}Y_\nu[b(ax-x^2)^{\frac{1}{2}}], \quad x<a$ $0, \quad x>a$ $ab\le 2\rho_{\nu,1}, \quad -1<\nu<1$ $N=\pi\{2\cos(\frac{1}{2}\pi\nu)J_{\frac{1}{2}\nu}(\frac{1}{4}ab)$ $\cdot Y_{\frac{1}{2}\nu}(\frac{1}{4}ab)-\sin(\frac{1}{2}\pi\nu)$ $\cdot[J^2_{\frac{1}{2}\nu}(\frac{1}{4}ab)+Y^2_{\frac{1}{2}\nu}(\frac{1}{4}ab)]\}$	$-\pi\cos(\frac{1}{2}ay)\{\cos(\frac{1}{2}\pi\nu)$ $\cdot[J_{\frac{1}{2}\nu}(\frac{1}{4}az+\frac{1}{4}ay)Y_{\frac{1}{2}\nu}(\frac{1}{4}az-\frac{1}{4}ay)$ $+Y_{\frac{1}{2}\nu}(\frac{1}{4}az+\frac{1}{4}ay)J_{\frac{1}{2}\nu}(\frac{1}{4}az-\frac{1}{4}ay)]$ $-\sin(\frac{1}{2}\pi\nu)[J_{\frac{1}{2}\nu}(\frac{1}{4}az+\frac{1}{4}ay)J_{\frac{1}{2}\nu}(\frac{1}{4}az-\frac{1}{4}ay)$ $+Y_{\frac{1}{2}\nu}(\frac{1}{4}az+\frac{1}{4}ay)Y_{\frac{1}{2}\nu}(\frac{1}{4}az-\frac{1}{4}ay)]\},$ $z=(b^2+y^2)^{\frac{1}{2}}$
385	$-\log(ax-x^2)J_0[b(ax-x^2)^{\frac{1}{2}}], \quad x<a$ $0, \quad x>a$ $ab\le 2\tau_{0,1}, \quad a\le 2$ $N=4b^{-1}\{\sin(\frac{1}{2}ab)$ $\cdot[\mathrm{Ci}(ab)-\mathrm{Ci}(\frac{1}{2}ab)+\log(\frac{1}{2}a)]$ $-\cos(\frac{1}{2}ab)[\mathrm{Si}(ab)-\mathrm{Si}(\frac{1}{2}ab)]\}$	$-4z^{-1}\{\sin(\frac{1}{2}az)[\mathrm{Ci}(az)-\frac{1}{2}\,\mathrm{Ci}(\frac{1}{2}az+\frac{1}{2}ay)$ $-\frac{1}{2}\,\mathrm{Ci}(\frac{1}{2}az-\frac{1}{2}ay)+\log(\frac{1}{2}ab)-\log z]$ $-\cos(\frac{1}{2}az)[\mathrm{Si}(az)-\frac{1}{2}\,\mathrm{Si}(\frac{1}{2}az+\frac{1}{2}ay)$ $-\frac{1}{2}\,\mathrm{Si}(\frac{1}{2}az-\frac{1}{2}ay)]\}\cos(\frac{1}{2}ay),$ $z=(b^2+y^2)^{\frac{1}{2}}$

	$2Nf(x)$	$Ng(y)$
386	$-Y_0[b(a^2-x^2)^{\frac{1}{2}}],\quad x<a$ $0,\quad x>a$ $ab\leq\rho_{0,1}$ $N=-2\pi^{-1}b^{-1}[\sin(ab)\ \mathrm{Ci}(ab)$ $-\cos(ab)\ \mathrm{Si}(ab)]$	$-(\pi z)^{-1}\{\sin(az)[\mathrm{Ci}(az+ay)$ $+\mathrm{Ci}(az-ay)]-\cos(az)$ $\cdot[\mathrm{Si}(az+ay)+\mathrm{Si}(az-ay)]\},$ $z=(b^2+y^2)^{\frac{1}{2}}$
387	$-(a^2-x^2)^{-\frac{1}{2}}Y_\nu[b(a^2-x^2)^{\frac{1}{2}}],\quad x<a$ $0,\quad x>a$ $ab\leq\rho_{\nu,1},\quad -1<\nu<1$ $N=-\frac{1}{2}\pi\{2\cos(\frac{1}{2}\pi\nu)J_{\frac{1}{2}\nu}(\frac{1}{2}ab)Y_{\frac{1}{2}\nu}(\frac{1}{2}ab)$ $-\sin(\frac{1}{2}\pi\nu)[J^2_{\frac{1}{2}\nu}(\frac{1}{2}ab)+Y^2_{\frac{1}{2}\nu}(\frac{1}{2}ab)]\}$	$-\frac{1}{2}\pi\{\cos(\frac{1}{2}\pi\nu)[J_{\frac{1}{2}\nu}(z_1)Y_{\frac{1}{2}\nu}(z_2)$ $+Y_{\frac{1}{2}\nu}(z_1)J_{\frac{1}{2}\nu}(z_2)]-\sin(\frac{1}{2}\pi\nu)$ $\cdot[J_{\frac{1}{2}\nu}(z_1)J_{\frac{1}{2}\nu}(z_2)+Y_{\frac{1}{2}\nu}(z_1)Y_{\frac{1}{2}\nu}(z_2)]\},$ $z_{1,2}=\frac{1}{2}a[(b^2+y^2)^{\frac{1}{2}}\pm y]$
388	$(a^2+x^2)^{-\nu}\{J_\nu[b(a^2+x^2)^{\frac{1}{2}}]\}^2,\quad \nu>0$ $N=(2b^2)^{\nu-\frac{1}{2}}\Gamma(\nu)[\Gamma(\frac{1}{2}+2\nu)]^{-1}$ $\cdot{}_1F_2(\nu;\frac{1}{2}+\nu,2\nu+\frac{1}{2};-a^2b^2)$	$(2a)^{\frac{1}{2}-\nu}[\Gamma(\frac{1}{2}+\nu)]^{-1}$ $\cdot\int_{\arcsin(y/2b)}^{\pi/2}\cos^{2\nu}t(4b^2\sin^2t-y^2)^{\frac{1}{2}\nu-\frac{1}{4}}$ $\cdot J_{\nu-\frac{1}{2}}[a(4b^2\sin^2t-y^2)^{\frac{1}{2}}]\,dt,\quad y<2b$ $0,\quad y>2b$
389	$J_\nu(a\cos x),\quad x<\pi$ $0,\quad x>\pi$ $a\leq\tau_{\nu,1},\quad \nu>-1$ $N=\pi[J_{\frac{1}{2}\nu}(\frac{1}{2}a)]^2$	$\pi J_{\frac{1}{2}\nu-y}(\frac{1}{2}a)J_{\frac{1}{2}\nu+y}(\frac{1}{2}a)$
390	$J_\nu(a\sin x),\quad x<\pi$ $0,\quad x>\pi$ $a\leq\tau_{\nu,1},\quad \nu>-1$ $N=\pi[J_{\frac{1}{2}\nu}(\frac{1}{2}a)]^2$	$\pi\cos(\frac{1}{2}\pi y)J_{\frac{1}{2}\nu-\frac{1}{2}y}(\frac{1}{2}a)J_{\frac{1}{2}\nu+\frac{1}{2}y}(\frac{1}{2}a)$
391	$\csc x\,J_\nu(a\sin x),\quad x<\pi$ $0,\quad x>\pi$ $a\leq\tau_{\nu,1},\quad \nu>0$ $N=\frac{1}{2}\pi a\nu^{-1}$ $\cdot\{[J_{\frac{1}{2}(\nu-1)}(\frac{1}{2}a)]^2+[J_{\frac{1}{2}(\nu+1)}(\frac{1}{2}a)]^2\}$	$\frac{1}{2}\pi a\nu^{-1}\cos(\frac{1}{2}\pi y)$ $\cdot[J_{\frac{1}{2}(\nu-1-y)}(\frac{1}{2}a)J_{\frac{1}{2}(\nu-1+y)}(\frac{1}{2}a)$ $+J_{\frac{1}{2}(\nu+1-y)}(\frac{1}{2}a)J_{\frac{1}{2}(\nu+1+y)}(\frac{1}{2}a)]$
392	$\sec(\frac{1}{2}x)J_\nu[a\cos(\frac{1}{2}x)],\quad x<\pi$ $0,\quad x>\pi$ $a\leq\tau_{\nu,1},\quad \nu>0$ $N=\frac{1}{2}\pi a\nu^{-1}$ $\cdot\{[J_{\frac{1}{2}(\nu-1)}(\frac{1}{2}a)]^2+[J_{\frac{1}{2}(\nu+1)}(\frac{1}{2}a)]^2\}$	$\frac{1}{2}\pi a\nu^{-1}[J_{\frac{1}{2}(\nu-1-y)}(\frac{1}{2}a)J_{\frac{1}{2}(\nu-1+y)}(\frac{1}{2}a)$ $+J_{\frac{1}{2}(\nu+1-y)}(\frac{1}{2}a)J_{\frac{1}{2}(\nu+1+y)}(\frac{1}{2}a)]$

	$2Nf(x)$	$Ng(y)$
393	$(\cos x)^{-m}J_m(a\cos x),\quad x<\pi/2$ $0,\quad x>\pi/2$ $a\leq\tau_{m,1},\quad m=0,1,2,\ldots$ $N=\frac{1}{2}(\frac{1}{2}\pi a)^{\frac{1}{2}}m^{-1}[\mathbf{H}_{m-\frac{3}{2}}(a)+\mathbf{H}_{m+\frac{1}{2}}(a)]$	$\frac{1}{2}\pi(\frac{1}{2}a)^m m!\sum_{n=0}^{m}\epsilon_n[(m+n)!(m-n)!]^{-1}$ $\cdot J_{n-\frac{1}{2}y}(\frac{1}{2}a)J_{n+\frac{1}{2}y}(\frac{1}{2}a)$
394	$(\sin x)^{-m}J_m(a\sin x),\quad x<\pi$ $0,\quad x>\pi$ $a\leq\tau_{m,1},\quad m=0,1,2,\ldots$ $N=(\frac{1}{2}\pi a)^{\frac{1}{2}}m^{-1}$ $\cdot[\mathbf{H}_{m-\frac{3}{2}}(a)+\mathbf{H}_{m+\frac{1}{2}}(a)]$	$\pi(\frac{1}{2}a)^m m!\cos(\frac{1}{2}\pi y)$ $\cdot\sum_{n=0}^{m}\epsilon_n[(m+n)!(m-n)!]^{-1}J_{n-\frac{1}{2}y}(\frac{1}{2}a)$ $\cdot J_{n+\frac{1}{2}y}(\frac{1}{2}a)$
395	$J_\nu(2a\cos x)J_\mu(2a\cos x),\quad x<\pi/2$ $0,\quad x>\pi/2$ $a\leq\mathrm{Min}(\tau_{\nu,1};\tau_{\mu,1}),\quad \nu+\mu>-1$ $N=\frac{1}{2}\pi(\frac{1}{2}a)^{\nu+\mu}\Gamma(\nu+\mu+1)$ $\cdot\{\Gamma(1+\nu)\Gamma(1+\mu)[\Gamma(1+\frac{1}{2}\nu+\frac{1}{2}\mu)]^2\}^{-1}$ $\cdot{}_2F_3[\frac{1}{2}(\nu+\mu+1),\frac{1}{2}(\nu+\mu+1);$ $\nu+1,\mu+1,\nu+\mu+1;-4a^2]$	$\frac{1}{2}\pi(\frac{1}{2}a)^{\nu+\mu}\Gamma(\nu+\mu+1)$ $\cdot\{\Gamma(1+\nu)\Gamma(1+\mu)\Gamma[1+\frac{1}{2}(\nu+\mu+y)]$ $\cdot\Gamma[1+\frac{1}{2}(\nu+\mu-y)]\}^{-1}$ $\cdot{}_4F_5[\frac{1}{2}(1+\nu+\mu),\frac{1}{2}(1+\nu+\mu),$ $1+\frac{1}{2}(\nu+\mu),1+\frac{1}{2}(\nu+\mu);\nu+1,\mu+1,$ $\nu+\mu+1,1+\frac{1}{2}(\nu+\mu+y),$ $1+\frac{1}{2}(\nu+\mu-y);-4a^2]$
396	$J_0[(b-a\cosh x)^{\frac{1}{2}}],\quad \cosh x\leq b/a$ $0,\quad \cosh x>b/a$ $a\leq b\leq\tau_{0,1}$ $N=\frac{1}{2}\pi[J_0(z_1)Y_0(z_2)-J_0(z_2)Y_0(z_1)]$	$\frac{1}{2}\pi[J_{iy}(z_1)Y_{iy}(z_2)-J_{iy}(z_2)Y_{iy}(z_1)]$ $\cdot z_{\substack{1\\2}}=\frac{1}{2}[(b+a)^{\frac{1}{2}}\pm(b-a)^{\frac{1}{2}}]$
397	$(\cosh x)^{-\mu}J_\nu(a\,\mathrm{sech}\,x),\quad \nu+\mu>0,$ $a\leq\tau_{\nu,1}$ $N=2^{\mu-2}a^\nu[\Gamma(\nu+1)]^{-1}$ $\cdot B(\frac{1}{2}\mu+\frac{1}{2}\nu,\frac{1}{2}\mu+\frac{1}{2}\nu)$ $\cdot{}_1F_2[\frac{1}{2}(\mu+\nu);\frac{1}{2}(1+\mu+\nu),$ $1+\nu;-\frac{1}{2}a^2]$	$2^{\mu-2}a^\nu[\Gamma(\nu+1)\Gamma(\mu+\nu)]^{-1}$ $\cdot\lvert\Gamma\frac{1}{2}(\mu+\nu+iy)\rvert^2$ $\cdot{}_2F_3[\frac{1}{2}(\mu+\nu+iy,\frac{1}{2}(\mu+\nu-iy);$ $\frac{1}{2}(\mu+\nu),\frac{1}{2}(1+\mu+\nu),1+\nu;-\frac{1}{4}a^2)]$
398	$\mathrm{sech}(J_\nu x)\{[J_{ix}(a)]^2+[Y_{ix}(a)]^2\}$ $N=\mathbf{H}_0(2a)-Y_0(2a)$	$\mathbf{H}_0(2a\cosh\frac{1}{2}y)-Y_0(2a\cosh\frac{1}{2}y)$

11. Modified Bessel Functions

	$2Nf(x)$	$Ng(y)$
399	$e^{-bx}I_0(ax),\quad b>a$ $N=(b^2-a^2)^{-\frac{1}{2}}$	$2^{-\frac{1}{2}}[(b^2-a^2-y^2)^2+4b^2y^2]^{-\frac{1}{2}}$ $\cdot\{[(b^2-a^2-y^2)^2+4b^2y^2]^{\frac{1}{2}}+b^2-a^2-y^2\}^{\frac{1}{2}}$

	$2Nf(x)$	$Ng(y)$
400	$K_0(ax)$ $N=\frac{1}{2}\pi a^{-1}$	$\frac{1}{2}\pi(a^2+y^2)^{-\frac{1}{2}}$
401	$x^{2n}K_0(ax), \quad n=0, 1, 2, \ldots$ $N=\pi(2a)^{-2n-1}[(2n)!/n!]^2$	$(-1)^n\frac{1}{2}\pi(2n)!(a^2+y^2)^{-n-\frac{1}{2}}$ $\cdot P_{2n}[y(a^2+y^2)^{-\frac{1}{2}}]$
402	$x^{2n+1}K_0(ax), \quad n=0, 1, 2, \ldots$ $N=a^{-2n-2}2^{2n}(n!)^2$	$(-1)^{n+1}(2n+1)!$ $\cdot(a^2+y^2)^{-n-1}Q_{2n}[y(a^2+y^2)^{-\frac{1}{2}}]$
403	$x^{\frac{1}{2}}K_0(ax)$ $N=(2a)^{-\frac{1}{2}}a^{-1}[\Gamma(\frac{3}{4})]^2$	$\frac{1}{4}(2\pi)^{\frac{1}{2}}z^{\frac{3}{2}}\{2E[(\frac{1}{2}-\frac{1}{2}yz)^{\frac{1}{2}}]$ $+2E[(\frac{1}{2}+\frac{1}{2}yz)^{\frac{1}{2}}]-K[(\frac{1}{2}-\frac{1}{2}yz)^{\frac{1}{2}}]$ $-K[(\frac{1}{2}+\frac{1}{2}yz)^{\frac{1}{2}}]\}, \quad z=(a^2+y^2)^{-\frac{1}{2}}$
404	$x^{-\frac{1}{2}}K_0(ax)$ $N=2^{-\frac{3}{2}}a^{-\frac{1}{2}}[\Gamma(\frac{1}{4})]^2$	$(\frac{1}{2}\pi z)^{\frac{1}{2}}\{K[(\frac{1}{2}+\frac{1}{2}yz)^{\frac{1}{2}}]+K[(\frac{1}{2}-\frac{1}{2}yz)^{\frac{1}{2}}]\},$ $z=(a^2+y^2)^{-\frac{1}{2}}$
405	$K_\nu(ax), \quad -1<\nu<1$ $N=\frac{1}{2}\pi a^{-1}\sec(\frac{1}{2}\pi\nu)$	$\frac{1}{4}\pi\sec(\frac{1}{2}\pi\nu)(a^2+y^2)^{-\frac{1}{2}}$ $\cdot\{a^{-\nu}[y+(a^2+y^2)^{\frac{1}{2}}]^\nu+a^\nu[y+(a^2+y^2)^{\frac{1}{2}}]^{-\nu}\}$
406	$x^\nu K_\nu(ax), \quad \nu>-\frac{1}{2},$ $N=2^{\nu-1}\pi^{\frac{1}{2}}a^{-\nu-1}\Gamma(\frac{1}{2}+\nu)$	$2^{\nu-1}\pi^{\frac{1}{2}}a^\nu(a^2+y^2)^{-\nu-\frac{1}{2}}\Gamma(\frac{1}{2}+\nu)$
407	$x^\mu K_\nu(ax), \quad \mu+\nu>-1$ $N=a^{-\mu-1}2^{\mu-1}$ $\cdot\Gamma(\frac{1}{2}+\frac{1}{2}\mu+\frac{1}{2}\nu)\Gamma(\frac{1}{2}+\frac{1}{2}\mu-\frac{1}{2}\nu)$	$\frac{1}{4}\pi\sec[\frac{1}{2}\pi(\mu-\nu)]\Gamma(1+\mu+\nu)$ $\cdot(a^2+y^2)^{-\frac{1}{2}\mu-\frac{1}{2}}$ $\cdot\{P_\mu^{-\nu}[y(a^2+y^2)^{-\frac{1}{2}}]+P_\mu^{-\nu}[-y(a^2+y^2)^{-\frac{1}{2}}]\}$
408	$x^{-\nu}(b^2+x^2)^{-1}K_\nu(ax), \quad \nu<\frac{1}{2}$ $N=\frac{1}{4}\pi^2b^{-\nu-1}\sec(\pi\nu)$ $\cdot[\mathbf{H}_\nu(ab)-Y_\nu(ab)]$	$\frac{1}{4}\pi^{\frac{1}{2}}b^{-1}(2a)^{-\nu}\Gamma(\frac{1}{2}-\nu)$ $\cdot\left[e^{by}\int_y^\infty e^{-bt}(a^2+t^2)^{\nu-\frac{1}{2}}\,dt\right.$ $\left.+e^{-by}\int_{-y}^\infty e^{-bt}(a^2+t^2)^{\nu-\frac{1}{2}}\,dt\right]$
409	$I_0(bx)K_0(ax), \quad a>b$ $N=(a+b)^{-1}K[2(ab)^{\frac{1}{2}}(a+b)^{-1}]$	$[(a+b)^2+y^2]^{-\frac{1}{2}}$ $\cdot K\{2(ab)^{\frac{1}{2}}[(a+b)^2+y^2]^{-\frac{1}{2}}\}$
410	$I_\nu(bx)K_\nu(ax), \quad a>b, \quad \nu>-\frac{1}{2}$ $N=\frac{1}{2}(ab)^{-\frac{1}{2}}\mathfrak{Q}_{\nu-\frac{1}{2}}(a/2b+b/2a)$	$\frac{1}{2}(ab)^{-\frac{1}{2}}\mathfrak{Q}_{\nu-\frac{1}{2}}[(2ab)^{-1}(a^2+b^2+y^2)]$

	$2Nf(x)$	$Ng(y)$
411	$x^{-\frac{1}{2}}I_\nu(ax)K_\nu(ax), \quad \nu>-\frac{1}{4}$ $N=2^{-\frac{3}{2}}\pi^{\frac{1}{2}}\Gamma(\frac{1}{4})\Gamma(\frac{1}{4}+\nu)$ $\cdot[\Gamma(\frac{3}{4})\Gamma(\frac{3}{4}+\nu)]^{-1}$	$\Gamma(\frac{1}{4}+\nu)[\Gamma(\frac{1}{4}-\nu)]^{-1}(\frac{1}{2}\pi/y)^{\frac{1}{2}}$ $\cdot e^{i\pi\nu}\mathfrak{Q}_{-\frac{3}{4}}^{-\nu}[(1+4a^2y^{-2})^{\frac{1}{2}}]\mathfrak{P}_{-\frac{3}{4}}^{-\nu}[(1+4a^2y^{-2})^{\frac{1}{2}}]$
412	$x^{-2\nu}I_\nu(ax)K_\nu(ax), \quad 0<\nu<\frac{1}{2}$ $N=\frac{1}{4}a^{2\nu-1}\Gamma(\nu)\Gamma(\frac{1}{2}-\nu)[\Gamma(\frac{1}{2}+2\nu)]^{-1}$	$2^{-3\nu-1}\pi^{\frac{1}{2}}\Gamma(\frac{1}{2}-\nu)(y/a)^\nu$ $\cdot(y^2+4a^2)^{\nu-\frac{1}{2}}P_{\nu-\frac{1}{2}}^{-\nu}[(y^2-4a^2)(y^2+4a^2)^{-1}]$
413	$x^{\nu-\mu}I_\mu(bx)K_\nu(ax), \quad \nu>-\frac{1}{2}, \quad a>b$ $N=2^{\nu-\mu-1}b^\mu a^{-\nu-1}\pi^{\frac{1}{2}}\Gamma(\frac{1}{2}+\nu)$ $\cdot[\Gamma(1+\mu)]^{-1}\,{}_2F_1(\frac{1}{2}+\nu,\frac{1}{2};\mu+1;$ $b^2/a^2)$	$\frac{1}{2}\pi\int_0^\infty t^{\nu-\mu}J_\nu(at)J_\mu(bt)e^{-yt}\,dt$
414	$[K_0(ax)]^2$ $N=\frac{1}{4}\pi^2a^{-1}$	$\pi(4a^2+y^2)^{-\frac{1}{2}}K[y(4a^2+y^2)^{-\frac{1}{2}}]$
415	$K_0(ax)K_0(bx)$ $N=\pi(a+b)^{-1}K[(a-b)^{\frac{1}{2}}(a+b)^{-\frac{1}{2}}]$	$\pi[(a+b)^2+y^2]^{-\frac{1}{2}}$ $\cdot K\{[y^2+(a-b)^2]^{\frac{1}{2}}[y^2+(a+b)^2]^{-\frac{1}{2}}\}$
416	$K_\nu(ax)K_\nu(bx), \quad -\frac{1}{2}<\nu<\frac{1}{2}$ $N=\frac{1}{4}\pi^2(ab)^{-\frac{1}{2}}\sec(\pi\nu)$ $\cdot\mathfrak{P}_{\nu-\frac{1}{2}}[(a/2b)+(b/2a)]$	$\frac{1}{4}\pi^2\sec(\pi\nu)(ab)^{-\frac{1}{2}}$ $\cdot\mathfrak{P}_{\nu-\frac{1}{2}}[(2ab)^{-1}(y^2+a^2+b^2)]$
417	$x^{\frac{1}{2}}[K_\nu(ax)]^2, \quad -\frac{3}{4}<\nu<\frac{3}{4}$ $N=a^{-\frac{3}{2}}(2\pi)^{-\frac{1}{2}}\Gamma^2(\frac{3}{4})$ $\cdot\Gamma(\frac{3}{4}+\nu)\Gamma(\frac{3}{4}-\nu)$	$\Gamma(\frac{3}{4}+\nu)[\Gamma(-\frac{1}{4}-\nu)]^{-1}\exp(i2\pi\nu)$ $\cdot(\frac{1}{2}\pi y)^{\frac{1}{2}}(4a^2+y^2)^{-\frac{1}{2}}$ $\cdot\mathfrak{Q}_{-\frac{1}{4}}^{-\nu}[(1+4a^2y^{-2})^{\frac{1}{2}}]\mathfrak{Q}_{-\frac{1}{4}}^{-\nu}[(1+4a^2y^{-2})^{\frac{1}{2}}]$
418	$x^{-\frac{1}{2}}[K_\nu(ax)]^2, \quad -\frac{1}{4}<\nu<\frac{1}{4}$ $N=2^{-\frac{5}{2}}(a\pi)^{-\frac{1}{2}}\Gamma^2(\frac{1}{4})\Gamma(\frac{1}{4}+\nu)\Gamma(\frac{1}{4}-\nu)$	$\Gamma(\frac{1}{4}+\nu)[\Gamma(\frac{1}{4}-\nu)]^{-1}\exp(i2\pi\nu)(\frac{1}{2}\pi/y)^{\frac{1}{2}}$ $\cdot\{\mathfrak{Q}_{-\frac{3}{4}}^{-\nu}[(1+4a^2y^{-2})^{\frac{1}{2}}]\}^2$
419	$x^{\zeta-1}K_\nu(x)K_\mu(x), \quad \zeta>\|\mu\|+\|\nu\|$ $N=2^{\zeta-3}[\Gamma(\zeta)]^{-1}\Gamma[\frac{1}{2}(\zeta+\mu+\nu)]$ $\cdot\Gamma[\frac{1}{2}(\zeta+\mu-\nu)]\Gamma[\frac{1}{2}(\zeta-\mu+\nu)]$ $\cdot\Gamma[\frac{1}{2}(\zeta-\mu-\nu)]$	$2^{\zeta-3}[\Gamma(\zeta)]^{-1}\Gamma[\frac{1}{2}(\zeta+\mu+\nu)]\Gamma[\frac{1}{2}(\zeta-\mu+\nu)]$ $\cdot\Gamma[\frac{1}{2}(\zeta+\mu-\nu)]\Gamma[\frac{1}{2}(\zeta-\mu-\nu)]$ $\cdot{}_4F_3[\frac{1}{2}(\zeta+\mu+\nu),\frac{1}{2}(\zeta+\mu-\nu),\frac{1}{2}(\zeta-\mu+\nu),$ $\cdot(\zeta-\mu-\nu);\frac{1}{2},\frac{1}{2}\zeta,\frac{1}{2}(1+\zeta);-\frac{1}{4}y^2]$
420	$x^{2\nu}\exp(-x^2)I_\nu(x^2), \quad -\frac{1}{4}<\nu<0$ $N=\pi^{-1}2^{-\nu-\frac{3}{2}}\Gamma(-\nu)\Gamma(\frac{1}{2}+2\nu)$	$\pi^{-1}2^{-\nu-\frac{3}{2}}\Gamma(-\nu)\Gamma(\frac{1}{2}+2\nu)\exp(-\frac{1}{8}y^2)$ $\cdot{}_1F_1(\frac{1}{2}-\nu;1+\nu;\frac{1}{8}y^2)$ $+2^{2\nu-\frac{3}{2}}\Gamma(\nu)[\Gamma(\frac{1}{2}-\nu)]^{-1}$ $\cdot y^{-2\nu}\exp(-\frac{1}{8}y^2)\,{}_1F_1(\frac{1}{2}-2\nu;1-\nu;\frac{1}{8}y^2)$

	$2Nf(x)$	$Ng(y)$
421	$x^{-2\nu}\exp(-x^2)I_\nu(x^2),\quad \nu>0$ $N=2^{\nu-\frac{3}{2}}\Gamma(\nu)[\Gamma(\frac{1}{2}+2\nu)]^{-1}$	$2^{-\frac{1}{2}\nu}y^{\nu-1}\exp(-y^2/16)W_{-\frac{3}{2}\nu,\frac{1}{2}\nu}(\frac{1}{8}y^2)$
422	$K_0(ax^2)$ $N=2^{-\frac{5}{2}}a^{-\frac{1}{2}}\Gamma^2(\frac{1}{4})$	$\frac{1}{8}\pi a^{-1}yK_{\frac{1}{4}}(\frac{1}{8}y^2/a)$ $\cdot[I_{\frac{1}{4}}(\frac{1}{8}y^2/a)+I_{-\frac{1}{4}}(\frac{1}{8}y^2/a)]$
423	$x^{\frac{1}{2}}K_{\frac{1}{4}}(ax^2)$ $N=2^{-\frac{7}{4}}(a/\pi)^{-\frac{3}{2}}[\Gamma(\frac{3}{4})]^{-1}$	$\frac{1}{4}\pi a^{-1}(\frac{1}{2}\pi y)^{\frac{1}{2}}$ $\cdot[I_{-\frac{1}{4}}(\frac{1}{4}y^2/a)-\mathbf{L}_{-\frac{1}{4}}(\frac{1}{4}y^2/a)]$
424	$x^{\frac{3}{2}}K_{\frac{1}{4}}(ax^2)$ $N=2^{-\frac{5}{4}}\pi^{\frac{3}{2}}a^{-\frac{5}{4}}[\Gamma(\frac{1}{4})]^{-1}$	$\frac{1}{4}a^{-2}(\frac{1}{2}\pi y)^{\frac{3}{2}}$ $\cdot[I_{-\frac{3}{4}}(\frac{1}{4}y^2/a)-\mathbf{L}_{-\frac{3}{4}}(\frac{1}{4}y^2/a)]$
425	$x^{2\nu}K_\nu(ax^2),\quad \nu>-\frac{1}{4}$ $N=2^{\nu-2}\pi a^{-\nu-\frac{1}{2}}\Gamma(\frac{1}{4}+\nu)[\Gamma(\frac{3}{4})]^{-1}$	$2^{\nu-2}\pi a^{-\nu-\frac{1}{2}}\{\Gamma(\frac{1}{4}+\nu)[\Gamma(\frac{3}{4})]^{-1}$ $\cdot{}_1F_2(\frac{1}{4}+\nu;\frac{1}{2},\frac{3}{4};\frac{1}{64}y^2/a^2)$ $-(\frac{1}{4}y^2/a)\Gamma(\frac{3}{4}+\nu)[\Gamma(\frac{5}{4})]^{-1}$ $\cdot{}_1F_2(\frac{3}{4}+\nu;\frac{3}{2},\frac{5}{2};\frac{1}{64}y^2/a^2)\}$
426	$\exp(-x^2)K_0(x^2)$ $N=(\frac{1}{2}\pi)^{\frac{3}{2}}$	$(\frac{1}{2}\pi)^{\frac{3}{2}}\exp(-y^2/16)I_0(y^2/16)$
427	$x^{2\nu}\exp(-ax^2)K_\nu(ax^2),\quad \nu>-\frac{1}{4}$ $N=\frac{1}{2}(2a)^{-\nu-\frac{1}{2}}\pi\Gamma(\frac{1}{2}+2\nu)[\Gamma(1+\nu)]^{-1}$	$\frac{1}{2}\pi(2a)^{-\nu-\frac{1}{2}}\Gamma(\frac{1}{2}+2\nu)[\Gamma(1+\nu)]^{-1}$ $\cdot{}_1F_1(\frac{1}{2}-\nu;1+\nu;\frac{1}{8}y^2/a)$
428	$x^{-2\nu}\exp(x^2)K_\nu(x^2),\quad 0<\nu<\frac{1}{4}$ $N=2^{\nu-\frac{3}{2}}\cos(\pi\nu)\Gamma(\nu)\Gamma(\frac{1}{2}-2\nu)$	$\pi 2^{-\frac{1}{2}\nu}\Gamma(\frac{1}{2}-2\nu)[\Gamma(\frac{1}{2}+\nu)]^{-1}$ $\cdot y^{\nu-1}\exp(y^2/16)W_{\frac{3}{2}\nu,-\frac{1}{2}\nu}(\frac{1}{8}y^2)$
429	$x^\mu\exp(-x^2)K_\nu(x^2),$ $-1-2\nu<\mu<-1+2\nu$ $N=2^{-\frac{1}{2}\mu-\frac{3}{2}}\pi^{\frac{1}{2}}\Gamma(\frac{1}{2}+\frac{1}{2}\mu-\nu)$ $\cdot\Gamma(\frac{1}{2}+\frac{1}{2}\mu+\nu)[\Gamma(1+\frac{1}{2}\mu)]^{-1}$	$2^{-\frac{1}{2}\mu-\frac{3}{2}}\pi^{\frac{1}{2}}\Gamma(\frac{1}{2}+\frac{1}{2}\mu-\nu)\Gamma(\frac{1}{2}+\frac{1}{2}\mu+\nu)$ $\cdot[\Gamma(1+\frac{1}{2}\mu)]^{-1}$ $\cdot{}_2F_2(\frac{1}{2}+\frac{1}{2}\mu+\nu,\frac{1}{2}+\frac{1}{2}\mu-\nu;$ $\frac{1}{2},1+\frac{1}{2}\mu;\frac{1}{8}y^2)$
430	$x[K_{\frac{1}{4}}(ax^2)]^2$ $N=2^{-\frac{5}{2}}\pi^2a^{-1}$	$2^{-\frac{5}{2}}\pi^2a^{-1}[I_0(\frac{1}{8}y^2/a)-\mathbf{L}_0(\frac{1}{8}y^2/a)]$
431	$x^{-3}K_0(ax^{-1})$ $N=a^{-2}$	$-\pi a^{-1}yY_1[(2ay)^{\frac{1}{2}}]K_1[(2ay)^{\frac{1}{2}}]$
432	$K_0(ax^{\frac{1}{2}})$ $N=2a^{-2}$	$\frac{1}{2}y^{-1}[\mathrm{Ci}(\frac{1}{4}a^2/y)\sin(\frac{1}{4}a^2/y)$ $-\mathrm{si}(\frac{1}{4}a^2/y)\cos(\frac{1}{4}a^2/y)]$

	$2Nf(x)$	$Ng(y)$
433	$x^{-\frac{1}{2}}K_{2\nu}(ax^{\frac{1}{2}}), \quad -\frac{1}{2}<\nu<\frac{1}{2}$ $N=\pi a^{-1}\sec(\pi\nu)$	$-\frac{1}{4}\pi^{\frac{3}{2}}\sec(\pi\nu)y^{-\frac{1}{2}}$ $\cdot[\sin(\frac{1}{2}\pi\nu-\frac{1}{4}\pi-\frac{1}{8}a^2/y)J_\nu(\frac{1}{8}a^2/y)$ $+\cos(\frac{1}{2}\pi\nu-\frac{1}{4}\pi-\frac{1}{8}a^2/y)Y_\nu(\frac{1}{8}a^2/y)]$
434	$(a-x)^{\frac{1}{2}\nu}I_\nu[b(a-x)^{\frac{1}{2}}], \quad x<a$ $0, \quad x>a$ $\nu>-1$ $N=2b^{-1}a^{\frac{1}{2}\nu+\frac{1}{2}}I_{\nu+1}(ba^{\frac{1}{2}})$	$(\frac{1}{2}b)^\nu a^{\nu+1}y^{-\nu-1}U_{\nu+1}(2ay, iba^{\frac{1}{2}})$
435	$I_0[b(a^2-x^2)^{\frac{1}{2}}], \quad x<a$ $0, \quad x>a$ $N=b^{-1}\sinh(ab)$	$(y^2-b^2)^{-\frac{1}{2}}\sin[a(y^2-b^2)^{\frac{1}{2}}], \quad y>b$ $(b^2-y^2)^{-\frac{1}{2}}\sinh[a(b^2-y^2)^{\frac{1}{2}}], \quad y<b$
436	$(a^2-x^2)^{-\frac{1}{2}}I_{2\nu}[b(a^2-x^2)^{\frac{1}{2}}], \quad x<a$ $0, \quad x>a$ $\nu>-1$ $N=\frac{1}{2}\pi[I_\nu(\frac{1}{2}ab)]^2$	$\frac{1}{2}\pi J_\nu(z_1)J_\nu(z_2)$ $z_{1,2}=\frac{1}{2}a[y\pm(y^2-b^2)^{\frac{1}{2}}]$
437	$[x(a-x)]^{-\frac{1}{2}}I_{2\nu}\{b[x(a-x)]^{\frac{1}{2}}\}, \quad x<a$ $0, \quad x>a$ $\nu>-\frac{1}{2}$ $N=\pi[I_\nu(\frac{1}{4}ab)]^2$	$\pi\cos(\frac{1}{2}ay)J_\nu(z_1)J_\nu(z_2)$ $z_{1,2}=\frac{1}{4}a[y\pm(y^2-b^2)^{\frac{1}{2}}]$
438	$(a^2-x^2)^{\frac{1}{2}\nu}I_\nu[b(a^2-x^2)^{\frac{1}{2}}], \quad x<a$ $0, \quad x>a$ $\nu>-1$ $N=(\frac{1}{2}\pi a/b)^{\frac{1}{2}}a^\nu I_{\nu+\frac{1}{2}}(ab)$	$(\frac{1}{2}\pi a)^{\frac{1}{2}}(ab)^\nu(b^2-y^2)^{-\frac{1}{2}\nu-\frac{1}{4}}$ $\cdot I_{\nu+\frac{1}{2}}[a(b^2-y^2)^{\frac{1}{2}}], \quad y<b$ $(\frac{1}{2}\pi a)^{\frac{1}{2}}(ab)^\nu(y^2-b^2)^{-\frac{1}{2}\nu-\frac{1}{4}}$ $\cdot J_{\nu+\frac{1}{2}}[a(y^2-b^2)^{\frac{1}{2}}], \quad y>b$
439	$K_0[b(a^2+x^2)^{\frac{1}{2}}]$ $N=\frac{1}{2}\pi b^{-1}e^{-ab}$	$\frac{1}{2}\pi(b^2+y^2)^{-\frac{1}{2}}\exp[-a(b^2+y^2)^{\frac{1}{2}}]$
440	$K_{2n}[b(a^2+x^2)^{\frac{1}{2}}], \quad n=1, 2, 3, \ldots$ $N=\pi^{\frac{1}{2}}b^{-1}n\sum_{k=0}^{n}(n+k-1)![k!(n-k)!]^{-1}$ $\cdot(\frac{1}{2}ab)^{\frac{1}{2}-k}K_{k-\frac{1}{2}}(ab)$	$(\frac{1}{2}\pi a)^{\frac{1}{2}}n\sum_{k=0}^{n}(n+k-1)![k!(n-k)!]^{-1}$ $\cdot(\frac{1}{2}b^2a)^{-k}(b^2+y^2)^{\frac{1}{2}k-\frac{1}{4}}K_{k-\frac{1}{2}}[a(b^2+y^2)^{\frac{1}{2}}]$
441	$\cos[\nu\arctan(x/b)]K_\nu[a(b^2+x^2)^{\frac{1}{2}}],$ $N=\frac{1}{2}\pi a^{-1}e^{-ab} \quad -1\le\nu\le 1$	$\frac{1}{4}\pi a^{-\nu}(a^2+y^2)^{-\frac{1}{2}}\exp[-b(a^2+y^2)^{\frac{1}{2}}]$ $\cdot\{[(a^2+y^2)^{\frac{1}{2}}+y]^\nu+[(a^2+y^2)^{\frac{1}{2}}-y]^\nu\}$

	$2Nf(x)$	$Ng(y)$
442	$I_0[b(ax-x^2)^{\frac{1}{2}}],\quad x<a$ $0,\quad x>a$ $N=2b^{-1}\sinh(\frac{1}{2}ab)$	$2\cos(\frac{1}{2}ay)$ $\cdot\begin{cases}(b^2-y^2)^{-\frac{1}{2}}\sinh[\frac{1}{2}a(b^2-y^2)^{\frac{1}{2}}], & y<b\\(y^2-b^2)^{-\frac{1}{2}}\sin[\frac{1}{2}a(y^2-b^2)^{\frac{1}{2}}], & y>b\end{cases}$
443	$(ax-x^2)^{\frac{1}{2}\nu}I_\nu[b(ax-x^2)^{\frac{1}{2}}],\quad x<a$ $0,\quad x>a$ $\nu>-1$ $N=(\pi a/b)^{\frac{1}{2}}(\frac{1}{2}a)^\nu I_{\nu+\frac{1}{2}}(\frac{1}{2}ab)$	$(\pi a)^{\frac{1}{2}}(\frac{1}{2}ab)^\nu\cos(\frac{1}{2}ay)$ $\cdot\begin{cases}(b^2-y^2)^{-\frac{1}{2}\nu-\frac{1}{4}}I_{\nu+\frac{1}{2}}[\frac{1}{2}a(b^2-y^2)^{\frac{1}{2}}], & y<b\\(y^2-b^2)^{-\frac{1}{2}\nu-\frac{1}{4}}J_{\nu+\frac{1}{2}}[\frac{1}{2}a(y^2-b^2)^{\frac{1}{2}}], & y>b\end{cases}$
444	$K_0[b(ax-x^2)^{\frac{1}{2}}],\quad x<a$ $0,\quad x>a$ $N=b^{-1}\{i\cosh(\frac{1}{2}ab)$ $\cdot[\mathrm{Si}(-i\frac{1}{2}ab)-\mathrm{Si}(i\frac{1}{2}ab)]$ $-\sinh(\frac{1}{2}ab)[\mathrm{Ci}(i\frac{1}{2}ab)$ $+\mathrm{Ci}(-i\frac{1}{2}ab)]\}$	$-\cos(\frac{1}{2}ay)(y^2-b^2)^{-\frac{1}{2}}$ $\cdot\{\sin\alpha[\mathrm{Ci}(z_1)+\mathrm{Ci}(z_2)]$ $-\cos\alpha[\mathrm{Si}(z_1)-\mathrm{Si}(z_2)]\}$ $z_{1,2}=\frac{1}{2}a[y\pm(y^2-b^2)^{\frac{1}{2}}]$ $\alpha=\frac{1}{2}a(y^2-b^2)^{\frac{1}{2}}$ $\arg\alpha=\begin{cases}0, & y>b\\\frac{1}{2}\pi, & y<b\end{cases}$
445	$(ax-x^2)^{-\frac{1}{2}}K_\nu[b(ax-x^2)^{\frac{1}{2}}],\quad x<a$ $0,\quad x>a$ $-1<\nu<1$ $N=\frac{1}{2}\pi\sec(\frac{1}{2}\pi\nu)K_{\frac{1}{2}\nu}(\frac{1}{4}ab)$ $\cdot[I_{\frac{1}{2}\nu}(\frac{1}{4}ab)+I_{-\frac{1}{2}\nu}(\frac{1}{4}ab)]$	$-\frac{1}{4}\pi^2\sec(\frac{1}{2}\pi\nu)\cos(\frac{1}{2}ay)$ $\cdot[J_{\frac{1}{2}\nu}(\frac{1}{2}z_1)Y_{-\frac{1}{2}\nu}(\frac{1}{2}z_2)$ $+Y_{\frac{1}{2}\nu}(\frac{1}{2}z_1)J_{-\frac{1}{2}\nu}(\frac{1}{2}z_2)],\quad y>b$ $z_{1,2}=\frac{1}{2}a[y\pm(y^2-b^2)^{\frac{1}{2}}]$
446	$[x(1+x)]^{-\frac{1}{2}}K_{2\nu}\{b[x(1+x)]^{\frac{1}{2}}\},$ $-\frac{1}{2}<\nu<\frac{1}{2}$ $N=\frac{1}{8}\pi^2\sec(\pi\nu)$ $\cdot\{[J_\nu(\frac{1}{4}b)]^2+[Y_\nu(\frac{1}{4}b)]^2\}$	$\frac{1}{8}\pi^2\sec(\pi\nu)\{\cos(\frac{1}{2}y)$ $\cdot[J_\nu(z_1)J_\nu(z_2)+Y_\nu(z_1)Y_\nu(z_2)]+\sin(\frac{1}{2}y)$ $\cdot[J_\nu(z_2)Y_\nu(z_1)-J_\nu(z_1)Y_\nu(z_2)]\}$ $z_{1,2}=\frac{1}{4}[(b^2+y^2)^{\frac{1}{2}}\pm y]$
447	$(a^2+x^2)^{-\frac{1}{2}}K_1[b(a^2+x^2)^{\frac{1}{2}}]$ $N=\frac{1}{2}\pi(ab)^{-1}e^{-ab}$	$\frac{1}{2}\pi(ab)^{-1}\exp[-a(b^2+y^2)^{\frac{1}{2}}]$
448	$(a^2+x^2)^{-\frac{1}{2}}K_{2\nu}[b(a^2+x^2)^{\frac{1}{2}}]$ $N=\frac{1}{2}[K_\nu(\frac{1}{2}ab)]^2$	$\frac{1}{2}K_\nu(z_1)K_\nu(z_2)$ $z_{1,2}=\frac{1}{2}a[(b^2+y^2)^{\frac{1}{2}}\pm y]$
449	$(a^2+x^2)^{-\frac{1}{2}\nu}K_\nu[b(a^2+x^2)^{\frac{1}{2}}]$ $N=\frac{1}{2}(2\pi a/b)^{\frac{1}{2}}a^{-\nu}K_{\nu-\frac{1}{2}}(ab)$	$(\frac{1}{2}\pi a)^{\frac{1}{2}}(ab)^{-\nu}$ $\cdot(b^2+y^2)^{\frac{1}{2}\nu-\frac{1}{4}}K_{\nu-\frac{1}{2}}[a(b^2+y^2)^{\frac{1}{2}}]$
450	$(a^2+x^2)^{-1}K_0[b(a^2+x^2)^{\frac{1}{2}}]$ $N=-\frac{1}{2}\pi a^{-1}\,\mathrm{Ei}(-ab)$	$-\frac{1}{4}\pi a^{-1}[e^{-ay}\,\mathrm{Ei}(-z_2)+e^{ay}\,\mathrm{Ei}(-z_1)]$ $z_{1,2}=a[(b^2+y^2)^{\frac{1}{2}}\pm y]$

	$2Nf(x)$	$Ng(y)$
451	$K_0[b(2ax+x^2)^{\frac{1}{2}}]$ $N=b^{-1}[\sin(ab)\ \mathrm{Ci}(ab)$ $\quad -\cos(ab)\ \mathrm{si}(ab)]$	$\frac{1}{2}a\alpha^{-1}\cos(ay)\{\sin\alpha[\mathrm{Ci}(z_1)+\mathrm{Ci}(z_2)]$ $\quad -\cos\alpha[\mathrm{si}(z_1)+\mathrm{si}(z_2)]\}+\frac{1}{2}a\alpha^{-1}\sin(ay)$ $\quad \cdot\{\cos\alpha[\mathrm{Ci}(z_1)-\mathrm{Ci}(z_2)]+\sin\alpha$ $\quad \cdot[\mathrm{si}(z_1)-\mathrm{si}(z_2)]\}$ $\alpha=a(b^2+y^2)^{\frac{1}{2}},\quad z_{1,2}=a[(b^2+y^2)^{\frac{1}{2}}\pm y]$
452	$\log(a^2+x^2)K_0[b(a^2+x^2)^{\frac{1}{2}}]$ $N=\pi b^{-1}[e^{-ab}\log a-e^{ab}\,\mathrm{Ei}(-2ab)]$	$\pi(b^2+y^2)^{-\frac{1}{2}}\{\exp[-a(b^2+y^2)^{\frac{1}{2}}][\log(ab)$ $\quad -\frac{1}{2}\log(b^2+y^2)]-\exp[a(b^2+y^2)^{\frac{1}{2}}]$ $\quad \cdot\mathrm{Ei}[-2a(b^2+y^2)^{\frac{1}{2}}]\}$
453	$K_0[b(a^2-x^2)^{\frac{1}{2}}],\quad x<a$ $0,\quad x>a$ $N=\frac{1}{2}b^{-1}\{i\cosh(ab)[\mathrm{Si}(-iab)-\mathrm{Si}(iab)]$ $\quad -\sinh(ab)[\mathrm{Ci}(iab)+\mathrm{Ci}(-iab)]\}$	$-\frac{1}{2}(y^2-b^2)^{-\frac{1}{2}}\{\sin\alpha[\mathrm{Ci}(z_1)+\mathrm{Ci}(z_2)]$ $\quad -\cos\alpha[\mathrm{Si}(z_1)-\mathrm{Si}(z_2)]\}$ $\alpha=a(y^2-b^2)^{\frac{1}{2}},\quad z_{1,2}=ay\pm\alpha$ $\arg(y^2-b^2)=\begin{cases}0, & y>b\\ \frac{1}{2}\pi, & y<b\end{cases}$
454	$(a^2-x^2)^{-\frac{1}{2}}K_\nu[b(a^2-x^2)^{\frac{1}{2}}],\quad x<a$ $0,\quad x>a$ $-1<\nu<1$ $N=\frac{1}{4}\pi\ \sec(\frac{1}{2}\pi\nu)K_{\frac{1}{2}\nu}(\frac{1}{2}ab)$ $\quad \cdot[I_{\frac{1}{2}\nu}(\frac{1}{2}ab)+I_{-\frac{1}{2}\nu}(\frac{1}{2}ab)]$	$-\frac{1}{8}\pi^2\sec(\frac{1}{2}\pi\nu)$ $\quad \cdot[J_{\frac{1}{2}\nu}(z_1)Y_{-\frac{1}{2}\nu}(z_2)+Y_{\frac{1}{2}\nu}(z_1)J_{-\frac{1}{2}\nu}(z_2)]$ $z_{1,2}=\frac{1}{2}a[y\pm(y^2-b^2)^{\frac{1}{2}}],\quad y>b$
455	$(a^2+x^2)^{-1}K_\nu[b(a^2+x^2)^{\frac{1}{2}}]$ $N=\frac{1}{4}b\nu^{-1}\{[K_{\frac{1}{2}+\frac{1}{2}\nu}(\frac{1}{2}ab)]^2$ $\quad -[K_{\frac{1}{2}-\frac{1}{2}\nu}(\frac{1}{2}ab)]^2\}$	$\frac{1}{4}b\nu^{-1}[K_{\frac{1}{2}+\frac{1}{2}\nu}(\frac{1}{2}az_1)K_{\frac{1}{2}+\frac{1}{2}\nu}(\frac{1}{2}az_2)$ $\quad -K_{\frac{1}{2}-\frac{1}{2}\nu}(\frac{1}{2}az_1)K_{\frac{1}{2}-\frac{1}{2}\nu}(\frac{1}{2}az_2)]$ $z_{1,2}=(b^2+y^2)^{\frac{1}{2}}\pm y$
456	$0,\quad x<a$ $K_0[b(x^2-a^2)^{\frac{1}{2}}],\quad x>a$ $N=a^{-1}[\sin(ab)\ \mathrm{Ci}(ab)$ $\quad -\cos(ab)\ \mathrm{si}(ab)]$	$\frac{1}{2}(b^2+y^2)^{-\frac{1}{2}}\{\sin\alpha[\mathrm{Ci}(z_1)+\mathrm{Ci}(z_2)]$ $\quad -\cos\alpha[\mathrm{si}(z_1)+\mathrm{si}(z_2)]\}$ $\alpha=a(b^2+y^2)^{\frac{1}{2}},\quad z_{1,2}=\alpha\pm ay$
457	$0,\quad x<a$ $(x^2-a^2)^{-\frac{1}{2}}K_\nu[b(x^2-a^2)^{\frac{1}{2}}],\quad x>a$ $-1<\nu<1$ $N=\frac{1}{8}\pi^2\sec(\frac{1}{2}\pi\nu)$ $\quad \cdot\{[J_{\frac{1}{2}\nu}(\frac{1}{2}ab)]^2+[Y_{\frac{1}{2}\nu}(\frac{1}{2}ab)]^2\}$	$\frac{1}{8}\pi^2\sec(\frac{1}{2}\pi\nu)$ $\quad \cdot[J_{\frac{1}{2}\nu}(z_1)J_{\frac{1}{2}\nu}(z_2)+Y_{\frac{1}{2}\nu}(z_1)Y_{\frac{1}{2}\nu}(z_2)]$ $z_{1,2}=\frac{1}{2}a[(b^2+y^2)^{\frac{1}{2}}\pm y]$
458	$I_\nu[a(c^2+x^2)^{\frac{1}{2}}]K_\mu[b(c^2+x^2)^{\frac{1}{2}}],\quad a<b$ $N=\frac{1}{2}\pi c^{\mu-\nu}I_\nu(ac)K_\mu(bc)$	$\frac{1}{2}\pi\int_0^\infty t^{\mu-\nu+1}J_\nu(at)J_\mu(bt)$ $\quad \cdot(t^2+c^2)^{-\frac{1}{2}}\exp[-y(c^2+t^2)^{\frac{1}{2}}]\,dt$

	$2Nf(x)$	$Ng(y)$
459	$I_0\{a[(b^2+x^2)^{\frac{1}{2}}-x]\}$ $\cdot K_0\{a[(b^2+x^2)^{\frac{1}{2}}+x]\}$ $N=\frac{1}{4}\pi a^{-1}[I_0(2ab)-\mathbf{L}_0(2ab)]$	$\frac{1}{4}\pi(4a^2+y^2)^{-\frac{1}{2}}$ $\cdot\{I_0[b(4a^2+y^2)^{\frac{1}{2}}]-\mathbf{L}_0[b(4a^2+y^2)^{\frac{1}{2}}]\}$
460	$(a^2+x^2)^{\frac{1}{2}\nu}K_\nu[b(a^2+x^2)^{\frac{1}{2}}]$ $N=(\frac{1}{2}\pi a/b)^{\frac{1}{2}}a^\nu K_{\nu+\frac{1}{2}}(ab)$	$(\frac{1}{2}\pi a)^{\frac{1}{2}}(ab)^\nu$ $\cdot(b^2+y^2)^{-\frac{1}{2}\nu-\frac{1}{4}}K_{\nu+\frac{1}{2}}[a(b^2+y^2)^{\frac{1}{2}}]$
461	$K_0\{a[x+(x^2-b^2)^{\frac{1}{2}}]\}$ $\cdot K_0\{a[x-(x^2-b^2)^{\frac{1}{2}}]\}$ $N=\frac{1}{4}\pi^2a^{-1}[\mathbf{H}_0(2ab)-Y_0(2ab)]$	$\frac{1}{2}\pi^2(4a^2+y^2)^{-\frac{1}{2}}$ $\cdot\{\mathbf{H}_0[b(4a^2+y^2)^{\frac{1}{2}}]-Y_0[b(4a^2+y^2)^{\frac{1}{2}}]\}$
462	$I_\nu\{a[(b^2+x^2)^{\frac{1}{2}}-x]\}$ $\cdot K_\nu\{a[(b^2+x^2)^{\frac{1}{2}}+x]\}$ $N=a^{-1}[\frac{1}{4}\pi\sec(\pi\nu)I_{2\nu}(2ab)$ $+\frac{1}{2}is_{0,2\nu}(i2ab)]$	$(4a^2+y^2)^{-\frac{1}{2}}\{\frac{1}{2}\pi\sec(\pi\nu)I_{2\nu}[b(4a^2+y^2)^{\frac{1}{2}}]$ $+is_{0,2\nu}[ib(4a^2+y^2)^{\frac{1}{2}}]\}$
463	$K_\nu\{a[x+(x^2-b^2)^{\frac{1}{2}}]\}$ $\cdot K_\nu\{a[x-(x^2-b^2)^{\frac{1}{2}}]\}$ $N=\frac{1}{2}\pi a^{-1}S_{0,2\nu}(2ab)$	$\pi(4a^2+y^2)^{-\frac{1}{2}}S_{0,2\nu}[b(4a^2+y^2)^{\frac{1}{2}}]$
464	$K_\nu\{a[(b^2+x^2)^{\frac{1}{2}}-x]\}$ $\cdot K_\nu\{a[(b^2+x^2)^{\frac{1}{2}}+x]\}$ $N=\frac{1}{2}\pi a^{-1}K_{2\nu}(2ab)$	$\pi(4a^2+y^2)^{-\frac{1}{2}}K_{2\nu}[b(4a^2+y^2)^{\frac{1}{2}}]$
465	$K_0[a(ix)^{\frac{1}{2}}]K_0[a(-ix)^{\frac{1}{2}}]$ $N=\frac{1}{2}\pi a^{-2}$	$\frac{1}{8}\pi^2y^{-1}[\mathbf{H}_0(\frac{1}{2}a^2y^{-1})-Y_0(\frac{1}{2}a^2y^{-1})]$
466	$K_\nu[a(ix)^{\frac{1}{2}}]K_\nu[a(-ix)^{\frac{1}{2}}],\quad -1<\nu<1$ $N=\frac{1}{2}\pi a^{-2}\sec(\frac{1}{2}\pi\nu)$	$\frac{1}{4}\pi^2y^{-1}\sec(\frac{1}{2}\pi\nu)S_{0,\nu}(\frac{1}{2}a^2y^{-1})$
467	$I_{2\nu}(2a\sin x),\quad x<\pi$ $0,\quad x>\pi$ $\nu>-\frac{1}{2}$ $N=\pi[I_\nu(a)]^2$	$\pi\cos(\frac{1}{2}\pi y)I_{\nu-\frac{1}{2}y}(a)I_{\nu+\frac{1}{2}y}(a)$
468	$I_{2\nu}(2a\cos\frac{1}{2}x),\quad x<\pi$ $0,\quad x>\pi$ $\nu>-\frac{1}{2}$ $N=\pi[I_\nu(a)]^2$	$\pi I_{\nu-y}(a)I_{\nu+y}(a)$

	$2Nf(x)$	$Ng(y)$
469	$(\sin x)^{-m} I_m(2a \sin x), \quad x<\pi$ $0, \quad x>\pi$ $m=0, 1, 2, \ldots$ $N=\pi a^m m! \sum_{n=0}^{m} (-1)^n \epsilon_n [(m+n)!$ $\cdot (m-n)!]^{-1} [I_n(a)]^2$	$\pi a^m m! \cos(\frac{1}{2}\pi y) \sum_{n=0}^{m} (-1)^n \epsilon_n$ $\cdot [(m+n)!(m-n)!]^{-1} I_{n-\frac{1}{2}y}(a) I_{n+\frac{1}{2}y}(a)$
470	$(\cos\frac{1}{2}x)^{-m} I_m(2a \cos\frac{1}{2}x), \quad x<\pi$ $0, \quad x>\pi$ $m=0, 1, 2, \ldots$ $N=\pi a^m m! \sum_{n=0}^{m} (-1)^n \epsilon_n [(m+n)!$ $\cdot (m-n)!]^{-1} [I_n(a)]^2$	$\pi a^m m! \sum_{n=0}^{m} (-1)^n \epsilon_n$ $\cdot [(m+n)!(m-n)!]^{-1} I_{n-y}(a) I_{n+y}(a)$
471	$\csc x\, I_{2\nu}(2a \sin x), \quad x<\pi$ $0, \quad x>\pi$ $\nu>0$ $N=\frac{1}{2}\pi a \nu^{-1} \{[I_{\nu-\frac{1}{2}}(a)]^2 + [I_{\nu+\frac{1}{2}}(a)]^2\}$	$\frac{1}{2}\pi a \nu^{-1} \cos(\frac{1}{2}\pi y)$ $\cdot [I_{\nu-\frac{1}{2}-\frac{1}{2}y}(a) I_{\nu-\frac{1}{2}+\frac{1}{2}y}(a)$ $- I_{\nu+\frac{1}{2}-\frac{1}{2}y}(a) I_{\nu+\frac{1}{2}+\frac{1}{2}y}(a)]$
472	$\sec(\frac{1}{2}x) I_{2\nu}(2a \cos\frac{1}{2}x), \quad x<\pi$ $0, \quad x>\pi$ $\nu>0$ $N=\frac{1}{2}\pi a \nu^{-1} \{[I_{\nu-\frac{1}{2}}(a)]^2 - [I_{\nu+\frac{1}{2}}(a)]^2\}$	$\frac{1}{2}\pi a \nu^{-1} [I_{\nu-\frac{1}{2}-y}(a) I_{\nu-\frac{1}{2}+y}(a)$ $- I_{\nu+\frac{1}{2}-y}(a) I_{\nu+\frac{1}{2}+y}(a)]$
473	$(\operatorname{sech} x)^{\mu} I_\nu(a \operatorname{sech} x), \quad \mu+\nu>0$ $N=2^{\mu-2} a^\nu [\Gamma(1+\nu)]^{-1} B[\frac{1}{2}(\mu+\nu),$ $\frac{1}{2}(\mu+\nu)]\, {}_1F_2[\frac{1}{2}(\mu+\nu);$ $\frac{1}{2}(1+\mu+\nu), \nu+1; \frac{1}{4}a^2)]$	$2^{\mu-2} a^\nu [\Gamma(1+\nu)\Gamma(\mu+\nu)]^{-1}$ $\cdot \lvert \Gamma(\frac{1}{2}\mu+\frac{1}{2}\nu+i\frac{1}{2}y) \rvert^2$ $\cdot {}_2F_3[\frac{1}{2}(\mu+\nu+iy), \frac{1}{2}(\mu+\nu-iy);$ $\frac{1}{2}(\mu+\nu), \frac{1}{2}(1+\mu+\nu), \nu+1; \frac{1}{4}a^2]$
474	$K_{2\nu}(2a \sin x), \quad x<\pi$ $0, \quad x>\pi$ $-\frac{1}{2}<\nu<\frac{1}{2}$ $N=\frac{1}{2}\pi^2 \csc(2\pi\nu) \{[I_{-\nu}(a)]^2 - [I_\nu(a)]^2\}$	$\frac{1}{2}\pi^2 \csc(2\pi\nu) \cos(\frac{1}{2}\pi y)$ $\cdot [I_{-\nu-\frac{1}{2}y}(a) I_{-\nu+\frac{1}{2}y}(a)$ $- I_{\nu-\frac{1}{2}y}(a) I_{\nu+\frac{1}{2}y}(a)]$
475	$K_{2\nu}(2a \cos\frac{1}{2}x), \quad x<\pi$ $0, \quad x>\pi$ $-\frac{1}{2}<\nu<\frac{1}{2}$ $N=\frac{1}{2}\pi^2 \csc(2\pi\nu) \{[I_{-\nu}(a)]^2 - [I_\nu(a)]^2\}$	$\frac{1}{2}\pi^2 \csc(2\pi\nu) [I_{-\nu-y}(a) I_{-\nu+y}(a)$ $- I_{\nu-y}(a) I_{\nu+y}(a)]$

	$2Nf(x)$	$Ng(y)$
476	$K_0[(a^2+b^2-2ab\cos x)^{\frac{1}{2}}]$, $x<\pi$ 0, $x>\pi$ $N=\pi I_0(b)K_0(a)$, $a\geq b$	$y\sin(\pi y)\sum_{n=0}^{\infty}(-1)^n\epsilon_n(y^2-n^2)^{-1}I_n(b)K_n(a)$
477	$(\sec x\cos\frac{1}{2}x)^{\frac{1}{2}}$ $\cdot\exp(-a^2\sec x)K_{\frac{1}{4}}[a^2(1+\sec x)]$, $x<\frac{1}{2}\pi$ 0, $x>\frac{1}{2}\pi$ $N=2^{-\frac{1}{2}}\pi a^{-\frac{1}{2}}\exp(-a^2)[D_{-\frac{3}{4}}(2a)]^2$	$(2a)^{-\frac{1}{2}}\pi\exp(-a^2)D_{-\frac{3}{4}+y}(2a)D_{-\frac{3}{4}-y}(2a)$
478	$K_{2\nu}(2a\sinh\frac{1}{2}x)$, $-\frac{1}{2}<\nu<\frac{1}{2}$ $N=\frac{1}{4}\pi^2\sec(\pi\nu)$ $\cdot\{[J_\nu(a)]^2+[Y_\nu(a)]^2\}$	$\frac{1}{4}\pi^2\{J_{iy-\nu}(a)J_{iy+\nu}(a)+Y_{iy-\nu}(a)Y_{iy+\nu}(a)$ $+\tan(\pi\nu)[J_{iy+\nu}(a)Y_{iy-\nu}(a)$ $-J_{iy-\nu}(a)Y_{iy+\nu}(a)]\}$
479	$K_{2\nu}(2a\cosh\frac{1}{2}x)$ $N=[K_\nu(a)]^2$	$K_{\nu+iy}(a)K_{\nu-iy}(a)$
480	$\operatorname{sech}(\frac{1}{2}x)V_{\frac{1}{2}\nu}(2a\cosh\frac{1}{2}x)$ $N=\frac{1}{2}a\nu^{-1}\{[K_{\frac{1}{2}+\nu}(a)]^2-[K_{\frac{1}{2}-\nu}(a)]^2\}$	$\frac{1}{2}a\nu^{-1}[K_{\frac{1}{2}+\nu+iy}(a)K_{\frac{1}{2}+\nu-iy}(a)$ $-K_{\frac{1}{2}-\nu+iy}(a)K_{\frac{1}{2}-\nu-iy}(a)]$
481	$(\cosh\frac{1}{2}x)^{-m}K_m(2a\cosh\frac{1}{2}x)$ $m=0, 1, 2, \ldots$ $N=(-1)^m m!a^m\sum_{n=0}^{m}(-1)^n\epsilon_n$ $\cdot[(m+n)!(m-n)!]^{-1}[K_n(a)]^2$	$(-1)^m m!a^m\sum_{n=0}^{m}(-1)^n\epsilon_n$ $\cdot[(m+n)!(m-n)!]^{-1}K_{n+iy}(a)K_{n-iy}(a)$
482	$(\cosh x)^{\frac{1}{2}}\exp[(a\sinh x)^2]$ $\cdot K_{\frac{1}{4}}[(a\cosh x)^2]$ $N=2^{-9/4}a^{-\frac{3}{4}}[\Gamma(\frac{1}{4})]^2W_{\frac{1}{4},0}(2a^2)$	$2^{-9/4}a^{-\frac{3}{4}}\,|\,\Gamma(\frac{1}{4}+\frac{1}{2}iy)\,|^2\,W_{\frac{1}{4},\frac{1}{2}iy}(2a^2)$
483	$(\cosh x)^{\frac{1}{2}}\exp[-(a\sinh x)^2]$ $\cdot K_{\frac{1}{4}}[(a\cosh x)^2]$ $N=2^{-\frac{7}{4}}a^{-\frac{3}{2}}\pi W_{-\frac{1}{4},0}(2a^2)$	$2^{-\frac{7}{4}}a^{-\frac{3}{2}}\pi W_{-\frac{1}{4},\frac{1}{2}iy}(2a^2)$
484	$K_0[a(2\cosh x)^{\frac{1}{2}}]$ $N=K_0(ae^{\frac{1}{4}i\pi})K_0(ae^{-\frac{1}{4}i\pi})$	$K_{iy}(ae^{\frac{1}{4}i\pi})K_{iy}(ae^{-\frac{1}{4}i\pi})$

	$2Nf(x)$	$Ng(y)$
485	$K_0[(b^2+a^2\sinh^2 x)^{\frac{1}{2}}]$ $N=\frac{1}{2}K_0\{\frac{1}{2}[b+(b^2-a^2)^{\frac{1}{2}}]\}$ $\cdot K_0\{\frac{1}{2}[b-(b^2-a^2)^{\frac{1}{2}}]\}$	$\frac{1}{2}K_{\frac{1}{2}iy}\{\frac{1}{2}[b+(b^2-a^2)^{\frac{1}{2}}]\}$ $\cdot K_{\frac{1}{2}iy}\{\frac{1}{2}[b-(b^2-a^2)^{\frac{1}{2}}]\}$
486	$K_0[(b^2+a^2\cosh^2 x)^{\frac{1}{2}}]$ $N=\frac{1}{2}K_0\{\frac{1}{2}[(a^2+b^2)^{\frac{1}{2}}+b]\}$ $\cdot K_0\{\frac{1}{2}[(a^2+b^2)^{\frac{1}{2}}-b]\}$	$\frac{1}{2}K_{\frac{1}{2}iy}\{\frac{1}{2}[(a^2+b^2)^{\frac{1}{2}}+b]\}$ $\cdot K_{\frac{1}{2}iy}\{\frac{1}{2}[(a^2+b^2)^{\frac{1}{2}}-b]\}$
487	$K_0[(a^2+b^2+2ab\cosh x)^{\frac{1}{2}}]$ $N=K_0(a)K_0(b)$	$K_{iy}(a)K_{iy}(b)$
488	$\{[(a+be^x)(b+ae^x)^{-1}]^{\frac{1}{2}\nu}$ $+[(b+ae^x)(a+be^x)^{-1}]^{\frac{1}{2}\nu}\}$ $\cdot K_\nu[(a^2+b^2+2ab\cosh x)^{\frac{1}{2}}]$ $N=2K_{\frac{1}{2}\nu}(a)K_{\frac{1}{2}\nu}(b)$	$K_{\frac{1}{2}\nu+iy}(a)K_{\frac{1}{2}\nu-iy}(b)$ $+K_{\frac{1}{2}\nu+iy}(b)K_{\frac{1}{2}\nu-iy}(a)$
489	$[K_{ix}(a)]^2$ $N=\frac{1}{2}\pi K_0(2a)$	$\frac{1}{2}\pi K_0(2a\cosh\frac{1}{2}y)$
490	$K_{ix}(ae^{\frac{1}{4}i\pi})K_{ix}(ae^{-\frac{1}{4}i\pi})$ $N=\frac{1}{2}\pi K_0(2^{\frac{1}{2}}a)$	$\frac{1}{2}\pi K_0[a(2\cosh y)^{\frac{1}{2}}]$
491	$K_{\nu+ix}(a)K_{\nu-ix}(a)$ $N=\frac{1}{2}\pi K_{2\nu}(2a)$	$\frac{1}{2}\pi K_{2\nu}(2a\cosh\frac{1}{2}y)$

12. Functions Related to Bessel Functions

	$2Nf(x)$	$Ng(y)$
492	$x^{-1}\mathbf{H}_0(ax)$ $N=\frac{1}{2}\pi$	$\arccos(y/a),\quad y<a$ $0,\quad y>a$
493	$x^{-\nu-1}\mathbf{H}_\nu(x),\quad \nu\geq\frac{1}{2}$ $N=2^{-\nu-1}\pi[\Gamma(1+\nu)]^{-1}$	$(2\pi)^{\frac{1}{2}}(1-y^2)^{\frac{1}{2}\nu+\frac{1}{4}}P_{\nu-\frac{1}{2}}^{-\nu-\frac{1}{2}}(y),\quad y<1$ $0,\quad y>1$
494	$x^\nu[\mathbf{H}_\nu(ax)-Y_\nu(ax)],\quad -\frac{1}{2}<\nu<0$ $N=-2^\nu\pi^{\frac{1}{2}}\csc(\pi\nu)a^{-\nu-1}[\Gamma(\frac{1}{2}-\nu)]^{-1}$	$(\frac{1}{2}\pi)^{-\frac{1}{2}}\cos(\pi\nu)a^\nu\Gamma(1+2\nu)y^{-\nu-\frac{1}{2}}$ $\cdot\begin{cases}(a^2-y^2)^{-\frac{1}{2}\nu-\frac{1}{4}}\mathfrak{P}_{-\nu-\frac{1}{2}}^{-\nu-\frac{1}{2}}(a/y), & y<a\\ (y^2-a^2)^{-\frac{1}{2}\nu-\frac{1}{4}}P_{-\nu-\frac{1}{2}}^{-\nu-\frac{1}{2}}(y/a), & y>a\end{cases}$

	$2Nf(x)$	$Ng(y)$
495	$x^{-\nu-1}[I_{-\nu}(ax)-\mathbf{L}_\nu(ax)]$, $\nu<0$ $N=2^{-\nu-1}a^\nu\cos(\pi\nu)\Gamma(-\nu)$	$-\pi^{-\frac{1}{2}}2^{-\nu-1}a^{-\nu}\cos(\pi\nu)\nu^{-1}\Gamma(\frac{1}{2}-\nu)$ $\cdot(a^2+y^2)^\nu {}_2F_1[-\nu,\frac{1}{2};1-\nu;a^2/(a^2+y^2)]$
496	$\mathbf{H}_0(ax^2)$ $N=\frac{1}{2}(2a)^{-\frac{1}{2}}\Gamma(\frac{1}{4})[\Gamma(\frac{3}{4})]^{-1}$	$\frac{1}{8}\pi a^{-1}y\{[J_{\frac{1}{4}}(y^2/8a)]^2-[Y_{-\frac{1}{4}}(y^2/8a)]^2\}$
497	$I_0(ax^2)-\mathbf{L}_0(ax^2)$ $N=\frac{1}{2}\pi^{-1}(2a)^{-\frac{1}{2}}[\Gamma(\frac{1}{4})]^2$	$2^{-\frac{1}{2}}(2\pi a)^{-1}y[K_{\frac{1}{4}}(y^2/8a)]^2$
498	$(a^2+x^2)^{-\frac{1}{2}}\mathbf{H}_0[b(a^2+x^2)^{\frac{1}{2}}]$ $N=\frac{1}{2}\pi[J_0(\frac{1}{2}ab)]^2$	$0,\quad y>b$
499	$(a^2+x^2)^{-\frac{1}{2}-\frac{1}{2}\nu}\mathbf{H}_\nu[b(a^2+x^2)^{\frac{1}{2}}]$, $\nu>-\frac{3}{2}$ $N=\pi2^{-\nu-1}b^{-\nu}[\Gamma(\nu+1)]^{-1}$ $\cdot{}_1F_2(\frac{1}{2};1,\nu+1;-\frac{1}{4}a^2b^2)$	$0,\quad y>b$
500	$(a^2+x^2)^{-\frac{1}{2}}\{\mathbf{H}_0[b(a^2+x^2)^{\frac{1}{2}}]$ $-Y_0[b(a^2+x^2)^{\frac{1}{2}}]\}$ $N=\frac{1}{4}\pi\{[J_0(\frac{1}{2}ab)]^2+[Y_0(\frac{1}{2}ab)]^2\}$	$\pi^{-1}K_0\{\frac{1}{2}a[y+(y^2-b^2)^{\frac{1}{2}}]\}$ $\cdot K_0\{\frac{1}{2}a[y-(y^2-b^2)^{\frac{1}{2}}]\}$
501	$(a^2+x^2)^{-\frac{1}{2}}\{I_0[b(a^2+x^2)^{\frac{1}{2}}]$ $-\mathbf{L}_0[b(a^2+x^2)^{\frac{1}{2}}]\}$ $N=I_0(\frac{1}{2}ab)-K_0(\frac{1}{2}ab)$	$I_0\{\frac{1}{2}a[(b^2+y^2)^{\frac{1}{2}}-y]\}K_0\{\frac{1}{2}a[(b^2+y^2)^{\frac{1}{2}}+y]\}$
502	$\mathbf{H}_0(2a\cosh\frac{1}{2}x)$ $N=\pi[J_0(a)]^2$	$\frac{1}{2}\pi\operatorname{sech}(\pi y)\{[J_{iy}(a)]^2+[J_{-iy}(a)]^2\}$
503	$\mathbf{H}_0(2a\cosh\frac{1}{2}x)-Y_0(2a\cosh\frac{1}{2}x)$ $N=\frac{1}{2}\pi\{[J_0(a)]^2+[Y_0(a)]^2\}$	$\frac{1}{2}\pi\operatorname{sech}(\pi y)\{[J_{iy}(a)]^2+[Y_{iy}(a)]^2\}$
504	$I_0(2a\cosh\frac{1}{2}x)-\mathbf{L}_0(2a\cosh\frac{1}{2}x)$ $N=2I_0(a)K_0(a)$	$\operatorname{sech}(\pi y)K_{iy}(a)[I_{iy}(a)+I_{-iy}(a)]$
505	$\mathbf{H}_0(ax^2)-Y_0(ax^2)$ $N=2^{-\frac{3}{2}}a^{-\frac{1}{2}}\Gamma(\frac{1}{4})[\Gamma(\frac{3}{4})]^{-1}$	$\frac{1}{8}\pi a^{-1}y\{[J_{\frac{1}{4}}(\frac{1}{8}y^2/a)]^2+[Y_{\frac{1}{4}}(\frac{1}{8}y^2/a)]^2\}$
506	$\csc(\pi\nu)[\mathbf{J}_\nu(ax^2)-J_\nu(ax^2)]$ $N=2^{-\frac{3}{2}}a^{-\frac{1}{2}}\Gamma(\frac{1}{4}+\frac{1}{2}\nu)$ $\cdot[\Gamma(\frac{3}{4}+\frac{1}{2}\nu)]^{-1}$, $\nu>-\frac{1}{2}$	$(2ra)^{-\frac{1}{2}}\Gamma(\frac{1}{2}+\nu)$ $\cdot D_{-\nu-\frac{1}{2}}[y(2ai)^{\frac{1}{2}}]D_{-\nu-\frac{1}{2}}[y(-2ai)^{\frac{1}{2}}]$

	$2Nf(x)$	$Ng(y)$
507	$-[Y_\nu(ax^2)+E_\nu(ax^2)], \quad -\frac{1}{2}<\nu<\frac{1}{2}$ $N=2^{-\frac{3}{2}}a^{-\frac{1}{2}}\{\Gamma(\frac{1}{4}-\frac{1}{2}\nu)$ $\cdot[\Gamma(\frac{3}{4}-\frac{1}{2}\nu)]^{-1}+\cos(\pi\nu)$ $\cdot\Gamma(\frac{1}{4}+\frac{1}{2}\nu)[\Gamma(\frac{3}{4}+\frac{1}{2}\nu)]^{-1}\}$	$(2\pi a)^{-\frac{1}{2}}\{\Gamma(\frac{1}{2}-\nu)D_{\nu-\frac{1}{2}}[y(2ai)^{-\frac{1}{2}}]$ $\cdot D_{\nu-\frac{1}{2}}[y(-2ai)^{-\frac{1}{2}}]+\cos(\pi\nu)\Gamma(\frac{1}{2}+\nu)$ $\cdot D_{-\nu-\frac{1}{2}}[y(2ai)^{-\frac{1}{2}}]D_{-\nu-\frac{1}{2}}[y(-2ai)^{-\frac{1}{2}}]\}$
508	$\csc(\pi\nu)$ $\cdot[\mathbf{J}_\nu(a\cosh x)-J_\nu(a\cosh x)]$ $N=-\frac{1}{4}if'(0)$	$-\frac{1}{4}i\pi\,\operatorname{csch}(\pi y)f(y)$ $f(y)=J_{\frac{1}{2}(\nu+iy)}(\frac{1}{2}a)Y_{\frac{1}{2}(\nu-iy)}(\frac{1}{2}a)$ $-J_{\frac{1}{2}(\nu-iy)}(\frac{1}{2}a)Y_{\frac{1}{2}(\nu+iy)}(\frac{1}{2}a)$
509	$-[Y_\nu(a\cosh x)$ $+E_\nu(a\cosh x)]$ $N=-\frac{1}{4}if'(0)$	$-\frac{1}{4}i\pi\,\operatorname{csch}(\pi y)f(y)$ $f(y)=J_{-\frac{1}{2}\nu+\frac{1}{2}iy}(\frac{1}{2}a)Y_{-\frac{1}{2}(\nu+\frac{1}{2}iy)}(\frac{1}{2}a)$ $-J_{-\frac{1}{2}(\nu+iy)}(\frac{1}{2}a)Y_{-(\frac{1}{2}\nu-iy)}(\frac{1}{2}a)+\cos(\pi\nu)$ $\cdot[J_{\frac{1}{2}(\nu+iy)}(\frac{1}{2}a)Y_{\frac{1}{2}(\nu-iy)}(\frac{1}{2}a)-J_{\frac{1}{2}(\nu-iy)}(\frac{1}{2}a)$ $\cdot Y_{\frac{1}{2}(\nu+iy)}(\frac{1}{2}a)]$
510	$x^{-\mu-1}S_{\mu,\nu}(ax), \quad \mu\pm\nu<0$ $N=\frac{1}{4}\pi a^\mu\Gamma(-\frac{1}{2}\mu-\frac{1}{2}\nu)\Gamma(-\frac{1}{2}\mu+\frac{1}{2}\nu)$ $\cdot[\Gamma(\frac{1}{2}-\frac{1}{2}\mu+\frac{1}{2}\nu)\Gamma(\frac{1}{2}-\frac{1}{2}\mu-\frac{1}{2}\nu)]^{-1}$	$\frac{1}{4}(\frac{1}{2}\pi/a)^{\frac{1}{2}}2^{-\mu}\Gamma(-\frac{1}{2}\nu-\frac{1}{2}\mu)\Gamma(\frac{1}{2}\nu-\frac{1}{2}\mu)$ $\cdot\begin{cases}(y^2-a^2)^{\frac{1}{2}\mu+\frac{1}{4}}\mathfrak{P}_{\nu-\frac{1}{2}}^{\frac{1}{2}+\mu}(y/a), & y>a\\(a^2-y^2)^{\frac{1}{2}\mu+\frac{1}{4}}P_{\nu-\frac{1}{2}}^{\frac{1}{2}+\mu}(y/a), & y<a\end{cases}$
511	$x^\nu S_{\mu,\nu}(ax), \quad \nu>-\frac{1}{2}, \quad \mu-\|\nu\|<1,$ $-2<\mu+\nu<0$ $N=-2^{\nu+\mu-1}\pi^{\frac{3}{2}}a^{-\nu-1}\csc[\frac{1}{2}\pi(\nu+\mu)]$ $\cdot\Gamma(\frac{1}{2}+\nu)[\Gamma(\frac{1}{2}-\frac{1}{2}\nu-\frac{1}{2}\mu)$ $\cdot\Gamma(\frac{1}{2}+\frac{1}{2}\nu-\frac{1}{2}\mu)]^{-1}$	$\pi^{\frac{1}{2}}(2\nu+1)^{-1}2^{\nu+\mu}a^{-\nu-1}$ $\Gamma(1+\frac{1}{2}\nu+\frac{1}{2}\mu)[\Gamma(\frac{1}{2}-\frac{1}{2}\nu-\frac{1}{2}\mu)]^{-1}$ $\cdot{}_2F_1(\frac{1}{2}+\nu, 1+\frac{1}{2}\nu+\frac{1}{2}\mu; \frac{3}{2}+\nu; 1-y^2/a^2)$
512	$S_{\mu,\nu}(\cosh x), \quad \mu<\|\nu\|+1$ $N=\frac{1}{2}(2a)^{\mu+1}\{\Gamma[\frac{1}{2}(1-\mu-\nu)]$ $\cdot\Gamma[\frac{1}{2}(1-\mu+\nu)]\}^{-1}$ $\cdot\int_0^\infty t^{-\mu}(a^2+t^2)^{-1}[K_{\frac{1}{2}\nu}(\frac{1}{2}t)]^2\,dt$	$2^\mu a^{\mu+1}\{\Gamma[\frac{1}{2}(1-\mu-\nu)]\Gamma[\frac{1}{2}(1-\mu+\nu)]\}^{-1}$ $\cdot\int_0^\infty t^{-\mu}(a^2+t^2)^{-1}K_{\frac{1}{2}\nu+i\frac{1}{2}y}(\frac{1}{2}t)K_{\frac{1}{2}\nu-i\frac{1}{2}y}(\frac{1}{2}t)\,dt$
513	$(\cosh x)^{\frac{1}{2}}S_{\mu,\frac{1}{2}}(a\cosh x), \quad \mu<\frac{1}{2}$ $N=(2a)^{-\frac{1}{2}}2^{-\mu-1}[\Gamma(\frac{1}{2}-\mu)]^{-1}[\Gamma(\frac{1}{4}-\frac{1}{2}\mu)]^{-1}$ $\cdot S_{\mu+\frac{1}{2},0}(a)$	$(2a)^{-\frac{1}{2}}2^{-\mu-1}[\Gamma(\frac{1}{2}-\mu)]^{-1}$ $\cdot\Gamma(\frac{1}{4}-\frac{1}{2}\mu-\frac{1}{2}iy)\Gamma(\frac{1}{4}-\frac{1}{2}\mu+\frac{1}{2}iy)$ $\cdot S_{\mu+\frac{1}{2},iy}(a)$
514	$(a^2+x^2)^{-\frac{1}{2}}\{\frac{1}{2}\pi\sec(\frac{1}{2}\pi\nu)I_\nu[b(a^2+x^2)^{\frac{1}{2}}]$ $+is_{0,\nu}[ib(a^2+x^2)^{\frac{1}{2}}]\}$ $N=\frac{1}{2}\pi I_{\frac{1}{2}\nu}(\frac{1}{2}ab)K_{\frac{1}{2}\nu}(\frac{1}{2}ab)$	$\frac{1}{2}\pi I_{\frac{1}{2}\nu}\{\frac{1}{2}a[(b^2+y^2)^{\frac{1}{2}}-y]\}$ $\cdot K_{\frac{1}{2}\nu}\{\frac{1}{2}a[(b^2+y^2)^{\frac{1}{2}}+y]\}$

	$2Nf(x)$	$Ng(y)$
515	$(a^2+x^2)^{-\frac{1}{2}}$ $\cdot S_{0,\nu}[b(a^2+x^2)^{\frac{1}{2}}]$	$\frac{1}{2}K_{\frac{1}{2}\nu}\{\frac{1}{2}a[y+(y^2-b^2)^{\frac{1}{2}}]\}$ $\cdot K_{\frac{1}{2}\nu}\{\frac{1}{2}a[y-(y^2-b^2)^{\frac{1}{2}}]\}$
516	$S_{0,ix}(a)$ $N=\frac{1}{2}\pi$	$\frac{1}{2}\pi\exp(-a\sinh y)$
517	$\operatorname{sech}(\frac{1}{2}\pi x)S_{0,ix}(a)$ $N=\sin a\operatorname{Ci}(a)-\cos a\operatorname{si}(a)$	$\sin(a\cosh y)\operatorname{Ci}(a\cosh y)$ $-\cos(a\cosh y)\operatorname{si}(a\cosh y)$
518	$x\operatorname{csch}(\pi x)S_{-1,ix}(a)$ $N=-[\cos a\operatorname{Ci}(a)+\sin a\operatorname{si}(a)]$	$-[\cos(a\cosh y)\operatorname{Ci}(a\cosh y)$ $+\sin(a\cosh y)\operatorname{si}(a\cosh y)]$
519	$xS_{\mu,\frac{1}{2}}(ax^2),\quad -2<\mu<0$ $N=-\frac{1}{2}\pi^{\frac{1}{2}}a^{-1}\cos(\frac{1}{2}\pi\mu)$ $\cdot\Gamma(-1-\mu)\Gamma(2+\mu)[\Gamma(\frac{1}{2}-\mu)]^{-1}$	$\frac{1}{4}\pi^{\frac{1}{2}}a^{-\frac{3}{2}}\Gamma(2+\mu)$ $\cdot[\Gamma(\frac{1}{2}-\mu)]^{-1}yS_{-\mu-\frac{3}{2},\frac{1}{2}}(\frac{1}{4}y^2/a)$

13. Parabolic Cylindrical Functions

	$2Nf(x)$	$Ng(y)$
520	$\exp(\frac{1}{4}a^2x^2)D_{-2}(ax)$ $N=a(\frac{1}{2}\pi)^{\frac{1}{2}}$	$(\frac{1}{2}\pi)^{\frac{1}{2}}a\exp(\frac{1}{4}y^2/a^2)D_{-2}(y/a)$
521	$\exp(\frac{1}{4}a^2x^2)D_\nu(ax),\quad \nu<-1$ $N=\pi^{\frac{1}{2}}2^{\frac{1}{2}\nu-\frac{1}{2}}\Gamma(-\frac{1}{2}-\frac{1}{2}\nu)a^{-1}$ $\cdot[\Gamma(-\frac{1}{2}\nu)\Gamma(\frac{1}{2}-\frac{1}{2}\nu)]^{-1}$	$\pi^{\frac{1}{2}}(2a)^\nu[\Gamma(-\frac{1}{2}\nu)]^{-1}y^{-\nu-1}$ $\cdot\exp(\frac{1}{2}y^2/a^2)\Gamma(\frac{1}{2}+\frac{1}{2}\nu,\frac{1}{2}y^2/a^2)$
522	$\exp(-\frac{1}{4}a^2x^2)D_\nu(ax)$ $N=2^{\frac{1}{2}\nu-\frac{1}{2}}\pi^{\frac{1}{2}}a^{-1}[\Gamma(1-\frac{1}{2}\nu)]^{-1},\quad \nu\leq 1$	$2^{\frac{1}{2}\nu-\frac{1}{2}}\pi^{\frac{1}{2}}[\Gamma(1-\frac{1}{2}\nu)]^{-1}a^{-1}$ $\cdot{}_1F_1(1;1-\frac{1}{2}\nu;-\frac{1}{2}y^2/a^2)$
523	$x^{-\nu-1}\exp(\frac{1}{4}a^2x)D_{2\nu-1}(ax^{\frac{1}{2}}),\quad \nu<0$ $N=-(\frac{1}{2}\pi)^{\frac{1}{2}}\nu^{-1}a^{2\nu}$	$-(\pi/2)^{\frac{1}{2}}\nu^{-1}[y+(a+y^{\frac{1}{2}})^2]^\nu$ $\cdot\cos\{2\nu\arctan[y^{\frac{1}{2}}/(a+y^{\frac{1}{2}})]\}$
524	$x^\mu\exp(-\frac{1}{4}a^2x^2)D_\nu(ax),\quad \mu>-1,\quad \nu\leq 1$ $N=2^{\frac{1}{2}(\nu-\mu-1)}\pi^{\frac{1}{2}}\Gamma(1+\mu)$ $\cdot[\Gamma(1+\frac{1}{2}\mu-\frac{1}{2}\nu)]^{-1}a^{-\mu-1}$	$2^{\frac{1}{2}(\nu-\mu-1)}\pi^{\frac{1}{2}}\Gamma(1+\mu)[\Gamma(1+\frac{1}{2}\mu-\frac{1}{2}\nu)]^{-1}$ $\cdot a^{-\mu-1}{}_2F_2(\frac{1}{2}+\frac{1}{2}\mu,1+\frac{1}{2}\mu;\frac{1}{2},1+\frac{1}{2}\mu-\frac{1}{2}\nu;$ $-\frac{1}{2}y^2/a^2)$

	$2Nf(x)$	$Ng(y)$
525	$(\cos x)^{-\frac{1}{2}\nu-\frac{1}{2}} \exp(-a^2 \sec x)$ $\cdot D_\nu[2a(1+\sec x)^{\frac{1}{2}}], \quad x<\frac{1}{2}\pi$ $0, \quad x>\frac{1}{2}\pi$ $\nu \leq 1$ $N=\pi^{\frac{1}{2}} 2^{\frac{1}{2}\nu} \exp(-a^2)[D_{\frac{1}{2}\nu-\frac{1}{2}}(2a)]^2$	$\pi^{\frac{1}{2}} 2^{\frac{1}{2}\nu} \exp(-a^2) D_{\frac{1}{2}\nu-\frac{1}{2}+y}(2a) D_{\frac{1}{2}\nu-\frac{1}{2}-y}(2a)$
526	$D_\nu[ax(i)^{\frac{1}{2}}] D_\nu[ax(-i)^{\frac{1}{2}}], \quad \nu<-\frac{1}{2}$ $N=-\frac{1}{2}\pi^{\frac{1}{2}} a^{-1}[(\nu+\frac{1}{2})\Gamma(-\nu)]^{-1}$	$\frac{1}{2}\pi^{\frac{3}{2}}[a\Gamma(-\nu)]^{-1} \sec(\pi\nu)$ $\cdot[J_{-\nu-\frac{1}{2}}(\frac{1}{2}y^2/a^2) - \mathbf{J}_{-\nu-\frac{1}{2}}(\frac{1}{2}y^2/a^2)]$
527	$\exp[(a \sinh x)^2] D_\nu(2a \cosh x), \quad \nu<0$ $N=2^{\frac{1}{2}\nu-\frac{3}{2}} a^{-1} \pi^{\frac{1}{2}} \Gamma(-\frac{1}{2}\nu)$ $\cdot[\Gamma(\frac{1}{2}-\frac{1}{2}\nu)]^{-1} W_{\frac{1}{2}\nu+\frac{1}{2},0}(2a^2)$	$2^{-\frac{1}{2}\nu-\frac{5}{2}} a^{-1}[\Gamma(-\nu)]^{-1} \Gamma(-\frac{1}{2}\nu+i\frac{1}{2}y)$ $\cdot\Gamma(-\frac{1}{2}\nu-i\frac{1}{2}y) W_{\frac{1}{2}\nu+\frac{1}{2},\frac{1}{2}iy}(2a^2)$
528	$\exp[-(a \sinh x)^2] D_\nu(2a \cosh x), \quad \nu \leq 1$ $N=2^{\frac{1}{2}\nu-\frac{3}{2}} \pi^{\frac{1}{2}} a^{-1} W_{\frac{1}{2}\nu,0}(2a^2)$	$2^{\frac{1}{2}\nu-\frac{3}{2}} \pi^{\frac{1}{2}} a^{-1} W_{\frac{1}{2}\nu,\frac{1}{2}iy}(2a^2)$
529	$\operatorname{sech}(\pi x) D_{-\frac{1}{2}+ix}(a) D_{-\frac{1}{2}-ix}(a)$ $N=\frac{1}{2}\pi^{\frac{1}{2}} \exp(\frac{1}{2}a^2) \operatorname{Erfc}(a)$	$\frac{1}{2}\pi^{\frac{1}{2}}(\operatorname{sech} y)^{\frac{1}{2}} \exp(\frac{1}{2}a^2 \operatorname{sech} y)$ $\cdot\operatorname{Erfc}[2^{-\frac{1}{2}} a(1+\operatorname{sech} y)^{\frac{1}{2}}]$

TABLE II

FUNCTIONS VANISHING IDENTICALLY FOR NEGATIVE VALUES OF THE ARGUMENT

Definitions

Here $f(x)$ is assumed to vanish identically for negative x. The Fourier transform $G(y)$ of $f(x)$ then becomes

$$G(y) = \int_{-\infty}^{\infty} f(x) e^{ixy}\, dx = \int_{0}^{\infty} f(x) e^{ixy}\, dx = g(y) + ih(y),$$

$$g(y) = \int_{0}^{\infty} f(x) \cos(xy)\, dx, \qquad h(y) = \int_{0}^{\infty} f(x) \sin(xy)\, dx,$$

$$g(0) = \int_{0}^{\infty} f(x)\, dx = 1.$$

Table II, which follows, gives $Nh(y)$, while the properties N and $g(y)$ are the same as those listed in Table I under the same number.

1. Algebraic Functions

	$Nf(x)$	$Nh(y)$
1	$1, \quad x<a$ $0, \quad x>a$	$2y^{-1}\sin^2(ya/2)$
2	$x, \quad x<1$ $2-x, \quad 1<x<2$ $0, \quad x>2$	$4y^{-2}\sin y \sin^2(\frac{1}{2}y)$
3	$x^{-\frac{1}{2}}, \quad x<1$ $0, \quad x>1$	$(2\pi/y)^{\frac{1}{2}}S(y)$
4	$(a+x)^{-1}, \quad x<b$ $0, \quad x>b$	$\sin(ay)[\mathrm{Ci}(ay+by)-\mathrm{Ci}(ay)]$ $+\cos(ay)[\mathrm{si}(ay+by)-\mathrm{si}(ay)]$
5	$0, \quad x<b$ $(a+x)^{-n}, \quad x>b$ $n=2, 3, 4, \ldots$	$\sum_{m=1}^{n-1}\frac{(m-1)!}{(n-1)!}(a+b)^{-m}(-y)^{n-m-1}$ $\cdot\cos[\frac{1}{2}\pi(n-m)-by]$ $-(-y)^{n-1}[(n-1)!]^{-1}[\cos(ay+\frac{1}{2}\pi n)$ $\cdot\mathrm{Ci}(ay+by)+\sin(ay+\frac{1}{2}\pi n)\,\mathrm{si}(ay+by)]$
6	$0, \quad x<b$ $[x(a+x)]^{-1}, \quad x>b$	$a^{-1}[\cos(ay)\,\mathrm{si}(ay+by)-\mathrm{si}(by)$ $-\sin(ay)\,\mathrm{Ci}(ay+by)]$
7	$(a^2+x^2)^{-1}$	$(2a)^{-1}[e^{-ay}\,\overline{\mathrm{Ei}}(ay)-e^{ay}\,\mathrm{Ei}(-ay)]$
13	$x^{-\frac{1}{2}}(a+x)^{-1}$	$\pi a^{-\frac{1}{2}}\{\cos(ay)[C(ay)-S(ay)]$ $-\sin(ay)[1-C(ay)-S(ay)]\}$
14	$(a+x)^{-\frac{3}{2}}$	$(2\pi y)^{\frac{1}{2}}\{\sin(ay)[1-2S(ay)^{\frac{1}{2}}]$ $+\cos(ay)[1-2C(ay)^{\frac{1}{2}}]\}$
15	$x^{-\frac{1}{2}}(a^2+x^2)^{-\frac{1}{2}}$	$(\frac{1}{2}\pi)^{\frac{1}{2}}y^{\frac{1}{2}}I_{\frac{1}{4}}(\frac{1}{2}ay)K_{\frac{1}{4}}(\frac{1}{2}ay)$
16	$(a-x)^{-\frac{1}{2}}, \quad x<a$ $0, \quad x>a$	$(2\pi)^{\frac{1}{2}}y^{-\frac{1}{2}}[\sin(ay)C(ay)$ $-\cos(ay)S(ay)]$
17	$(a-x)^{-1}, \quad x<b$ $0, \quad x>b, \quad b<a$	$\sin(ay)[\mathrm{Ci}(ay)-\mathrm{Ci}(ay-by)]$ $-\cos(ay)[\mathrm{si}(ay)-\mathrm{si}(ay-by)]$

	$Nf(x)$	$Nh(y)$
18	$(a^2-x^2)^{-1}, \quad x<b$ $0, \quad x>b, \quad b<a$	$\frac{1}{2}a^{-1}\{\cos(ay)[\mathrm{si}(ay+by)-2\,\mathrm{si}(ay)$ $+\mathrm{si}(ay-by)]$ $-\sin(ay)[\mathrm{Ci}(ay+by)-2\,\mathrm{Ci}(ay)$ $+\mathrm{Ci}(ay-by)]\}$
19	$0, \quad x<b$ $(x^2-a^2)^{-1}, \quad x>b, \quad b>a$	$\cos(ay)[\mathrm{si}(by-ay)-\mathrm{si}(by+ay)]$ $-\sin(ay)[\mathrm{Ci}(by-ay)+\mathrm{Ci}(by-ay)]$
20	$0, \quad x<b$ $x^{-1}(x-b)^{-\frac{1}{2}}, \quad x>b$	$b^{-\frac{1}{2}}\pi[C(by)-S(by)]$
22	$0, \quad x<b$ $(x-b)^{-\frac{1}{2}}(x+b)^{-1}, \quad x>b$	$(2b)^{-\frac{1}{2}}\pi\{[C(2by)-S(2by)]\cos(by)$ $-[1-C(2by)-S(2by)]\sin(by)\}$
23	$0, \quad x<b$ $(x-b)^{-\frac{1}{2}}(x+a)^{-1}, \quad x>b$	$\pi(a+b)^{-\frac{1}{2}}\{\cos(ay)[C(ay+by)-S(ay+by)]$ $-\sin(ay)[1-C(ay+by)-S(ay+by)]\}$
24	$(a^2-x^2)^{-\frac{1}{2}}, \quad x<a$ $0, \quad x>a$	$\frac{1}{2}\pi\mathbf{H}_0(ay)$
25	$x(a^2-x^2)^{-\frac{1}{2}}, \quad x<a$ $0, \quad x>a$	$\frac{1}{2}a\pi J_1(ay)$
26	$x^{-\frac{1}{2}}(a^2-x^2)^{-\frac{1}{2}}, \quad x<a$ $0, \quad x>a$	$(\frac{1}{2}\pi)^{\frac{3}{2}}y^{\frac{1}{2}}[J_{\frac{1}{4}}(\frac{1}{2}ay)]^2$
29	$x^{-\frac{1}{2}}(a^2+x^2)^{-\frac{1}{2}}[x+(a^2+x^2)^{\frac{1}{2}}]^{-\frac{3}{2}}$	$2^{-\frac{1}{2}}a^{-2}\pi e^{-\frac{1}{2}ay}I_1(\frac{1}{2}ay)$
30	$0, \quad x<a$ $x^{-\frac{1}{2}}(x^2-a^2)^{-\frac{1}{2}}, \quad x>a$	$-(\frac{1}{2}\pi)^{\frac{3}{2}}y^{\frac{1}{2}}J_{\frac{1}{4}}(\frac{1}{2}ay)\,Y_{\frac{1}{4}}(\frac{1}{2}ay)$
31	$0, \quad x<a$ $x^{-1}(x^2-a^2)^{-\frac{1}{2}}, \quad x>a$	$-\frac{1}{4}\pi^2y[\mathbf{H}_0(ay)\,Y_1(ay)$ $+Y_0(ay)\mathbf{H}_{-1}(ay)]$
32	$(a^2+x^2)^{-\frac{1}{2}}[a+(a^2+x^2)^{\frac{1}{2}}]^{-\frac{1}{2}}$	$-i\pi(2a)^{-\frac{1}{2}}\,\mathrm{Erf}[i(ay)^{\frac{1}{2}}]\,\mathrm{Erfc}[(ay)^{\frac{1}{2}}]$
35	$x^{-\frac{1}{2}}(a^2+x^2)^{-\frac{1}{2}}[x+(a^2+x^2)^{\frac{1}{2}}]^{-\frac{1}{2}}$	$2^{\frac{1}{2}}a^{-1}\sinh(\frac{1}{2}ay)\,K_0(\frac{1}{2}ay)$

	$Nf(x)$	$Nh(y)$
36	$0, \quad x<a$ $(x^2-a^2)^{-\frac{1}{2}}[x+(x^2-a^2)^{\frac{1}{2}}]^{-n}, \quad x>a$ $n=1, 2, 3, \ldots$	$\frac{1}{2}\pi a^{-n}[\cos(\frac{1}{2}n\pi)J_n(ay)$ $-\sin(\frac{1}{2}n\pi)Y_n(ay)]$ $-a^{-n}\sum_{k=1}^{n} k!(n+k-1)![(2k)!(n-k)!]^{-1}$ $\cdot(\frac{1}{2}ay)^{-k}\sin(ay+\frac{1}{2}\pi k)$

2. Arbitrary Powers

	$Nf(x)$	$Nh(y)$
37	$x^{\nu-1}, \quad x<1$ $0, \quad x>1, \quad \nu>0$	$-\frac{1}{2}i\nu^{-1}[{}_1F_1(\nu;\nu+1;iy)$ $-{}_1F_1(\nu;\nu+1;-iy)]$
38	$(b-x)^{\nu}, \quad x<b$ $0, \quad x>b, \quad \nu>-1$	$-\frac{1}{2}y^{-\nu-1}\{\exp[-i(\frac{1}{2}\nu\pi-by)]\gamma(\nu+1, iby)$ $-\exp[i(\frac{1}{2}\nu\pi-by)]\gamma(\nu+1, -iby)\}$
39	$x^{\nu-1}(b-x)^{\mu-1}, \quad x<b$ $0, \quad x>b, \quad \nu, \mu>0$	$\frac{1}{2}ib^{\nu+\mu-1}B(\nu, \mu)$ $\cdot[{}_1F_1(\nu;\nu+\mu;-iby)-{}_1F_1(\nu;\nu+\mu;iby)]$
40	$x^{\nu}(a+x)^{-1}, \quad -1<\nu<0$	$2^{\nu}\pi^{\frac{1}{2}}a^{\nu+\frac{1}{2}}y^{\frac{1}{2}}\left[2\frac{\Gamma(\nu+\frac{3}{2})}{\Gamma(-\frac{1}{2}\nu)}S_{-\nu-\frac{3}{2},\frac{1}{2}}(ay)\right.$ $\left.-\frac{\Gamma(1+\frac{1}{2}\nu)}{\Gamma(\frac{1}{2}-\frac{1}{2}\nu)}S_{-\nu-\frac{1}{2},\frac{1}{2}}(ay)\right]$
41	$x^{\nu}(a^2+x^2)^{-1}, \quad -1<\nu<1$	$2^{\nu-2}\pi^{\frac{1}{2}}\Gamma(\frac{1}{2}\nu)[\Gamma(\frac{3}{2}-\frac{1}{2}\nu)]^{-1}y^{1-\nu}$ $\cdot{}_1F_2(1;\frac{3}{2}-\frac{1}{2}\nu, 1-\frac{1}{2}\nu;\frac{1}{4}a^2y^2)$ $-\frac{1}{2}\pi a^{\nu-1}\csc(\frac{1}{2}\pi\nu)\sinh(ay)$
42	$(a^2+x^2)^{-\nu-\frac{1}{2}}, \quad \nu>0$	$2^{-\nu-1}\pi^{\frac{1}{2}}\Gamma(\frac{1}{2}-\nu)(y/a)^{\nu}[I_{\nu}(ay)-\mathbf{L}_{-\nu}(ay)]$
43	$(x^2+2ax)^{-\nu-\frac{1}{2}}, \quad 0<\nu<\frac{1}{2}$	$2^{-\nu-1}a^{-\nu}\pi^{\frac{1}{2}}\Gamma(\frac{1}{2}-\nu)y^{\nu}$ $\cdot[J_{\nu}(ay)\cos(ay)+Y_{\nu}(ay)\sin(ay)]$
44	$x^{-\nu-\frac{1}{2}}(a^2+x^2)^{-\frac{1}{2}}[(a^2+x^2)^{\frac{1}{2}}+a]^{\nu}, \quad \nu<\frac{3}{2}$	$2^{\frac{1}{2}}a^{-1}\Gamma(\frac{3}{4}-\frac{1}{2}\nu)y^{-\frac{1}{2}}W_{\frac{1}{2}\nu,\frac{1}{4}}(ay)M_{-\frac{1}{2}\nu,\frac{1}{4}}(ay)$
45	$x^{-\frac{1}{2}}(a^2+x^2)^{-\frac{1}{2}}[(a^2+x^2)^{\frac{1}{2}}-x]^{\nu}, \quad \nu>-\frac{1}{2}$	$a^{\nu}(\frac{1}{2}\pi)^{\frac{1}{2}}y^{\frac{1}{2}}I_{\frac{1}{4}+\frac{1}{2}\nu}(\frac{1}{2}ay)K_{\frac{1}{4}-\frac{1}{2}\nu}(\frac{1}{2}ay)$

	$Nf(x)$	$Nh(y)$
46	$x^{-\frac{1}{2}}(a^2+x^2)^{-\frac{1}{2}}[x+(a^2+x^2)^{\frac{1}{2}}]^{\nu}, \quad \nu<\frac{1}{2}$	$a^{\nu}(\frac{1}{2}\pi)^{\frac{1}{2}}y^{\frac{1}{2}}I_{\frac{1}{4}-\frac{1}{2}\nu}(\frac{1}{2}ay)K_{\frac{1}{4}+\frac{1}{2}\nu}(\frac{1}{2}ay)$
47	$(a^2+x^2)^{-\frac{1}{2}}$ $\cdot[(a^2+x^2)^{\frac{1}{2}}+x]^{-\nu}, \quad \nu>0$	$a^{-\nu}\pi \csc(\nu\pi)[I_{\nu}(ay) \sin(\frac{1}{2}\nu\pi)$ $+\frac{1}{2}i\mathbf{J}_{\nu}(iay)-\frac{1}{2}i\mathbf{J}(-iay)]$
48	$[x+(a^2+x^2)^{\frac{1}{2}}]^{-\nu}, \quad \nu>1$	$y^{-1}a^{-\nu}\{1+\nu\pi \csc(\pi\nu)[I_{\nu}(ay) \cos(\frac{1}{2}\pi\nu)$ $-\frac{1}{2}\mathbf{J}_{\nu}(iay)-\frac{1}{2}\mathbf{J}_{\nu}(-iay)]\}$
49	$x^{\nu}(a^2+x^2)^{-\mu-1}, \quad \nu>-1, \quad \nu-2\mu<1$	$\frac{1}{2}a^{\nu-2\mu}B(1+\frac{1}{2}\nu, \mu-\frac{1}{2}\nu)$ $\cdot y\,{}_1F_2(1+\frac{1}{2}\nu; 1+\frac{1}{2}\nu-\mu, \frac{3}{2}; \frac{1}{4}a^2y^2)$ $+2^{\nu-2\mu-2}\pi^{\frac{1}{2}}\Gamma(\frac{1}{2}\nu-\mu)[\Gamma(\mu-\frac{1}{2}\nu+\frac{3}{2})]^{-1}$ $\cdot y^{2\mu-\nu+1}{}_1F_2(\mu+1; \mu-\frac{1}{2}\nu+\frac{3}{2}, \mu-\frac{1}{2}\nu+1;$ $\frac{1}{4}a^2y^2)$
50	$(a^2-x^2)^{\nu-\frac{1}{2}}, \quad x<a$ $0, \quad x>a$ $\nu>-\frac{1}{2}$	$2^{\nu-1}a^{\nu}\Gamma(\frac{1}{2}+\nu)\pi^{\frac{1}{2}}y^{-\nu}\mathbf{H}_{\nu}(ay)$
51	$x(a^2-x^2)^{\nu-\frac{1}{2}}, \quad x<a$ $0, \quad x>a$ $\nu>-\frac{1}{2}$	$2^{\nu-1}a^{\nu}\Gamma(\frac{1}{2}+\nu)\pi^{\frac{1}{2}}y^{-\nu}J_{\nu+1}(ay)$
52	$(2ax-x^2)^{\nu-\frac{1}{2}}, \quad x<2a$ $0, \quad x>2a$ $\nu>-\frac{1}{2}$	$(2a)^{\nu}\pi^{\frac{1}{2}}\Gamma(\frac{1}{2}+\nu)y^{-\nu}\sin(ay)$ $\cdot J_{\nu}(ay)$
53	$x^{\nu}(a^2-x^2)^{\mu}, \quad x<a$ $0, \quad x>a, \quad \nu,\mu>-1$	$\frac{1}{2}a^{\nu+2\mu+2}B(\mu+1, 1+\frac{1}{2}\nu)y$ $\cdot{}_1F_2(1+\frac{1}{2}\nu; \frac{3}{2}, 2+\frac{1}{2}\nu+\mu; -\frac{1}{4}a^2y^2)$
54	$x^{-\nu-\frac{1}{2}}(a^2-x^2)^{-\frac{1}{2}}$ $\cdot\{[a+(a^2-x^2)^{\frac{1}{2}}]^{\nu}+[a-(a^2-x^2)^{\frac{1}{2}}]^{\nu}\},$ $x<a$ $0, \quad x>a$ $-\frac{1}{2}<\nu<\frac{1}{2}$	$(2a)^{\frac{1}{2}}B(\frac{3}{4}+\frac{1}{2}\nu, \frac{3}{4}-\frac{1}{2}\nu)y$ $\cdot{}_1F_1(\frac{3}{4}-\frac{1}{2}\nu; \frac{3}{2}; -iay)$ $\cdot{}_1F_1(\frac{3}{4}-\frac{1}{2}\nu; \frac{3}{2}; iay)$
55	$x^{-\frac{1}{2}}(b^2-x^2)^{-\frac{1}{2}}\{[(b+x)^{\frac{1}{2}}+i(b-x)^{\frac{1}{2}}]^{4\nu}$ $+[(b+x)^{\frac{1}{2}}-i(b-x)^{\frac{1}{2}}]^{4\nu}\}, \quad x<b$ $0, \quad x>b$	$2^{2\nu-\frac{1}{2}}b^{2\nu}\pi^{\frac{3}{2}}y^{\frac{1}{2}}$ $\cdot J_{\nu+\frac{1}{4}}(\frac{1}{2}by)J_{-\nu+\frac{1}{4}}(\frac{1}{2}by)$

	$Nf(x)$	$Nh(y)$
56	$(x^2-a^2)^{-\nu-\frac{1}{2}}, \quad x>a$ $0, \quad x<a$ $0<\nu<\frac{1}{2}$	$2^{-\nu-1}a^{-\nu}\pi^{\frac{1}{2}}\Gamma(\frac{1}{2}-\nu)y^{\nu}$ $\cdot J_{\nu}(ay)$
57	$0, \quad x<a$ $(x^2-2ax)^{-\nu-\frac{1}{2}}, \quad x>a$ $0<\nu<\frac{1}{2}$	$2^{-\nu-1}a^{-\nu}\pi^{\frac{1}{2}}\Gamma(\frac{1}{2}-\nu)$ $\cdot[J_{\nu}(ay)\cos ay-Y_{\nu}(ay)\sin(ay)]$
58	$0, \quad x<a$ $x^{-1}(x^2-a^2)^{-\nu-\frac{1}{2}}, \quad x>a$ $-\frac{1}{2}<\nu<\frac{1}{2}$	$\frac{1}{4}\pi^2\sec(\nu\pi)a^{-2\nu}y$ $\cdot[\mathbf{H}_{\nu}(ay)Y_{\nu-1}(ay)-Y_{\nu}(ay)\mathbf{H}_{\nu-1}(ay)]$
59	$0, \quad x<a$ $x^{-\frac{1}{2}}(x^2-a^2)^{-\frac{1}{2}}\{[x+(x^2-a^2)^{\frac{1}{2}}]^{\nu}$ $+[x-(x^2-a^2)^{\frac{1}{2}}]^{\nu}\}, \quad x>a$ $-\frac{1}{2}<\nu<\frac{1}{2}$	$-\frac{1}{2}a^{\nu}\pi(\frac{1}{2}\pi y)^{\frac{1}{2}}$ $\cdot[J_{\frac{1}{4}+\frac{1}{2}\nu}(\frac{1}{2}ay)Y_{\frac{1}{4}-\frac{1}{2}\nu}(\frac{1}{2}ay)$ $+J_{\frac{1}{4}-\frac{1}{2}\nu}(\frac{1}{2}ay)Y_{\frac{1}{4}+\frac{1}{2}\nu}(\frac{1}{2}ay)]$

3. Exponential Functions

	$Nf(x)$	$Nh(y)$
60	e^{-ax}	$y(a^2+y^2)^{-1}$
61	$x^{-1}(e^{-bx}-e^{-ax}), \quad a>b$	$\arctan[(a-b)y/(y^2+ab)]$
62	$x^{\frac{1}{2}}e^{-ax}$	$\frac{1}{2}\pi^{\frac{1}{2}}(a^2+y^2)^{-\frac{3}{4}}\sin[\frac{3}{2}\arctan(y/a)]$
63	$x^{-\frac{1}{2}}e^{-ax}$	$(\frac{1}{2}\pi)^{\frac{1}{2}}(a^2+y^2)^{-\frac{1}{2}}[(a^2+y^2)^{\frac{1}{2}}-a]^{\frac{1}{2}}$
65	$x^{\nu-1}e^{-ax}, \quad \nu>0$	$\Gamma(\nu)(a^2+y^2)^{-\nu/2}\sin[\nu\arctan(y/a)]$
66	$(e^{ax}+1)^{-1}$	$\frac{1}{2}y^{-1}-\frac{1}{2}\pi a^{-1}\operatorname{csch}(\pi y/a)$
68	$x^{-2}(1-e^{-ax})^2$	$2a\arctan[ay/(y^2+2a^2)]$ $-\frac{1}{2}y\log[y^2(y^2+4a^2)/(a^2+y^2)^2]$
69	$0, \quad x<b$ $(x-b)^{\nu}e^{-ax}, \quad x>b, \quad \nu>-1$	$e^{-ab}\Gamma(\nu+1)(a^2+y^2)^{-\frac{1}{2}(\nu+1)}$ $\cdot\sin[by+(\nu+1)\arctan(y/a)]$

	$Nf(x)$	$Nh(y)$
70	$e^{-ax}(1-e^{-bx})^{\nu-1}, \quad \nu>0$	$-\frac{1}{2}ib^{-1}\{B[\nu, (a-iy)/b]-B[\nu, (a+iy)/b]\}$
71	$x^{\nu-1}(e^{ax}+1)^{-1}, \quad \nu>0$	$\Gamma(\nu)\{y^{-\nu}\sin(\frac{1}{2}\nu\pi)$ $+\frac{1}{2}i(2a)^{-\nu}[\zeta(\nu, \frac{1}{2}+\frac{1}{2}iy/a)$ $-\zeta(\nu, \frac{1}{2}-\frac{1}{2}iy/a)-\zeta(\nu, \frac{1}{2}iy/a)$ $+\zeta(\nu, -\frac{1}{2}iy/a)]\}$
72	$x^{\nu-1}(e^{ax}-1)^{-1}, \quad \nu>1$	$\frac{1}{2}ia^{-\nu}\Gamma(\nu)\{\zeta[\nu, 1+(iy/a)]$ $-\zeta(\nu, 1-iy/a)\}$
73	$\exp(-ax^2)$	$-\frac{1}{2}ia^{-\frac{1}{2}}\pi^{\frac{1}{2}}\exp(-y^2/4a)\ \mathrm{Erf}(\frac{1}{2}iy/a^{\frac{1}{2}})$
74	$x^{-\frac{1}{2}}\exp(-ax^2)$	$2^{-\frac{3}{2}}a^{-\frac{1}{2}}\pi y^{\frac{1}{2}}\exp(-\frac{1}{8}a^{-1}y^2)I_{\frac{1}{4}}(\frac{1}{8}a^{-1}y^2)$
75	$x^{\frac{1}{2}}\exp(-ax^2)$	$2^{-7/2}a^{-\frac{3}{2}}\pi y^{\frac{3}{2}}\exp(-\frac{1}{8}a^{-1}y^2)$ $\cdot[I_{-\frac{1}{4}}(\frac{1}{8}a^{-1}y^2)-I_{\frac{3}{4}}(\frac{1}{8}a^{-1}y^2)]$
77	$x^{\nu}\exp(-ax^2), \quad \nu>-1$	$\frac{1}{2}a^{-\frac{1}{2}\nu-1}\Gamma(1+\frac{1}{2}\nu)y$ $\cdot{}_1F_1(\frac{1}{2}\nu+1; \frac{3}{2}; -\frac{1}{4}y^2/a^2)$
79	$\exp(-ax-b^2x^2)$	$-\dfrac{1}{4}i\pi^{\frac{1}{2}}b^{-1}\left\{\exp\left[\left(\dfrac{a-iy}{2b}\right)^2\right]\mathrm{Erfc}\left(\dfrac{a-iy}{2b}\right)\right.$ $\left.-\exp\left[\left(\dfrac{a+iy}{2b}\right)^2\right]\mathrm{Erfc}\left(\dfrac{a+iy}{2b}\right)\right\}$
80	$x^{\nu-1}\exp(-ax-bx^2), \quad \nu>0$	$\dfrac{1}{2}i(2b)^{-\frac{1}{2}\nu}\exp\left(\dfrac{a^2-y^2}{8b}\right)\Gamma(\nu)$ $\cdot\left\{\exp\left(i\dfrac{ay}{4b}\right)D_{-\nu}[(a+iy)(2b)^{-\frac{1}{2}}]\right.$ $\left.-\exp\left(-i\dfrac{ay}{4b}\right)D_{-\nu}[(a-iy)(2b)^{-\frac{1}{2}}]\right\}$
81	$\exp[-(ax)^3]$	$i(3a)^{-\frac{3}{2}}y^{\frac{1}{2}}\{e^{i\frac{1}{4}\pi}S_{0,\frac{1}{3}}[2(\frac{1}{3}y/a)^{\frac{3}{2}}e^{\frac{3}{4}i\pi}]$ $-e^{-\frac{1}{4}i\pi}S_{0,\frac{1}{3}}[2(\frac{1}{3}y/a)^{\frac{3}{2}}e^{-\frac{3}{4}i\pi}]\}$

	$Nf(x)$	$Nh(y)$
82	$x^{\mu}\exp(-ax^{c}),\quad \mu>-1,\quad 0<c\leq 1$	$\sum_{n=0}^{\infty}(-a)^{n}\Gamma(\mu+1+nc)(n!)^{-1}$ $\cdot\cos[\frac{1}{2}\pi(\mu+nc)]y^{-\mu-1-nc}$
83	$x^{\mu}\exp(-ax^{c}),\quad \mu>-1,\quad c\geq 1$	$c^{-1}\sum_{n=0}^{\infty}(-1)^{n}a^{-(\mu+2+2n)/c}[(2n+1)!]^{-1}$ $\cdot\Gamma[(\mu+2n+2)/c]y^{2n+1}$
84	$x^{-\frac{3}{2}}\exp(-a/x)$	$(\pi/a)^{\frac{1}{2}}\exp[-(2ay)^{\frac{1}{2}}]\sin[(2ay)^{\frac{1}{2}}]$
85	$x^{-\nu-1}\exp(-\frac{1}{4}a^{2}/x),\quad \nu>0$	$i2^{\nu}a^{-\nu}y^{\frac{1}{2}\nu}\{e^{\frac{1}{4}i\pi\nu}K_{\nu}[a(iy)^{\frac{1}{2}}]$ $-e^{-i\frac{1}{4}\pi\nu}K_{\nu}[a(-iy)^{\frac{1}{2}}]\}$
86	$x^{-\frac{1}{2}}\exp[-ax-(b/x)]$	$\pi^{\frac{1}{2}}(a^{2}+y^{1})^{-\frac{1}{2}}\exp(-2b^{\frac{1}{2}}u)$ $\cdot[u\sin(2b^{\frac{1}{2}}v)+v\cos(2b^{\frac{1}{2}}v)]$, $u=2^{-\frac{1}{2}}[(a^{2}+y^{2})^{\frac{1}{2}}+a]^{\frac{1}{2}}$, $v=2^{-\frac{1}{2}}[(a^{2}+y^{2})^{\frac{1}{2}}-a]^{\frac{1}{2}}$
87	$x^{-\frac{3}{2}}\exp[-ax-(b/x)]$	$b^{-\frac{1}{2}}\pi^{\frac{1}{2}}\exp(-2b^{\frac{1}{2}}u)\sin(2b^{\frac{1}{2}}v)$, $u=2^{-\frac{1}{2}}[(a^{2}+y^{2})^{\frac{1}{2}}+a]^{\frac{1}{2}}$ $v=2^{-\frac{1}{2}}[(a^{2}+y^{2})^{\frac{1}{2}}-a]^{\frac{1}{2}}$
88	$x^{\nu-1}\exp[-ax-(b^{2}/x)]$	$ib^{\nu}\{(a+iy)^{-\frac{1}{2}\nu}K_{\nu}[2b(a+iy)^{\frac{1}{2}}]$ $-(a-iy)^{-\frac{1}{2}\nu}K_{\nu}[2b(a-iy)^{\frac{1}{2}}]\}$
89	$x^{-2}\exp(-a^{2}x^{-2})$	$a^{-1}\sum_{m=0}^{\infty}\frac{(a^{2}y^{2})^{m+\frac{1}{2}}}{m!(2m+1)!}[\psi(2m+2)+\frac{1}{2}\psi(m+1)$ $-\log(ay)]$
90	$\exp(-ax^{\frac{1}{2}})$	$y^{-1}+a(\frac{1}{2}\pi)^{\frac{1}{2}}y^{-\frac{3}{2}}\{\sin(a/4y)$ $\cdot[\frac{1}{2}-C(a/4y)]$ $-\cos(a^{2}/4y)[\frac{1}{2}-S(a^{2}/4y)]\}$
91	$x^{-\frac{1}{2}}\exp(-ax^{\frac{1}{2}})$	$(2\pi)^{\frac{1}{2}}y^{-\frac{1}{2}}\{\cos(a^{2}/4y)[\frac{1}{2}-C(a^{2}/4y)]$ $+\sin(a^{2}/4y)[\frac{1}{2}-S(a^{2}/4y)]\}$
92	$x^{-\frac{3}{4}}\exp(-ax^{\frac{1}{2}})$	$-\frac{1}{2}a^{\frac{1}{2}}\pi y^{-\frac{1}{2}}\{J_{\frac{1}{4}}(a^{2}/8y)\cos[\frac{1}{8}\pi+(a^{2}/8y)]$ $+Y_{\frac{1}{4}}(a^{2}/8y)\sin[\frac{1}{8}\pi+(a^{2}/8y)]\}$

	$Nf(x)$	$Nh(y)$
93	$x^{\nu-1}\exp(-ax^{\frac{1}{2}}),\quad \nu>0$	$i\Gamma(2\nu)(2y)^{-\nu}\{\exp[-i(\frac{1}{2}\nu\pi+(a^2/8y))]$ $\cdot D_{-2\nu}[a(-i/2y)^{\frac{1}{2}}]$ $-\exp[i(\frac{1}{2}\nu\pi+(a^2/8y))]D_{-2\nu}[a(i/2y)^{\frac{1}{2}}]\}$
95	$(a^2+x^2)^{-\frac{1}{2}}$ $\cdot\exp[-b(a^2+x^2)^{\frac{1}{2}}]$	$\arctan(y/b)\, I_0[a(b^2+y^2)^{\frac{1}{2}}]$ $-(1/\pi)\int_0^{\pi}\exp(ab\cos t)\,t$ $\cdot\sinh(ay\sin t)\;dt$
96	$x^{-\frac{1}{2}}(b^2+x^2)^{-\frac{1}{2}}$ $\cdot\exp[-a(b^2+x^2)^{\frac{1}{2}}]$	$2^{-\frac{1}{2}}\pi^{\frac{1}{2}}y^{\frac{1}{2}}I_{\frac{1}{4}}\{\frac{1}{2}b[(a^2+y^2)^{\frac{1}{2}}-a]\}$ $\cdot K_{\frac{1}{4}}\{\frac{1}{2}b[(a^2+y^2)^{\frac{1}{2}}+a]\}$
101	$x^{\nu-\frac{1}{2}}(a^2+x^2)^{-\frac{1}{2}}$ $\cdot[(a^2+x^2)^{\frac{1}{2}}+a]^{-\nu}\exp[-b(a^2+x^2)^{\frac{1}{2}}],$ $\nu>-\frac{1}{2}$	$2^{\frac{3}{4}+\frac{1}{2}\nu}a^{-\frac{3}{4}}\Gamma(\frac{3}{4}+\frac{1}{2}\nu)\,y^{-\frac{1}{2}}$ $\cdot[(b^2+y^2)^{\frac{1}{2}}+b]^{\frac{1}{4}}$ $\cdot D_{-\nu-\frac{1}{2}}\{(2a)^{\frac{1}{2}}[(b^2+y^2)^{\frac{1}{2}}+b]^{\frac{1}{2}}\}$ $\cdot M_{\frac{1}{2}\nu,\frac{1}{4}}\{a[(b^2+y^2)^{\frac{1}{2}}-b]\}$
102	$(a^2+x^2)^{-\frac{1}{2}}[x+(a^2+x^2)^{\frac{1}{2}}]^{\nu}$ $\cdot\exp[-b(a^2+x^2)^{\frac{1}{2}}]$	$a^{\nu}\csc(\nu\pi)\{\pi\sin[\nu\arctan(y/b)]$ $\cdot I_{-\nu}[a(b^2+y^2)^{\frac{1}{2}}]$ $-\int_0^{\pi}\exp(ab\cos t)\sinh(ay\sin t)$ $\cdot\sin(\nu t)\;dt\}$
103	$(a^2+x^2)^{-\frac{1}{2}}[(a^2+x^2)^{\frac{1}{2}}+a]^{-\frac{1}{2}}$ $\cdot\exp[-b(a^2+x^2)^{\frac{1}{2}}]$	$-i(2a)^{-\frac{1}{2}}\pi e^{ab}\,\mathrm{Erfc}\{a^{\frac{1}{2}}[(b^2+y^2)^{\frac{1}{2}}+b]^{\frac{1}{2}}\}$ $\cdot\mathrm{Erf}\{ia^{\frac{1}{2}}[(b^2+y^2)^{\frac{1}{2}}-b]^{\frac{1}{2}}\}$
105	$x^{-2\nu-\frac{1}{2}}(a^2-x^2)^{-\frac{1}{2}}$ $\cdot\{[a-(a^2-x^2)^{\frac{1}{2}}]^{2\nu}\exp[b(a^2-x^2)^{\frac{1}{2}}]$ $+[a+(a^2-x^2)^{\frac{1}{2}}]^{2\nu}\exp[-b(a^2-x^2)^{\frac{1}{2}}]\},$ $x<a$ $0,\quad x>a$ $\frac{1}{4}\pm\nu>0$	$2^{\frac{3}{2}}a^{-1}\pi^{-\frac{1}{2}}\Gamma(\frac{3}{4}+\nu)$ $\cdot\Gamma(\frac{3}{4}-\nu)\,y^{-\frac{1}{2}}.$ $\cdot M_{\nu,\frac{1}{4}}\{a[b+(b^2-y^2)^{\frac{1}{2}}]\}$ $\cdot M_{-\nu,\frac{1}{4}}\{a[b-(b^2-y^2)^{\frac{1}{2}}]\},\quad b>y$

4. Logarithmic Functions

	$Nf(x)$	$Nh(y)$
106	$-\log x,\quad x<1$ $0,\quad x>1$	$y^{-1}[\gamma+\log y-\mathrm{Ci}(y)]$

	$Nf(x)$	$Nh(y)$
107	$\log(a+x), \quad x<b$ $0, \quad x>b$ $a\geq 1$	$y^{-1}\{\log a-\cos(by)\log(a+b)$ $+\sin(ay)[\mathrm{si}(ay+by)-\mathrm{si}(ay)]$ $+\cos(ay)[\mathrm{Ci}(ay+by)-\mathrm{Ci}(ay)]\}$
111	$\pm\log[(a^2+x^2)/(b^2+x^2)]$ $\pm$ according as $a\gtrless b$	$\pm y^{-1}[2\log(a/b)+e^{by}\,\mathrm{Ei}(-by)$ $-e^{ay}\,\mathrm{Ei}(-ay)+e^{-by}\,\mathrm{Ei}(by)$ $-e^{-ay}\,\mathrm{Ei}(ay)]$
113	$\log\{[x+(a^2+x^2)^{\frac{1}{2}}]/2x\}$	$y^{-1}[K_0(ay)+\gamma+\log(\frac{1}{2}ay)]$
115	$\log(1+a^2x^{-2})$	$y^{-1}[2\gamma+2\log(ay)-e^{ay}\,\mathrm{Ei}(-ay)$ $-e^{-ay}\,\mathrm{Ei}(ay)]$
116	$-\log(a-x), \quad x<a$ $0, \quad x>a$ $a\leq 1$	$-y^{-1}\{\log a-\sin(ay)\,\mathrm{Si}(ay)$ $-\cos(ay)[\mathrm{Ci}(ay)-\gamma-\log y]\}$
117	$-\log(a-x), \quad x<b$ $0, \quad x>b$ $b<a\leq 1$	$-y^{-1}\{\log a-\cos(by)\log(a-b)$ $+\cos(ay)[\mathrm{Ci}(ay-by)-\mathrm{Ci}(ay)]$ $+\sin(ay)[\mathrm{Si}(ay-by)-\mathrm{Si}(ay)]\}$
118	$-\log(a^2-x^2), \quad x<a$ $0, \quad x>a$ $a\leq 1$	$-y^{-1}\{\cos(ay)[\mathrm{Ci}(2ay)+\gamma+\log(y/2a)$ $-2\,\mathrm{Ci}(ay)]+\sin(ay)[\mathrm{Si}(2ay)$ $-2\,\mathrm{Si}(ay)]+2\log a\}$
120	$-\log(a^2-x^2), \quad x<b$ $0, \quad x>b$ $b<a\leq 1$	$-y^{-1}\{2\log a-\cos(by)\log(a^2-b^2)$ $-\cos(ay)[2\,\mathrm{Ci}(ay)-\mathrm{Ci}(ay-by)$ $-\mathrm{Ci}(ay+by)]-\sin(ay)$ $\cdot[2\,\mathrm{si}(ay)-\mathrm{si}(ay-by)-\mathrm{si}(ay+by)]\}$
121	$-[x(1-x)]^{-\frac{1}{2}}\log[x(1-x)], \quad x<1$ $0, \quad x>1$	$-\pi\sin(\frac{1}{2}y)[\frac{1}{2}\pi Y_0(\frac{1}{2}y)$ $-(\gamma+\log 4y)J_0(\frac{1}{2}y)]$
123	$-(2ax-x^2)^{n-\frac{1}{2}}\log(2ax-x^2), \quad x<2a$ $0, \quad x>2a$ $a\leq 1$ $n=0, 1, 2, \ldots$	$-\pi(2n)!(n!)^{-1}\sin(ay)(2y/a)^{-n}\{\frac{1}{2}\pi Y_n(ay)$ $+\frac{1}{2}n!\sum_{m=0}^{n-1}(\frac{1}{2}ay)^{m-n}[m!(n-m)]^{-1}J_m(ay)$ $+J_n(ay)[2\sum_{m=0}^{n-1}(2m+1)^{-1}-\gamma$ $-\log(2y/a)]\}$ For $n=0$, $\sum_{m=0}^{n-1}(\ )=0$

	$Nf(x)$	$Nh(y)$
124	$0, \quad x<a$ $-\log\{[(x+a)^{\frac{1}{2}}+(x-a)^{\frac{1}{2}}]/2x^{\frac{1}{2}}\}, \quad x>a$	$\frac{1}{2}y^{-1}[\frac{1}{2}\pi Y_0(ay)+\log 2-\mathrm{Ci}(ay)]$
125	$e^{-ax}(\log x)^2$	$(a^2+y^2)^{-1}\{\frac{1}{6}\pi^2 y-y[\arctan(y/a)]^2$ $+[\gamma+\frac{1}{2}\log(a^2+y^2)][\gamma y+\frac{1}{2}y$ $\cdot\log(a^2+y^2)-2a\arctan(y/a)]\}$
127	$\log(1+e^{-ax})$	$y^{-1}\{\log 2-\frac{1}{4}\psi[1+(iy/2a)]$ $-\frac{1}{4}\psi[1-i(y/2a)]+\frac{1}{4}\psi[\frac{1}{2}+(iy/2a)]$ $+\frac{1}{4}\psi[\frac{1}{2}-(iy/2a)]\}$
128	$-\log(1-e^{-ax})$	$y^{-1}\{\gamma+\frac{1}{2}\psi[1+(iy/a)]+\frac{1}{2}\psi[1-(iy/a)]\}$

5. Trigonometric Functions

	$Nf(x)$	$Nh(y)$
129	$x^{-1}(\sin ax)^2$	$\frac{1}{4}\{(y+2a)\log(y+2a)+(y-2a)$ $\cdot\log(y-2a)-2y\log y\}$
130	$x^{-2m}(\sin ax)^{2m}, \quad m=1,2,3,\ldots$	$(m!)^2 y\log\left(\frac{y}{a}\right)$ $+\sum_{n=1}^{m}(-1)^n\frac{(y-2an)^{2m-1}\log[2n-(y/a)]}{(m-n)!(m+n)!}$ $+\sum_{n=1}^{m}(-1)^n\frac{(y+2an)^{2m-1}\log[2n+(y/a)]}{(m-n)!(m+n)!}$
131	$e^{-ax}(\sin x)^{2n}, \quad n=0,1,2,\ldots$	$\frac{-(-y)^{-n-1}}{2n+1}\left\{\left[\binom{n+\frac{1}{2}y+i\frac{1}{2}a}{2n+1}\right]^{-1}\right.$ $\left.+\left[\binom{n+\frac{1}{2}-i\frac{1}{2}a}{2n+1}\right]^{-1}\right\}$
132	$(\cosh a-\cos x)^{-1}, \quad x<\pi$ $0, \quad x>\pi$	$y[\cos(\pi y)/\sinh a][\pi y^{-1}\csc(\pi y)]$ $+\sum_{n=0}^{\infty}(-1)^n\epsilon_n(n^2-y^2)^{-1}e^{-na}$

	$Nf(x)$	$Nh(y)$
137	$(\sin x)^{\alpha}, \quad x<\pi$ $0, \quad x>\pi$ $\alpha>-1$	$\dfrac{2^{-\alpha}\pi \sin(\frac{1}{2}y)\Gamma(1+\alpha)}{\Gamma(1+\frac{1}{2}\alpha+\frac{1}{2}y)\Gamma(1+\frac{1}{2}\alpha-\frac{1}{2}y)}$
144	$x^{-\frac{1}{2}}(a^2-x^2)^{-\frac{1}{2}}\cos[b(a^2-x^2)^{\frac{1}{2}}], \quad x<a$ $0, \quad x>\mathrm{a}$ $ab\leq\frac{1}{2}\pi$	$2^{-\frac{3}{2}}\pi^{\frac{3}{2}}y^{\frac{1}{2}}$ $\cdot J_{\frac{1}{4}}\{\frac{1}{2}a[(b^2+y^2)^{\frac{1}{2}}-b]\}$ $\cdot J_{\frac{1}{4}}\{\frac{1}{2}a[(b^2+y^2)^{\frac{1}{2}}+b]\}$
148	$(\sin x)^{-\frac{1}{2}}\exp(-a\sin x), \quad x<\pi$ $0, \quad x>\pi$	$(\frac{1}{2}a)^{\frac{1}{2}}\pi^{\frac{3}{2}}\sin(\frac{1}{2}\pi y)[I_{-\frac{1}{4}-\frac{1}{2}y}(\frac{1}{2}a)I_{-\frac{1}{4}+\frac{1}{2}y}(\frac{1}{2}a)$ $-I_{\frac{1}{4}-\frac{1}{2}y}(\frac{1}{2}a)I_{\frac{1}{4}+\frac{1}{2}y}(\frac{1}{2}a)]$
149	$(\sin x)^{-\frac{1}{2}}\exp(a\sin x), \quad x<\pi$ $0, \quad x>\pi$	$(\frac{1}{2}a)^{\frac{1}{2}}\pi^{\frac{3}{2}}\sin(\frac{1}{2}\pi y)[I_{-\frac{1}{4}-\frac{1}{2}y}(\frac{1}{2}a)I_{-\frac{1}{4}+\frac{1}{2}y}(\frac{1}{2}a)$ $+I_{\frac{1}{4}-\frac{1}{2}y}(\frac{1}{2}a)I_{\frac{1}{4}+\frac{1}{2}y}(\frac{1}{2}a)]$
151	$\log[\csc(\pi x)], \quad x<1$ $0, \quad x>1$	$2y^{-1}\sin^2(\frac{1}{2}y)\{\gamma+\log 2+\frac{1}{2}\psi[1+(y/2\pi)]$ $+\frac{1}{2}\psi[1-\frac{1}{2}(y/\pi)]\}$
153	$(\sin\pi x)^{\nu-1}\log(\csc\pi x), \quad x<1$ $0, \quad x>1$ $\nu>0$	$2^{1-\nu}\Gamma(\nu)\sin(\frac{1}{2}y)\{\Gamma[\frac{1}{2}+\frac{1}{2}\nu+(y/2\pi)]$ $\cdot\Gamma[\frac{1}{2}+\frac{1}{2}\nu-(y/2\pi)]\}^{-1}$ $\cdot\{\log 2+\frac{1}{2}\psi[\frac{1}{2}+\frac{1}{2}\nu+(y/2\pi)]$ $+\frac{1}{2}\psi[\frac{1}{2}+\frac{1}{2}\nu-(y/2\pi)]-\psi(\nu)\}$
161	$\sin(a\sin x), \quad x<\pi, \quad a\leq\pi$ $0, \quad x>\pi$	$\frac{1}{2}\pi[\mathbf{J}_y(a)-\mathbf{J}_{-y}(a)]$
164	$\cos(a\sin x), \quad x<\frac{1}{2}\pi, \quad a\leq\frac{1}{2}\pi$ $0, \quad x>\frac{1}{2}\pi$	$\frac{1}{2}\pi\tan(\frac{1}{2}\pi y)$ $\cdot[\mathbf{J}_y(a)+\mathbf{J}_{-y}(a)]$
165	$(\sin x)^{-\frac{1}{2}}\sin(2a\sin x), \quad x<\pi$ $0, \quad x>\pi$ $a\leq\frac{1}{2}\pi$	$\pi(a\pi)^{\frac{1}{2}}\sin(\frac{1}{2}\pi y)$ $\cdot J_{\frac{1}{4}-\frac{1}{2}y}(a)J_{\frac{1}{4}+\frac{1}{2}y}(a)$
166	$(\sin x)^{-\frac{1}{2}}\cos(2a\sin x), \quad x<\pi$ $0, \quad x>\pi$ $a\leq\frac{1}{2}\pi$	$\pi(a\pi)^{\frac{1}{2}}\sin(\frac{1}{2}\pi y)$ $\cdot J_{-\frac{1}{4}-\frac{1}{2}y}(a)J_{-\frac{1}{4}+\frac{1}{2}y}(a)$
169	$(\sin x)^{-\frac{3}{2}}\sin(2a\sin x), \quad x<\pi$ $0, \quad x>\pi$ $a\leq\frac{1}{2}\pi$	$2(a\pi)^{\frac{3}{2}}\sin(\frac{1}{2}\pi y)$ $\cdot[J_{-\frac{1}{4}+\frac{1}{2}y}(a)J_{-\frac{1}{4}-\frac{1}{2}y}(a)$ $+J_{\frac{3}{4}+\frac{1}{2}y}(a)J_{\frac{3}{4}-\frac{1}{2}y}(a)]$

6. Inverse Trigonometric Functions

	$Nf(x)$	$Nh(y)$
173	$\arcsin x, \quad x<1$ $0, \quad x>1$	$\frac{1}{2}\pi y^{-1}[J_0(y)-\cos y]$
174	$\arccos x, \quad x<1$ $0, \quad x>1$	$\frac{1}{2}\pi y^{-1}[1-J_0(y)]$
175	$x^{-1}\arcsin x, \quad x<1$ $0, \quad x>1$	$\frac{1}{2}\pi\left[\mathrm{si}(y)+\int_y^\infty t^{-1}\mathbf{H}_0(t)\,dt\right]$

7. Hyperbolic Functions

	$Nf(x)$	$Nh(y)$
177	$\operatorname{sech}(ax)$	$-\frac{1}{2}a^{-1}\{\pi\tanh(\frac{1}{2}a^{-1}\pi y)$ $+i[\psi(\frac{1}{4}+\frac{1}{4}a^{-1}y)-\psi(\frac{1}{4}-\frac{1}{4}a^{-1}y)]\}$
178	$[\operatorname{sech}(ax)]^2$	$\frac{1}{4}a^{-2}y[\psi(\frac{1}{4}ia^{-1}y)+\psi(-\frac{1}{4}ia^{-1}y)$ $-\psi(\frac{1}{2}+\frac{1}{4}ia^{-1}y)-\psi(\frac{1}{2}-\frac{1}{4}ia^{-1}y)]$
182	$x\operatorname{csch}(ax)$	$i\frac{1}{4}a^{-2}[\psi'(\frac{1}{2}+i\frac{1}{2}a^{-1}y)-\psi'(\frac{1}{2}-i\frac{1}{2}a^{-1}y)]$
183	$\cosh(ax)\operatorname{sech}(bx), \quad a<b$	$i\frac{1}{4}b^{-1}\left[\psi\left(\frac{3b+a+iy}{4b}\right)-\psi\left(\frac{3b+a-iy}{4b}\right)\right.$ $+\psi\left(\frac{3b-a+iy}{4b}\right)-\psi\left(\frac{3b-a-iy}{4b}\right)$ $\left.-i\frac{2\pi\sinh(b^{-1}\pi y)}{\cosh(b^{-1}\pi y)+\cos(ab^{-1}\pi)}\right]$
184	$\sinh(ax)\operatorname{sech}(bx), \quad a<b$	$\pi b^{-1}\sin(\frac{1}{2}ab^{-1}\pi)\sinh(\frac{1}{2}b^{-1}\pi y)$ $\cdot[\cosh(b^{-1}\pi y)+\cos(ab^{-1}\pi)]^{-1}$

	$Nf(x)$	$Nh(y)$
185	$\sinh(ax)\ \mathrm{csch}(bx), \quad a<b$	$\frac{1}{2}b^{-1}\pi\left\{\sinh\left(\frac{\pi y}{b}\right)\left[\cosh\left(\frac{\pi y}{b}\right)+\cos\left(\frac{a\pi}{b}\right)\right]^{-1}+i\pi^{-1}\left[\psi\left(\frac{a+b+iy}{2b}\right)-\psi\left(\frac{a+b+iy}{2b}\right)\right]\right\}$
186	$1-\tanh(ax)$	$y^{-1}-\frac{1}{2}a^{-1}\pi\ \mathrm{csch}(\frac{1}{2}a^{-1}\pi y)$
187	$x^{s-1}(x^{-1}-\mathrm{csch}x), \quad -1<s<1$	$i2^{-s}\Gamma(s)[\zeta(s,\frac{1}{2}-i\frac{1}{2}y)-\zeta(s,\frac{1}{2}+i\frac{1}{2}y)]$ $-\cos(\frac{1}{2}s\pi)\Gamma(s-1)y^{-s+1}$
188	$x^{-1}\sinh(ax)\ \mathrm{sech}(bx), \quad a<b$	$\arctan[\tan(\frac{1}{2}ab^{-1}\pi)\ \tanh(\frac{1}{2}b^{-1}\pi y)]$ $-i\left[\log\Gamma\left(\frac{3b-a+iy}{4b}\right)-\log\Gamma\left(\frac{3b-a-iy}{4b}\right)\right]$ $-i\left[\log\Gamma\left(\frac{3b+a-iy}{4b}\right)-\log\Gamma\left(\frac{3b+a+iy}{4b}\right)\right]$
190	$x^{-1}(x^{-1}-\mathrm{csch}x)$	$i\log\Gamma(\frac{1}{2}-i\frac{1}{2}y)-i\log\Gamma(\frac{1}{2}+i\frac{1}{2}y)$ $-y[\log(\frac{1}{2}y)-1]$
201	$[\cosh(ax)+\cos b]^{-1}, \quad -\pi<b<\pi$	$(\csc b)2y\sum_{n=0}^{\infty}(-1)^{n+1}\epsilon_n(y^2+a^2n^2)^{-1}\sin(bn)$
202	$[\cosh(ax)+\cosh b]^{-1}$	$\mathrm{csch}b\{y\sum_{n=0}^{\infty}(-1)^n\epsilon_n(y^2+n^2a^2)^{-1}e^{-nb}$ $-a^{-1}\pi\ \mathrm{csch}(a^{-1}\pi y)\cos(a^{-1}by)\}$

	$Nf(x)$	$Nh(y)$
210	$0, \quad x<a$ $(\cosh x-\cosh a)^{-\frac{1}{2}}, \quad x>a$	$2^{-\frac{1}{2}}\pi \tanh(\pi y) P_{-\frac{1}{2}+iy}(\cosh a)$
212	$(\operatorname{csch} ax)^{\nu}, \quad 0<\nu<1$	$2^{\nu}a^{-1}\pi \cos(\frac{1}{2}\nu\pi)\Gamma(1-\nu)$ $\cdot\left[\Gamma\left(1-\frac{1}{2}\nu+i\frac{y}{2a}\right)\right.$ $\left.\cdot\Gamma\left(1-\frac{1}{2}\nu-i\frac{y}{2a}\right)\right]^{-1} \sinh\left(\frac{\pi y}{2a}\right)$ $\cdot\left[\cosh\left(\frac{\pi y}{2a}\right)-\cos(\nu\pi)\right]^{-1}$
215	$0, \quad x<a$ $(\cosh x-\cosh a)^{-\nu}, \quad x>a$ $0<\nu<1$	$2^{-\frac{1}{2}}\pi^{-\frac{1}{2}}\Gamma(1-\nu)(\operatorname{csch} a)^{\nu-\frac{1}{2}}$ $\cdot\sinh(\pi y)\Gamma(\nu+iy)\Gamma(\nu-iy)$ $\cdot P^{\frac{1}{2}-\nu}_{-\frac{1}{2}+iy}(\cosh a)$
223	$x^{-1}e^{-ax}\sinh(bx)$	$\frac{1}{2}\arctan[2by(y^2+a^2-b^2)^{-1}]$
225	$(e^{bx}-1)^{-1}\sinh(ax), \quad a<b$	$-\frac{1}{2}y(a^2+y^2)^{-1}+\frac{1}{2}b^{-1}\pi\sinh(2b^{-1}\pi y)$ $\cdot[\cosh(2b^{-1}\pi y)-\cos(2ab^{-1}\pi)]^{-1}$ $+i\frac{1}{2}b^{-1}\{\psi[1+(a+iy)/b]$ $-\psi[1+(a-iy)/b]\}$
226	$e^{-ax}[\sinh(bx)]^{\nu}, \quad \nu>-1, \quad b\nu<a$	$-i2^{-\nu-2}b^{-1}\Gamma(\nu+1)$ $\cdot\left\{\frac{\Gamma[\frac{1}{2}b^{-1}(a-b\nu-iy)]}{\Gamma[\frac{1}{2}b^{-1}(a+b\nu-iy)+1]}\right.$ $\left.-\frac{\Gamma[\frac{1}{2}b^{-1}(a-b\nu+iy)]}{\Gamma[\frac{1}{2}b^{-1}(a+b\nu+iy)+1]}\right\}$
231	$e^{-ax}\begin{cases}\operatorname{ctnh}(bx^{\frac{1}{2}})\\ \tanh(bx^{\frac{1}{2}})\end{cases}$	See Mordell, L. J. (1920). *Mess. Math.* **49**, 65–72.
240	$\log[\operatorname{ctnh}(ax)]$	$y^{-1}\left[\gamma+2\log 2+\frac{1}{2}\psi\left(\frac{1}{2}+i\frac{y}{4a}\right)\right.$ $\left.+\frac{1}{2}\psi\left(\frac{1}{2}-i\frac{y}{2a}\right)\right]$

	$Nf(x)$	$Nh(y)$
250	$\exp(-a \sinh x)$	$y S_{-1,iy}(a) = \frac{1}{2}\pi \operatorname{csch}(\pi y)[\mathbf{J}_{iy}(a) + \mathbf{J}_{-iy}(a) - J_{iy}(a) - J_{-iy}(a)]$
251	$\exp(-a \cosh x)$	$\operatorname{csch}(\pi y)\left\{\int_a^{\pi} \exp(a \cos t) \cosh(yt)\, dt - \frac{1}{2}\pi[I_{iy}(a) + I_{-iy}(a)]\right\}$
252	$(\sinh x)^{-\frac{1}{2}} \exp(-2a \sinh x)$	$\frac{1}{4} i a^{\frac{1}{2}} \pi^{\frac{3}{2}}[J_{\frac{1}{4}-\frac{1}{2}iy}(a) Y_{-\frac{1}{4}-i\frac{1}{2}y}(a) - J_{-\frac{1}{4}-i\frac{1}{2}y}(a) Y_{\frac{1}{4}-\frac{1}{2}iy}(a) - J_{\frac{1}{4}+i\frac{1}{2}y}(a) Y_{-\frac{1}{4}+i\frac{1}{2}y}(a) + J_{-\frac{1}{4}+i\frac{1}{2}y}(a) Y_{\frac{1}{4}+i\frac{1}{2}y}(a)]$
256	$(\sinh x)^{-\frac{1}{2}} \exp(-a \operatorname{csch} x)$	$-2^{\frac{1}{2}} \operatorname{Im}\{\Gamma(\frac{1}{2}+iy) \cdot D_{-\frac{1}{2}-iy}[(2ai)^{\frac{1}{2}}] D_{-\frac{1}{2}-iy}[(-2ai)^{\frac{1}{2}}]\}$
261	$(\csc x)^{\frac{1}{2}} \sinh(a \sin x), \quad x<\pi$ $0, \quad x>\pi$	$2^{-\frac{1}{2}} a^{\frac{1}{2}} \pi^{\frac{3}{2}} \sin(\frac{1}{2}\pi y) \cdot I_{\frac{1}{4}-\frac{1}{2}y}(\frac{1}{2}a) I_{\frac{1}{4}+\frac{1}{2}y}(\frac{1}{2}a)$
262	$(\csc x)^{\frac{1}{2}} \cosh(a \sin x), \quad x<\pi$ $0, \quad x>\pi$	$(\frac{1}{2}a)^{\frac{1}{2}} \pi^{\frac{3}{2}} \sin(\frac{1}{2}\pi y) \cdot I_{-\frac{1}{4}-\frac{1}{2}y}(\frac{1}{2}a) I_{-\frac{1}{4}+\frac{1}{2}}(\frac{1}{2}a)$
265	$(\csc x)^{\frac{3}{2}} \sinh(2a \sin x), \quad x<\pi$ $0, \quad x>\pi$	$2(a\pi)^{\frac{3}{2}} \sin[(\pi/2)y][I_{-\frac{1}{4}-\frac{1}{2}y}(a) I_{-\frac{1}{4}+\frac{1}{2}y}(a) - I_{\frac{3}{4}-\frac{1}{2}y}(a) I_{\frac{3}{4}+\frac{1}{2}y}(a)]$

8. Gamma and Related Functions

	$Nf(x)$	$Nh(y)$
277	$x \operatorname{Erfc}(ax)$	$\frac{4}{3} a^{-3} \pi^{-\frac{1}{2}} y\, {}_1F_1(2; \frac{5}{2}; -y^2/a^2)$
278	$\operatorname{Erfc}(ax)$	$y^{-1}[1 - \exp(-\frac{1}{4} a^{-2} y^2)]$
280	$x^{\nu-1} \operatorname{Erfc}(ax)$	$\frac{1}{2} a^{-\nu-1} \Gamma(1+\frac{1}{2}\nu) \Gamma(\frac{1}{2}+\frac{1}{2}\nu) \pi^{-\frac{1}{2}} [\Gamma(\frac{3}{2}+\frac{1}{2}\nu)]^{-1} y \cdot {}_2F_2(\frac{1}{2}+\frac{1}{2}\nu, 1+\frac{1}{2}\nu; \frac{3}{2}, \frac{3}{2}+\frac{1}{2}\nu; -y^2/4a^2)$
282	$\operatorname{Erfc}[(ax)^{\frac{1}{2}}]$	$y^{-1} - (\frac{1}{2}a)^{\frac{1}{2}} (a^2+y^2)^{-\frac{1}{2}} [(a^2+y^2)^{\frac{1}{2}} - a]^{-\frac{1}{2}}$

	$Nf(x)$	$Nh(y)$
291	$-\mathrm{Ei}(-ax)$	$\frac{1}{2}y^{-1}\log(1+y^2a^{-2})$
292	$-\mathrm{Ei}(-bx),\quad x<a$ $0,\quad x>a$	$-\frac{1}{2}y^{-1}\{\mathrm{Ei}[-a(b+iy)]+\mathrm{Ei}[-a(b-iy)]$ $-\log(1+y^2b^{-2})-2\cos(ay)\,\mathrm{Ei}(-ab)\}$
293	$-e^{-ax}\,\mathrm{Ei}(-bx),\quad a>-b$	$-(a^2+y)^{-1}\{a\arctan[y(a+b)^{-1}]$ $-\frac{1}{2}y\log[b^{-2}y^2+(1+ab^{-1})^2]\}$
302	$\mathrm{Si}(bx),\quad x<a$ $0,\quad x>a$	$\frac{1}{2}y^{-1}[\mathrm{Si}(ab+ay)+\mathrm{Si}(ab-ay)$ $-2\cos(ay)\,\mathrm{Si}(ab)]$
303	$e^{-bx}\,\mathrm{Si}(ax)$	$\dfrac{1}{2}(b^2+y^2)^{-1}\left\{\dfrac{1}{2}b\log\left[\dfrac{b^2+(a+y)^2}{b^2+(a-y)^2}\right]\right.$ $\left.+y\arctan\left(\dfrac{a+y}{b}\right)-y\arctan\left(\dfrac{y-a}{b}\right)\right\}$
311	$[\frac{1}{2}-C(ax^2)]\cos(ax^2)$ $+[\frac{1}{2}-S(ax^2)]\sin(ax^2)$	$2^{-\frac{3}{2}}a^{-\frac{1}{2}}\pi^{\frac{1}{2}}\left\{\left[C\left(\dfrac{y^2}{4a}\right)+S\left(\dfrac{y^2}{4a}\right)-1\right]\sin\left(\dfrac{y^2}{4a}\right)\right.$ $\left.+\left[C\left(\dfrac{y^2}{4a}\right)-S\left(\dfrac{y^2}{4a}\right)\right]\cos\left(\dfrac{y^2}{4a}\right)\right\}$
312	$S(ax^{-1})$	$\frac{1}{4}y^{-1}\{2-\exp[-2(ay)^{\frac{1}{2}}]$ $-\cos[(2ay)^{\frac{1}{2}}]-\sin[(2ay)^{\frac{1}{2}}]\}$

9. Elliptic Integrals and Legendre Functions

	$Nf(x)$	$Nh(y)$
320	$K[(\frac{1}{2}-\frac{1}{2}x)^{\frac{1}{2}}],\quad x<1$ $0,\quad x>1$	$\frac{9}{32}\pi^{\frac{3}{2}}[\Gamma(\frac{7}{4})]^{-2}y^{-\frac{1}{2}}s_{\frac{1}{2},0}(y)$
328	$0,\quad \sinh x<a^{-1}$ $\operatorname{csch} x\,K[(1-a^{-2}\operatorname{csch}^2 x)^{\frac{1}{2}}],\quad \sinh x>a^{-1}$	$-\frac{1}{2}\pi\tanh(\frac{1}{2}\pi y)$ $\cdot\mathfrak{P}_{-\frac{1}{2}+i\frac{1}{2}y}(P)\,\mathrm{Re}[\mathfrak{Q}_{-\frac{1}{2}+i\frac{1}{2}y}(P)],$ $P=(1+a^{-2})^{\frac{1}{2}}$

	$Nf(x)$	$Nh(y)$
333	$(1-x^2)^{-\frac{1}{2}\mu}P_\nu^\mu(x),\quad x<1$ $0,\quad x>1$ $\mu-\nu<3,\quad \mu+\nu<2,$ $\mu<\frac{1}{2}$	$2^{\mu-2}\pi^{\frac{1}{2}}(1-\mu-\nu)(2+\nu-\mu)$ $\cdot\{\Gamma[\frac{1}{2}(3-\mu-\nu)]\Gamma[\frac{1}{2}(4-\mu+\nu)]\}^{-1}$ $\cdot y^{\mu-\frac{1}{2}}s_{\frac{1}{2}-\mu,\frac{1}{2}+\nu}(y)$
344	$\mathfrak{P}_\nu(1+2a^2\sinh^2 x),\quad 0<a<1,$ $-1<\nu<0$	$-\frac{1}{2}\operatorname{sech}(\frac{1}{2}\pi y)\operatorname{Im}\{P_\nu^{-i\frac{1}{2}y}(r)$ $\cdot[Q_\nu^{i\frac{1}{2}y}(r)+Q_{-\nu-1}^{i\frac{1}{2}y}(r)]\},\quad r=(1-a^2)^{\frac{1}{2}}$
345	$\mathfrak{P}_\nu(1+2a^2\sinh^2 x),\quad a>1,$ $-1<\nu<0$	$[2a\sin(\nu\pi)]^{-1}\tanh(\frac{1}{2}\pi y)$ $\cdot\{P_{-\frac{1}{2}+i\frac{1}{2}y}^{-\nu-\frac{1}{2}}(s)\operatorname{Re}[Q_{-\frac{1}{2}+j\frac{1}{2}y}^{\nu+\frac{1}{2}}(s)]$ $+P_{-\frac{1}{2}+i\frac{1}{2}y}^{\nu+\frac{1}{2}}(s)\operatorname{Re}[Q_{-\frac{1}{2}+i\frac{1}{2}y}^{-\nu-\frac{1}{2}}(s)]\},$ $s=(1-a^{-2})^{\frac{1}{2}}$
348	$0,\quad \sinh x<1/a$ $\mathfrak{P}(2a^2\sinh^2 x-1),\quad \sinh x<1/a,$ $-1<\nu<0$	$(a\pi)^{-1}\operatorname{Im}\{\mathfrak{Q}_{-\frac{1}{2}+i\frac{1}{2}y}^{-\nu-\frac{1}{2}}[(1+a^{-2})^{\frac{1}{2}}]$ $\cdot\mathfrak{Q}_{-\frac{1}{2}+i\frac{1}{2}y}^{\nu+\frac{1}{2}}[(1+a^{-2})^{\frac{1}{2}}]\}$

10. Bessel Functions

	$Nf(x)$	$Nh(y)$
371	$x^{-\frac{1}{2}}[J_0(ax)]^2$	$(\frac{1}{2}\pi/y)^{\frac{1}{2}}\{P_{-\frac{1}{4}}[(1-4a^2y^{-2})^{\frac{1}{2}}]\}^2,\quad y>2a$
372	$x^{-\frac{1}{2}}[J_\nu(ax)]^2,\quad \nu>-\frac{1}{4}$	$(\frac{1}{2}\pi/y)^{\frac{1}{2}}\Gamma(\frac{3}{4}+\nu)[\Gamma(\frac{3}{4}-\nu)]^{-1}$ $\cdot\{P_{-\frac{1}{4}}^{-\nu}[(1-4a^2y^{-2})^{\frac{1}{2}}]\}^2,\quad y>2a$
373	$[x^\nu J_\nu(ax)]^2,\quad -\frac{1}{4}<\nu<0$	$2^{2\nu}\pi^{-1}\Gamma(\frac{1}{2}+\nu)[\Gamma(\frac{1}{2}-2\nu)]^{-1}(2a)^{\nu-\frac{1}{2}}$ $\cdot y^{-\nu-\frac{1}{2}}(4a^2-y^2)^{-\nu}e^{i\pi\nu}\mathfrak{Q}_{-\nu-\frac{1}{2}}^{-\nu}$ $\cdot[(y^2+4a^2)/4ay],\quad y<2a$ $2^{2\nu}\pi^{-\frac{3}{2}}\cos(2\pi\nu)\Gamma(\frac{1}{2}+\nu)(2a)^{\nu-\frac{1}{2}}$ $\cdot y^{-\nu-\frac{1}{2}}(y^2-4a^2)^{-\nu}e^{-i\pi\nu}\mathfrak{Q}_{\nu-\frac{1}{2}}^{\nu}$ $\cdot[(y^2+4a^2)/4ay],\quad y>2a$
380	$J_0[b(ax-x^2)^{\frac{1}{2}}],\quad x<a$ $0,\quad x>a$ $ab\leq 2\tau_{0,1}$	$2\sin(\frac{1}{2}ay)z^{-1}\sin(\frac{1}{2}az),\quad z=(b^2+y^2)^{\frac{1}{2}}$
381	$(ax-x^2)^{-\frac{1}{2}}J_\nu[b(ax-x^2)^{\frac{1}{2}}],\quad x<a$ $0,\quad x>a$ $ab\leq 2\tau_{\nu,1},\quad \nu>-1$	$\pi\sin(\frac{1}{2}ay)J_{\frac{1}{2}\nu}[\frac{1}{4}a(z+y)]J_{\frac{1}{2}\nu}[\frac{1}{4}a(z-y)],$ $z=(b^2+y^2)^{\frac{1}{2}}$

	$Nf(x)$	$Nh(y)$
382	$(ax-x^2)^{\frac{1}{2}\nu}J_\nu[b(ax-x^2)^{\frac{1}{2}}],\quad x<a$ $0,\quad x>a$ $ab\leq 2\tau_{\nu,1},\quad \nu>-1$	$(\pi a)^{\frac{1}{2}}(\frac{1}{2}ab)^\nu \sin(\frac{1}{2}ay)$ $\cdot(b^2+y^2)^{-\frac{1}{2}\nu-\frac{1}{4}}J_{\nu+\frac{1}{2}}[\frac{1}{2}a(b^2+y^2)^{\frac{1}{2}}]$
383	$-Y_0[b(ax-x^2)^{\frac{1}{2}}],\quad x<a$ $0,\quad x>a$ $ab\leq 2\zeta_{0,1}$	$-2\pi^{-1}\sin(\frac{1}{2}ay)z^{-1}$ $\cdot\{\sin(\frac{1}{2}az)[\mathrm{Ci}(\frac{1}{2}az+\frac{1}{2}ay)$ $+\mathrm{Ci}(\frac{1}{2}az-\frac{1}{2}ay)]-\cos(\frac{1}{2}az)$ $\cdot[\mathrm{Si}(\frac{1}{2}az+\frac{1}{2}ay)+\mathrm{Si}(\frac{1}{2}az-\frac{1}{2}ay)]\},$ $z=(b^2+y^2)^{\frac{1}{2}}$
384	$-(ax-x^2)^{-\frac{1}{2}}Y_\nu[b(ax-x^2)^{\frac{1}{2}}],\quad x<a$ $0,\quad x>a$ $ab\leq 2\zeta_{\nu,1},\quad -1<\nu<1$	$-\pi\sin(\frac{1}{2}ay)\{\cos(\frac{1}{2}\pi\nu)$ $\cdot[J_{\frac{1}{2}\nu}(\frac{1}{4}az+\frac{1}{4}ay)Y_{\frac{1}{2}\nu}(\frac{1}{4}az-\frac{1}{4}ay)$ $+Y_{\frac{1}{2}\nu}(\frac{1}{4}az+\frac{1}{4}ay)J_{\frac{1}{2}\nu}(\frac{1}{4}az-\frac{1}{4}ay)]$ $-\sin(\frac{1}{2}\pi\nu)[J_{\frac{1}{2}\nu}(\frac{1}{4}az+\frac{1}{4}ay)J_{\frac{1}{2}\nu}(\frac{1}{4}az-\frac{1}{4}ay)$ $+Y_{\frac{1}{2}\nu}(\frac{1}{4}az+\frac{1}{4}ay)Y_{\frac{1}{2}\nu}(\frac{1}{4}az-\frac{1}{4}ay)]\},$ $z=(b^2+y^2)^{\frac{1}{2}}$
385	$-\log(ax-x^2)J_0[b(ax-x^2)^{\frac{1}{2}}],\quad x<a$ $0,\quad x>a$ $ab\leq 2\tau_{0,1},\quad a\leq 2$	$-4z^{-1}\{\sin(\frac{1}{2}az)[\mathrm{Ci}(az)-\frac{1}{2}\mathrm{Ci}(\frac{1}{2}az+\frac{1}{2}ay)$ $-\frac{1}{2}\mathrm{Ci}(\frac{1}{2}az-\frac{1}{2}ay)+\log(\frac{1}{2}ab)-\log z]$ $-\cos(\frac{1}{2}az)[\mathrm{Si}(az)-\frac{1}{2}\mathrm{Si}(\frac{1}{2}az+\frac{1}{2}ay)$ $-\frac{1}{2}\mathrm{Si}(\frac{1}{2}az-\frac{1}{2}ay)]\}\sin(\frac{1}{2}ay),$ $z=(b^2+y^2)^{\frac{1}{2}}$
390	$J_\nu(a\sin x),\quad x<\pi$ $0,\quad x>\pi$ $\nu>-1,\quad a\leq\tau_{\nu,1}$	$\pi\sin(\frac{1}{2}\pi y)J_{\frac{1}{2}\nu-\frac{1}{2}y}(\frac{1}{2}a)J_{\frac{1}{2}\nu+\frac{1}{2}y}(\frac{1}{2}a)$
391	$\csc x\, J_\nu(a\sin x),\quad x<\pi$ $0,\quad x>\pi$ $\nu>0,\quad a\leq\tau_{\nu,1}$	$\frac{1}{2}\pi a\nu^{-1}[J_{\frac{1}{2}(\nu-1+y)}(\frac{1}{2}a)J_{\frac{1}{2}(\nu-1-y)}(\frac{1}{2}a)$ $+J_{\frac{1}{2}(\nu+1+y)}(\frac{1}{2}a)J_{\frac{1}{2}(\nu+1-y)}(\frac{1}{2}a)]$
394	$(\sin x)^{-m}J_m(a\sin x),\quad x<\pi$ $0,\quad x>\pi$ $m=0,1,2,\ldots,\quad a\leq\tau_{m,1}$	$\pi(\frac{1}{2}a)^m m!\sin(\frac{1}{2}\pi y)$ $\cdot\sum_{n=0}^{n}\epsilon_n[(m+n)!(m-n)!]^{-1}J_{n-\frac{1}{2}y}(\frac{1}{2}a)$ $\cdot J_{n+\frac{1}{2}y}(\frac{1}{2}a)$

11. Modified Bessel Functions

	$Nf(x)$	$Nh(y)$
399	$e^{-bx}I_0(ax), \quad b<a$	$2^{-\frac{1}{2}}[(b^2-a^2-y^2)^2+4b^2y^2]^{-\frac{1}{2}}$ $\cdot\{[(b^2-a^2-y^2)^2+4b^2y^2]^{\frac{1}{2}}-(b^2-a^2-y^2)\}^{\frac{1}{2}}$
400	$K_0(ax)$	$(a^2+y^2)^{-\frac{1}{2}}\log\{a^{-1}[y+(a^2+y^2)^{\frac{1}{2}}]\}$
401	$x^{2n}K_0(ax), \quad n=0, 1, 2, \ldots$	$(-1)^n(2n)!(a^2+y^2)^{-n-\frac{1}{2}}Q_{2n}[y(a^2+y^2)^{-\frac{1}{2}}]$
402	$x^{2n+1}K_0(ax), \quad n=0, 1, 2, \ldots$	$(-1)^n\frac{1}{2}\pi(2n+1)!(a^2+y^2)^{-n-1}$ $\cdot P_{2n+1}[y(a^2+y^2)^{-\frac{1}{2}}]$
403	$x^{\frac{1}{2}}K_0(ax)$	$\frac{1}{4}(2\pi)^{\frac{1}{2}}z^{\frac{3}{2}}\{2E[(\frac{1}{2}-\frac{1}{2}yz)^{\frac{1}{2}}]$ $-2E[(\frac{1}{2}+\frac{1}{2}yz)]^{\frac{1}{2}}+K[(\frac{1}{2}+\frac{1}{2}yz)^{\frac{1}{2}}]$ $-K[(\frac{1}{2}-\frac{1}{2}yz)^{\frac{1}{2}}]\}, \quad z=(a^2+y^2)^{-\frac{1}{2}}$
404	$x^{-\frac{1}{2}}K_0(ax)$	$(\frac{1}{2}\pi z)^{\frac{1}{2}}\{K[(\frac{1}{2}+\frac{1}{2}yz)^{\frac{1}{2}}]$ $-K[(\frac{1}{2}-\frac{1}{2}yz)^{\frac{1}{2}}]\}, \quad z=(a^2+y^2)^{-\frac{1}{2}}$
405	$K_\nu(ax), \quad -1<\nu<1$	$\frac{1}{4}\pi\csc(\frac{1}{2}\pi\nu)(a^2+y^2)^{-\frac{1}{2}}\{a^{-\nu}[y+(a^2+y^2)^{\frac{1}{2}}]^\nu$ $-a^\nu[y+(a^2+y^2)^{\frac{1}{2}}]^{-\nu}\}$
407	$x^\mu K_\nu(ax), \quad \mu\pm\nu>-1$	$\frac{1}{4}\pi\csc[\frac{1}{2}\pi(\mu-\nu)]\Gamma(1+\mu+\nu)(a^2+y^2)^{-\frac{1}{2}\mu-\frac{1}{2}}$ $\cdot\{P_\mu^{-\nu}[y(a^2+y^2)^{-\frac{1}{2}}]-P_\mu^{-\nu}[-y(a^2+y^2)^{-\frac{1}{2}}]\}$
411	$x^{-\frac{1}{2}}I_\nu(ax)K_\nu(ax), \quad \nu>-\frac{1}{4}$	$\Gamma(\frac{3}{4}+\nu)[\Gamma(\frac{3}{4}-\nu)]^{-1}(\frac{1}{2}\pi/y)^{\frac{1}{2}}$ $\cdot e^{i\pi\nu}\mathfrak{Q}_{-\frac{1}{4}}^{-\nu}[(1+4a^2y^{-2})^{\frac{1}{2}}]\mathfrak{P}_{-\frac{1}{4}}^{-\nu}[(1+4a^2y^{-2})^{\frac{1}{2}}]$
418	$x^{-\frac{1}{2}}[K_\nu(ax)]^2, \quad -\frac{1}{4}<\nu<\frac{1}{4}$	$\Gamma(\frac{3}{4}+\nu)[\Gamma(\frac{3}{4}-\nu)]^{-1}$ $\cdot e^{i2\pi\nu}(\frac{1}{2}\pi/y)^{\frac{1}{2}}\{\mathfrak{Q}_{-\frac{1}{4}}^{-\nu}[(1+4a^2y^{-2})^{\frac{1}{2}}]\}^2$
419	$x^{\zeta-1}K_\nu(x)K_\mu(x), \quad \zeta>\|\mu\|+\|\nu\|$	$2^{\zeta-2}[\Gamma(1+\zeta)]^{-1}\Gamma[\frac{1}{2}(\zeta+\mu+\nu+1)]$ $\cdot\Gamma[\frac{1}{2}(\zeta+\mu-\nu+1)]\Gamma[\frac{1}{2}(\zeta-\mu+\nu+1)]$ $\cdot\Gamma[\frac{1}{2}(\zeta-\mu-\nu-1)]y_4F_3[\frac{1}{2}(\zeta+\mu+\nu+1),$ $\frac{1}{2}(\zeta+\mu-\nu+1), \frac{1}{2}(\zeta-\mu+\nu+1),$ $\frac{1}{2}(\zeta-\mu-\nu+1); \frac{3}{2}, \frac{1}{2}(\zeta+1), 1+\zeta/2;$ $-\frac{1}{4}y^2]$

	$Nf(x)$	$Nh(y)$
420	$x^{2\nu}\exp(-x^2)I_\nu(x^2), \quad -\frac{1}{4}<\nu<0$	$2^{-\frac{3}{2}}\Gamma(\frac{1}{2}+\nu)[\Gamma(1-\nu)]^{-1}(\frac{1}{2}y)^{-2\nu}$ $\cdot y^{-2\nu}\exp(-y^2/8)\,{}_1F_1(\frac{1}{2}-2\nu;\,1-\nu;\,\frac{1}{8}y^2)$
423	$x^{\frac{1}{2}}K_{\frac{1}{4}}(ax^2)$	$\frac{1}{4}\pi a^{-1}(\frac{1}{2}\pi y)^{\frac{1}{2}}[I_{\frac{1}{4}}(\frac{1}{4}y^2/a)-\mathbf{L}_{\frac{1}{4}}(\frac{1}{4}y^2/a)]$
429	$x^\mu\exp(-x^2)K_\nu(x^2),$ $-1-2\nu<\mu<-1+2\nu$	$2^{-\frac{1}{2}\mu-2}\pi^{\frac{1}{2}}\Gamma(1+\frac{1}{2}\mu-\nu)\Gamma(1+\frac{1}{2}\mu+\nu)$ $\cdot[\Gamma(\frac{3}{2}+\frac{1}{2}\mu)]^{-1}y$ $\cdot{}_2F_2(1+\frac{1}{2}\mu+\nu,\,1+\frac{1}{2}\mu-\nu;\,\frac{3}{2}+\frac{1}{2}\mu_1\frac{3}{2};\,-\frac{1}{8}y^2)$
431	$x^{-3}K_0(ax^{-1})$	$\pi a^{-1}yJ_1[(2ay)^{\frac{1}{2}}]K_1[(2ay)^{\frac{1}{2}}]$
432	$K_0(ax^{\frac{1}{2}})$	$-\frac{1}{2}y^{-1}[\mathrm{Ci}(\frac{1}{4}a^2/y)\cos(\frac{1}{4}a^2/y)$ $+\mathrm{si}(\frac{1}{4}a^2/y)\sin(\frac{1}{4}a^2/y)]$
433	$x^{-\frac{1}{2}}K_{2\nu}(ax^{\frac{1}{2}}), \quad -\frac{1}{2}<\nu<\frac{1}{2}$	$-\frac{1}{4}\pi^{\frac{3}{2}}\sec(\pi\nu)y^{-\frac{1}{2}}$ $\cdot[\cos(\frac{1}{2}\pi\nu-\frac{1}{4}\pi-\frac{1}{8}a^2/y)J_\nu(\frac{1}{8}a^2/y)$ $-\sin(\frac{1}{2}\pi\nu-\frac{1}{4}\pi-\frac{1}{8}a^2/y)Y_\nu(\frac{1}{8}a^2/y)]$
437	$[x(a-x)]^{-\frac{1}{2}}I_{2\nu}\{b[x(a-x)]^{\frac{1}{2}}\}, \quad x<a$ $0, \quad x>a$ $\nu>-\frac{1}{2}$	$\pi\sin(\frac{1}{2}ay)J_\nu(z_1)J_\nu(z_2),$ $z_{1,2}=\frac{1}{4}a[y\pm(y^2-b^2)^{\frac{1}{2}}]$
442	$I_0[b(ax-x^2)^{\frac{1}{2}}], \quad x<a$ $0, \quad x>a$	$2\sin(\frac{1}{2}ay)$ $\cdot\begin{cases}(b^2-y^2)^{-\frac{1}{2}}\sinh[\frac{1}{2}a(b^2-y^2)^{\frac{1}{2}}], & y<b\\(y^2-b^2)^{-\frac{1}{2}}\sin[\frac{1}{2}a(y^2-b^2)^{\frac{1}{2}}], & y>b\end{cases}$
443	$(ax-x^2)^{\frac{1}{2}\nu}I_\nu[b(ax-x^2)^{\frac{1}{2}}], \quad x<a$ $0, \quad x>a$ $\nu>-1$	$(\pi a)^{\frac{1}{2}}(\frac{1}{2}ab)^\nu\sin(\frac{1}{2}ay)$ $\cdot\begin{cases}(b^2-y^2)^{-\frac{1}{2}\nu-\frac{1}{4}}I_{\nu+\frac{1}{2}}[\frac{1}{2}a(b^2-y^2)^{\frac{1}{2}}], & y<b\\(y^2-b^2)^{-\frac{1}{2}\nu-\frac{1}{4}}J_{\nu+\frac{1}{2}}[\frac{1}{2}a(y^2-b^2)^{\frac{1}{2}}], & y>b\end{cases}$
444	$K_0[b(ax-x^2)^{\frac{1}{2}}], \quad x<a$ $0, \quad x>a$	$-\sin(\frac{1}{2}ay)(y^2-b^2)^{-\frac{1}{2}}$ $\cdot\{\sin\alpha[\mathrm{Ci}(z_1)+\mathrm{Ci}(z_2)]$ $-\cos\alpha[\mathrm{Si}(z_1)-\mathrm{Si}(z_2)]\}$ $z_{1,2}=\frac{1}{2}a[y\pm(y^2-b^2)^{\frac{1}{2}}], \quad \alpha=\frac{1}{2}a(y^2-b^2)^{\frac{1}{2}}$ $\arg\alpha=\begin{cases}0, & y>b\\\frac{1}{2}\pi, & y<b\end{cases}$
445	$(ax-x^2)^{-\frac{1}{2}}K_\nu[b(ax-x^2)^{\frac{1}{2}}], \quad x<a$ $0, \quad x>a$ $-1<\nu<1$	$-\frac{1}{4}\pi^2\sec(\frac{1}{2}\pi\nu)\sin(\frac{1}{2}ay)[J_{\frac{1}{2}\nu}(\frac{1}{2}z_1)Y_{-\frac{1}{2}\nu}(\frac{1}{2}z_2)$ $+Y_{\frac{1}{2}\nu}(\frac{1}{2}z_1)J_{-\frac{1}{2}\nu}(\frac{1}{2}z_2)], \quad y>b$ $z_{1,2}=\frac{1}{2}a[y\pm(y^2-b^2)^{\frac{1}{2}}]$

	$Nf(x)$	$Nh(y)$
446	$[x(1+x)]^{-\frac{1}{2}}K_{2\nu}\{b[x(1+x)]^{\frac{1}{2}}\}$, $-\frac{1}{2}<\nu<\frac{1}{2}$	$\frac{1}{8}\pi^2 \sec(\pi\nu)\{\cos(\frac{1}{2}y)$ $\cdot[J_\nu(z_2)Y_\nu(z_1)-J_\nu(z_1)Y_\nu(z_2)]-\sin(\frac{1}{2}y)$ $\cdot[J_\nu(z_1)J_\nu(z_2)+Y_\nu(z_1)Y_\nu(z_2)]\}$, $z_{1,2}=\frac{1}{4}[(b^2+y^2)^{\frac{1}{2}}\pm y]$
451	$K_0[b(2ax+x^2)^{\frac{1}{2}}]$	$\frac{1}{2}a\alpha^{-1}\cos(ay)\{\cos\alpha[\mathrm{Ci}(z_1)-\mathrm{Ci}(z_2)]$ $+\sin\alpha[\mathrm{si}(z_1)-\mathrm{si}(z_2)]\}$ $-\frac{1}{2}a\alpha^{-1}\sin(ay)\{\sin\alpha[\mathrm{Ci}(z_1)+\mathrm{Ci}(z_2)]$ $-\cos\alpha[\mathrm{si}(z_1)+\mathrm{si}(z_2)]\}$ $\alpha=a(b^2+y^2)^{\frac{1}{2}},\quad z_{1,2}=\alpha\pm ay$
456	$K_0[b(x^2-a^2)^{\frac{1}{2}}],\quad x>a$ $0,\quad x<a$	$\frac{1}{2}(b^2+y^2)^{-\frac{1}{2}}\{\cos\alpha[\mathrm{Ci}(z_1)-\mathrm{Ci}(z_2)]$ $+\sin\alpha[\mathrm{si}(z_1)-\mathrm{si}(z_2)]\}$ $\alpha=a(b^2+y^2)^{\frac{1}{2}},\quad z_{1,2}=\alpha\pm ay$
457	$(x^2-a^2)^{-\frac{1}{2}}K_\nu[b(x^2-a^2)^{\frac{1}{2}}],\quad x>a$ $0,\quad x<a$ $-1<\nu<1$	$\frac{1}{8}\pi^2\sec(\frac{1}{2}\pi\nu)$ $\cdot[J_{\frac{1}{2}\nu}(z_2)Y_{\frac{1}{2}\nu}(z_1)-J_{\frac{1}{2}\nu}(z_1)Y_{\frac{1}{2}\nu}(z_2)]$ $z_{1,2}=\frac{1}{2}a[(b^2+y^2)^{\frac{1}{2}}\pm y]$
465	$K_0[a(ix)^{\frac{1}{2}}]K_0[a(-ix)^{\frac{1}{2}}]$	$\frac{1}{2}y^{-1}S_{-1,0}(\frac{1}{2}a^2y^{-1})$
466	$K_\nu[a(ix)^{\frac{1}{2}}]K_\nu[a(-ix)^{\frac{1}{2}}],\quad -1<\nu<1$	$\frac{1}{4}\pi\nu\csc(\frac{1}{2}\pi v)y^{-1}S_{-1,\nu}(\frac{1}{2}a^2y^{-1})$
467	$I_{2\nu}(2a\sin x),\quad x<\pi$ $0,\quad x>\pi$ $\nu>-\frac{1}{2}$	$\pi\sin(\frac{1}{2}\pi y)I_{\nu-\frac{1}{2}y}(a)I_{\nu+\frac{1}{2}y}(a)$
469	$(\sin x)^{-m}I_m(2a\sin x),\quad x<\pi$ $0,\quad x>\pi$ $m=0,1,2,\ldots$	$\pi a^m m!\sin(\frac{1}{2}\pi y)\sum_{n=0}^{m}(-1)^n\epsilon_n$ $\cdot[(m+n)!(m-n)!]^{-1}I_{n-\frac{1}{2}y}(a)I_{n+\frac{1}{2}y}(a)$
471	$\csc x\, I_{2\nu}(2a\sin x),\quad x<\pi$ $0,\quad x>\pi$ $\nu>0$	$\frac{1}{2}\pi a\nu^{-1}\sin(\frac{1}{2}\pi y)$ $\cdot[I_{\nu-\frac{1}{2}-\frac{1}{2}y}(a)I_{\nu-\frac{1}{2}+\frac{1}{2}y}(a)$ $-I_{\nu+\frac{1}{2}-\frac{1}{2}y}(a)I_{\nu+\frac{1}{2}+\frac{1}{2}y}(a)]$
474	$K_{2\nu}(2a\sin x),\quad x<\pi$ $0,\quad x>\pi$ $-\frac{1}{2}<\nu<\frac{1}{2}$	$\frac{1}{2}\pi^2\csc(2\pi\nu)\sin(\frac{1}{2}\pi y)$ $\cdot[I_{-\nu-\frac{1}{2}y}(a)I_{-\nu+\frac{1}{2}y}(a)$ $-I_{\nu-\frac{1}{2}y}(a)I_{\nu+\frac{1}{2}y}(a)]$

	$Nf(x)$	$Nh(y)$
476	$K_0[(a^2+b^2-2ab\cos x)^{\frac{1}{2}}], \quad x<\pi$ $0, \quad x>\pi$	$y\sum_{n=0}^{\infty}\epsilon_n(y^2-n^2)^{-1}[1-(-1)^n\cos(\pi y)]$ $\cdot I_n(b)K_n(a), \quad b\leq a$
478	$K_{2\nu}(2a\sinh\frac{1}{2}x)$	$\frac{1}{4}i\pi^2\csc(2\pi\nu)[J_{\nu-iy}(a)Y_{-\nu-iy}(a)$ $-J_{-\nu-iy}(a)Y_{\nu-iy}(a)-J_{\nu+iy}(a)Y_{-\nu+iy}(a)$ $+J_{-\nu+iy}(a)Y_{\nu+iy}(a)]$
487	$K_0[(a^2+b^2+2ab\cosh x)^{\frac{1}{2}}]$	$\sum_{n=0}^{\infty}(-1)^n\epsilon_n(n^2+y^2)^{-1}I_n(b)K_n(a)$ $-\frac{1}{2}\pi\operatorname{csch}(\pi y)K_{iy}(a)$ $\cdot[I_{iy}(b)+I_{-iy}(b)], \quad a\geq b$

12. Parabolic Cylindrical Functions

	$Nf(x)$	$Nh(y)$
520	$\exp(\frac{1}{4}a^2x^2)D_{-2}(ax)$	$-\frac{1}{2}a^{-2}y\exp(\frac{1}{2}y^2/a^2)\operatorname{Ei}(-\frac{1}{2}y^2/a^2)$
521	$\exp(\frac{1}{4}a^2x^2)D_\nu(ax), \quad \nu<-1$	$\pi^{\frac{1}{2}}[\Gamma(\frac{1}{2}-\frac{1}{2}\nu)]^{-1}(2a)^\nu$ $\cdot y^{-\nu-1}\exp(\frac{1}{2}y^2/a^2)\Gamma(1+\frac{1}{2}\nu, \frac{1}{2}y^2/a^2)$
524	$x^\mu\exp(-\frac{1}{4}a^2x^2)D_\nu(ax), \quad \nu>-1$	$2^{\frac{1}{2}(\nu-\mu-2)}\pi^{\frac{1}{2}}\Gamma(\mu+2)[\Gamma(\frac{1}{2}\mu-\frac{1}{2}\nu+\frac{3}{2})]^{-1}$ $a^{-\mu-2}y\,{}_2F_2(1+\frac{1}{2}\mu, \frac{3}{2}+\frac{1}{2}\mu; \frac{3}{2}, \frac{1}{2}\mu-\frac{1}{2}\nu+\frac{3}{2};$ $-\frac{1}{2}y^2/a^2)$

TABLE III

FUNCTIONS NOT BELONGING TO EITHER OF THESE CLASSES

Definition

Here

$$G(y) = \int_{-\infty}^{\infty} f(x) e^{ixy} \, dx,$$

$$G(0) = \int_{-\infty}^{\infty} f(x) \, dx = 1.$$

The following tables give $NG(y)$.

	$Nf(x)$	$NG(y)$
1	$1, \quad 0<x<b$ $0, \quad$ otherwise $N=b-a$	$iy^{-1}(e^{-iby}-e^{-iay})$
2	$x^n, \quad 0<x<b$ $0, \quad$ otherwise $n=0, 1, 2, \ldots$ $N=b^{n+1}(n+1)^{-1}$	$n!(iy)^{-n-1}-e^{-iby}\sum_{m=0}^{n}(n!/m!)b^m(iy)^{m-n-1}$
3	$x^\nu, \quad 0<x<b$ $0, \quad$ otherwise $\nu>-1$ $N=b^{\nu+1}(\nu+1)^{-1}$	$(iy)^{-\nu-1}\gamma(\nu+1, iby)$
4	$x^{-\nu}, \quad b<x<\infty$ $0, \quad -\infty<x<b$ $\nu>1$ $N=(\nu-1)^{-1}b^{1-\nu}$	$(iy)^{\nu-1}\Gamma(1-\nu, iby)$
5	$(a+x)^{-\nu}, \quad 0<x<\infty$ $0, \quad -\infty<x<0$ $\nu>1$ $N=(\nu-1)^{-1}a^{1-\nu}$	$(iy)^{\nu-1}e^{iay}\Gamma(1-\nu, iay)$
6	$(b-x)^\nu, \quad 0<x<b$ $0, \quad$ otherwise $\nu>-1$ $N=(\nu+1)^{-1}b^{\nu+1}$	$(iy)^{-\nu-1}e^{-iby}\gamma(\nu+1, -iby)$
7	$x^{-\nu}(a+x)^{-1}, \quad 0<x<\infty$ $0, \quad -\infty<x<0$ $0<\nu<1$ $N=\pi a^{-\nu}\csc(\pi\nu)$	$\pi a^{-\nu}\csc(\pi\nu)[\Gamma(\nu)]^{-1}$ $\cdot e^{iay}\Gamma(\nu, iay)$
8	$x^{-1}(x-b)^\nu, \quad b<x<\infty$ $0, \quad$ otherwise $0<\nu<1$ $N=\pi b^{-\nu}\csc(\pi\nu)$	$\pi b^{-\nu}\csc(\pi\nu)[\Gamma(\nu)]^{-1}\Gamma(\nu, iby)$

	$Nf(x)$	$NG(y)$
9	$x^{\nu-1}(1+x^2)^{-1}, \quad 0<x<\infty$ $0, \quad -\infty<x<0$ $0<\nu<2$ $N=\frac{1}{2}\pi\csc(\frac{1}{2}\pi\nu)$	$\pi\csc(\pi\nu)\,V_\nu(2iy,0)$
10	$(x-b)^{\nu-1}(x+b)^{-\nu-\frac{1}{2}}, \quad b<x$ $0, \quad \text{otherwise}$ $\nu>0$ $N=(2b/\pi)^{-\frac{1}{2}}\Gamma(\nu)[\Gamma(\frac{1}{2}+\nu)]^{-1}$	$2^{\nu-\frac{1}{2}}b^{-\frac{1}{2}}\Gamma(\nu)D_{-2\nu}[2(iby)^{\frac{1}{2}}]$
11	$x^{\nu-1}(a+x)^{-\nu-\frac{1}{2}}, \quad 0<x<\infty$ $0, \quad -\infty<x<0$ $\nu>0$ $N=(\pi/a)^{\frac{1}{2}}\Gamma(\nu)[\Gamma(\frac{1}{2}+\nu)]^{-1}$	$2^{\nu}a^{-\frac{1}{2}}\Gamma(\nu)e^{i\frac{1}{2}ay}D_{-2\nu}[(2iay)^{\frac{1}{2}}]$
12	$x^{\nu}(a+x)^{\mu}, \quad 0<x<\infty$ $0, \quad -\infty<x<0$ $-1<\nu<-\mu-1$ $N=a^{\nu+\mu+1}\Gamma(1+\nu)\Gamma(-1-\nu-\mu)$ $\cdot[\Gamma(-\mu)]^{-1}$	$a^{\frac{1}{2}(\nu+\mu)}\Gamma(1+\nu)(iy)^{-\frac{1}{2}(\nu+\mu+2)}$ $\cdot e^{i\frac{1}{2}ay}W_{\frac{1}{2}\mu-\frac{1}{2}\nu,\frac{1}{2}+\frac{1}{2}\nu+\frac{1}{2}\mu}(iay)$
13	$(x+a)^{2\mu-1}(x-b)^{2\nu-1}, \quad b<x<\infty$ $0, \quad \text{otherwise}$ $0<\nu<\frac{1}{2}-\mu$ $N=(a+b)^{2\mu+2\nu-1}\Gamma(2\nu)$ $\cdot\Gamma(1-2\mu-2\nu)[\Gamma(1-2\mu)]^{-1}$	$(a+b)^{\nu+\mu-1}\Gamma(2\nu)(iy)^{-\mu-\nu}$ $\cdot\exp[i\frac{1}{2}y(a-b)]W_{\mu-\nu,\mu+\nu-\frac{1}{2}}[iy(a+b)]$
14	$(x-a)^{2\mu-1}(b-x)^{2\nu-1}, \quad a<x<b$ $0, \quad \text{otherwise}$ $\mu,\nu>0$ $N=(b-a)^{2\mu+2\nu-1}\Gamma(2\mu)$ $\cdot\Gamma(2\nu)[\Gamma(2\mu+2\nu)]^{-1}$	$(b-a)^{\nu+\mu-1}(iy)^{-\mu-\nu}\exp[-i\frac{1}{2}y(a+b)]$ $\cdot M_{\mu-\nu,\mu+\nu-\frac{1}{2}}[iy(b-a)]$
15	$(1-x)^{\nu-1}(1+x)^{\mu-1}, \quad -1<x<1$ $0, \quad \text{otherwise}$ $\nu,\mu>0$ $N=2^{\nu+\mu-1}B(\mu,\nu)$	$2^{\nu+\mu-1}B(\mu,\nu)e^{iy}$ $\cdot{}_1F_1(\mu;\nu+\mu;-2iy)$
16	$[a^2+(x\pm b)^2]^{-\nu}$ $\nu>\frac{1}{2}$ $N=\frac{1}{2}\pi^{\frac{1}{2}}a^{1-2\nu}\Gamma(\nu-\frac{1}{2})[\Gamma(\nu)]^{-1}$	$\pi^{\frac{1}{2}}2^{\mp iby}[\Gamma(\nu)]^{-1}$ $\cdot(\frac{1}{2}\mid y\mid/a)^{\nu-\frac{1}{2}}K_{\nu-\frac{1}{2}}(a\mid y\mid)$

	$Nf(x)$	$NG(y)$
17	$[(a^2+x^2)^{\frac{1}{2}}+x]^{-\nu}, \quad 0<x<\infty$ $0, \quad -\infty<x<0$ $\nu>1$ $N=a^{\nu+1}\nu(\nu^2-1)^{-1}$	$ia^{-\nu}y^{-1}\{\nu\pi \csc(\pi\nu)$ $\cdot[\mathbf{J}_\nu(iay)-J_\nu(iay)]-1\}$
18	$(a^2+x^2)^{-\frac{1}{2}}[x+(a^2+x^2)^{\frac{1}{2}}]^{-\nu}, \quad 0<x<\infty$ $0, \quad -\infty<x<0$ $\nu>0$ $N=\nu^{-1}a^{-\nu}$	$\pi a^{-\nu} \csc(\pi\nu)$ $\cdot[\mathbf{J}_\nu(iay)-J_\nu(iay)]$
19	$e^{-cx}, \quad a<x<b$ $0, \quad$ otherwise $N=c^{-1}(e^{-ac}-e^{-bc})$	$(c-iy)^{-1}\{\exp[-a(c-iy)]$ $-\exp[-b(c-iy)]\}$
20	$(a+e^{-x})^{-1}e^{-\lambda x}$ $0<\lambda<1$ $N=\pi a^{\lambda-1} \csc(\pi\lambda)$	$\pi a^{\lambda-1+iy} \csc(\pi\lambda+i\pi y)$
21	$x(a+e^{-x})^{-1}e^{-\lambda x}$ $0<\lambda<1$ $N=\pi a^{\lambda-1} \csc(\pi\lambda)(\log a-\pi \operatorname{ctn}\pi\lambda)$	$\pi a^{\lambda-1+iy} \csc(\pi\lambda+i\pi y)$ $\cdot[\log a-\pi \operatorname{ctn}(\pi\lambda+i\pi y)]$
22	$(a+e^{-x})^{-1}(b+e^{-x})^{-1}e^{-\lambda x}$ $0<\lambda<2$ $N=\pi(b-a)^{-1} \csc(\pi\lambda)(a^{\lambda-1}-b^{\lambda-1})$	$\pi(b-c)^{-1} \csc(\pi\lambda+i\pi y)$ $\cdot(a^{\lambda-1+iy}-b^{\lambda-1+iy})$
23	$[e^{b/c}+e^{-x/c}]^{-\nu}e^{-ax}$ $0<a<\nu/c$ $N=c\exp[b(a-\nu/c)]B(ac, \nu-ac)$	$c\exp[b(a-\nu/c)]$ $\cdot\exp(iby)B[c(a+iy), \nu-c(a+iy)]$
24	$e^{-\lambda x}\log(1+e^{-x})$ $-1<\lambda<0$ $N=\pi\lambda^{-1} \csc(\pi\lambda)$	$\pi(\lambda+iy)^{-1} \csc(\pi\lambda+i\pi y)$
25	$(1+e^x)^{-1}\log(1+e^x)$ $N=\pi^2/6$	$i\pi \operatorname{csch}(\pi y)[\gamma+\psi(1-iy)]$
26	$(1+e^x)^{-\nu}\log(1+e^x)$ $\nu>0$ $N=\psi'(\nu)$	$-i\pi \operatorname{csch}(\pi y)[\psi(\nu)-\psi(\nu-iy)]$

	$Nf(x)$	$NG(y)$
27	$e^{-\lambda x}(a+e^{-x})^{-\nu}\log(a+e^{-x})$ $a\geq 1,\quad \nu>\lambda>0$ $N=a^{\lambda-\nu}B(\lambda,\nu-\lambda)$ $\cdot[\psi(\nu)-\psi(\nu-\lambda)+\log a]$	$a^{\lambda-\nu+iy}B(\lambda+iy,\nu-\lambda-iy)$ $\cdot[\psi(\nu)-\psi(\nu-\lambda-iy)+\log a]$
28	$e^{\nu x}\exp(-ae^{x})$ $N=a^{-\nu}\Gamma(\nu)$	$a^{-\nu-iy}\Gamma(\nu+iy)$
29	$\nu e^{\nu x}\exp(-ae^{x})$ $N=a^{-\nu}\Gamma(\nu)[\psi(\nu)-\log a]$	$a^{-\nu-iy}\Gamma(\nu+iy)[\psi(\nu+iy)-\log a]$
30	$[\exp(e^{-x})-1]^{-1}e^{-\lambda x}$ $\lambda>1$ $N=\Gamma(\lambda)\zeta(\lambda)$	$\Gamma(\lambda+iy)\zeta(\lambda+iy)$
31	$[\exp(e^{-x})+1]^{-1}e^{-\lambda x}$ $\lambda>0$ $N=(1-2^{1-\lambda})\Gamma(\lambda)\zeta(\lambda)$	$(1-2^{1-\lambda-iy})\Gamma(\lambda+iy)\tau(\lambda+iy)$
32	$\exp(-a^2e^{x}-b^2e^{-x})$ $N=2K_0(2ab)$	$2(b/a)^{iy}K_{iy}(2ab)$
33	$\operatorname{sech}(ax+b)$ $N=\pi a^{-1}$	$\pi a^{-1}e^{-iby/a}\operatorname{sech}(\frac{1}{2}\pi y/a)$
34	$[\operatorname{sech}(ax+b)]^{\nu}$ $\nu>0$ $N=2^{\nu-1}a^{-1}[\Gamma(\frac{1}{2}\nu)]^2[\Gamma(\nu)]^{-1}$	$2^{\nu-1}a^{-1}[\Gamma(\nu)]^{-1}\exp(iby/a)$ $\cdot\Gamma[\frac{1}{2}\nu-i(y/2a)]\Gamma[\frac{1}{2}\nu+i(y/2a)]$
35	$\exp[-b\tanh(ax)][\operatorname{sech}(ax)]^{\nu}$, $\quad 0<x<\infty$ 0, $\quad -\infty<x<0$ $\nu>0$ $N=\pi^{\frac{1}{2}}a^{-1}(\frac{1}{2}b)^{\frac{1}{2}-\frac{1}{2}\nu}\Gamma(\frac{1}{2}\nu)I_{\frac{1}{2}\nu-\frac{1}{2}}(b)$	$a^{-1}2^{\frac{1}{2}\nu-1}b^{-\frac{1}{2}\nu}[\Gamma(\nu)]^{-1}$ $\cdot\Gamma[\frac{1}{2}\nu+i(y/2a)]\Gamma[\frac{1}{2}\nu-i(y/2a)]$ $\cdot M_{iy/2a,\frac{1}{2}\nu-\frac{1}{2}}(2b)$
36	$\exp(a^2e^{x}+b^2e^{-x})\operatorname{Erfc}(ae^{\frac{1}{2}x}+be^{-\frac{1}{2}x})$ $N=2K_0(2ab)$	$2(b/a)^{iy}\operatorname{sech}(\pi y)K_{iy}(2ab)$

	$Nf(x)$	$NG(y)$
37	$[(a+x)(a-x)^{-1}]^{\frac{1}{2}}$ $\cdot I_1[b(a^2-x^2)^{\frac{1}{2}}], \quad \lvert x\rvert<a$ $0, \quad \lvert x\rvert>a$ $N=4b^{-1}\sinh^2(\frac{1}{2}ab)$	$2b^{-1}\{\cos[a(y^2-b^2)^{\frac{1}{2}}]-\cos(ay)$ $+iy(y^2-b^2)^{-\frac{1}{2}}\sin[a(y^2-b^2)^{\frac{1}{2}}]$ $-i\sin(ay)\}$
38	$\exp(-ae^x)I_\nu(be^x), \quad a>b, \quad \nu>0$ $N=\nu^{-1}b^\nu[a+(a^2-b^2)^{\frac{1}{2}}]^{-\nu}$	$\Gamma(\nu+iy)(a^2-b^2)^{-\frac{1}{2}iy}$ $\cdot \mathfrak{P}_{-iy}^{-\nu}[a(a^2-b^2)^{-\frac{1}{2}}]$
39	$x^{\frac{1}{2}}K_{\frac{1}{3}}(ax^{\frac{3}{2}})$ $N=2\pi 3^{-\frac{1}{2}}a^{-1}$	$2\pi 3^{-\frac{1}{2}}a^{-1}\exp(-i\frac{4}{27}y^3/a^2)$
40	$[(a+be^x)(ae^x+b)^{-1}]^{\frac{1}{2}\nu}$ $\cdot K_\nu[(a^2+b^2+2ab\cosh x)^{\frac{1}{2}}]$ $N=2K_{\frac{1}{2}\nu}(a)K_{\frac{1}{2}\nu}(b)$	$2K_{\frac{1}{2}\nu-iy}(a)K_{\frac{1}{2}\nu+iy}(b)$
41	$\exp[-(a\sinh x)^2]D_\nu(2a\sinh x), \quad 0>\nu>-1$ $N=2^{\frac{1}{2}\nu}(2\pi a^2)^{-\frac{1}{2}}\exp(a^2)$ $\cdot\cos(\frac{1}{2}\pi\nu)[\Gamma\frac{1}{2}(1+\nu)]^2W_{-\frac{1}{2}\nu,0}(2a^2)$	$2^{\frac{1}{2}\nu}(2\pi a^2)^{-\frac{1}{2}}$ $\cdot\exp(a^2)\Gamma[\frac{1}{2}(1+\nu+iy)]\Gamma[\frac{1}{2}(1+\nu-iy)]$ $\cdot\cos[\frac{1}{2}\pi(\nu-iy)]W_{-\frac{1}{2}\nu,i\frac{1}{2}y}(2a^2)$

PART II

TABLES OF THE INVERSE TRANSFORMS OF PART I

TABLE IA

EVEN FUNCTIONS

Definition

These tables contain the inverse transforms of the tables from pp. 15–73. The numbers at the beginning of each formula coincide with those of the corresponding formula pair of Part I.

1. Algebraic Functions

	$Ng(y)$	$2Nf(x)$
60	$a(a^2+y^2)^{-1}$ $N=a^{-1}$	e^{-ax}
129	$\frac{1}{2}\pi(a-\frac{1}{2}y), \quad y<2a$ $0, \quad y>2a$ $N=\frac{1}{2}\pi a$	$x^{-2}[\sin(ax)]^2$
306	$\frac{1}{2}\pi(a+y)^{-1}$ $N=\frac{1}{2}\pi a^{-1}$	$-\sin(ax)\ \mathrm{si}(ax)-\cos(ax)\ \mathrm{Ci}(ax)$
400	$\frac{1}{2}\pi(a^2+y^2)^{-\frac{1}{2}}$ $N=\frac{1}{2}\pi a^{-1}$	$K_0(ax)$
282	$(\frac{1}{2}a)^{\frac{1}{2}}(a^2+y^2)^{-\frac{1}{2}}[(a^2+y^2)^{\frac{1}{2}}+a]^{-\frac{1}{2}}$ $N=\frac{1}{2}a^{-1}$	$\mathrm{Erfc}[(ax)^{\frac{1}{2}}]$
63	$(\frac{1}{2}\pi)^{\frac{1}{2}}(a^2+y^2)^{-\frac{1}{2}}[(a^2+y^2)^{\frac{1}{2}}+a]^{\frac{1}{2}}$ $N=(\pi/a)^{\frac{1}{2}}$	$x^{-\frac{1}{2}}e^{-ax}$
399	$2^{-\frac{1}{2}}[(b^2-a^2-y^2)^2+4b^2y^2]^{-\frac{1}{2}}$ $\cdot\{[(b^2-a^2-y^2)^2+4b^2y^2]^{\frac{1}{2}}$ $+b^2-a^2-y^2\}^{\frac{1}{2}}$ $N=(b^2-a^2)^{-\frac{1}{2}}$	$e^{-bx}I_0(ax), \quad b>a$
131	$i(-1)^n2^{-2n-2}(2n+1)^{-1}$ $\cdot\left\{\left[\binom{n+\frac{1}{2}y+i\frac{1}{2}a}{2n+1}\right]^{-1}\right.$ $\left.-\left[\binom{n+\frac{1}{2}y-i\frac{1}{2}a}{2n+1}\right]^{-1}\right\}$ $N=i(-1)^n2^{-2n-2}(2n+1)^{-1}$ $\cdot\left\{\left[\binom{n+i\frac{1}{2}a}{2n+1}\right]^{-1}-\left[\binom{n-i\frac{1}{2}a}{2n+1}\right]^{-1}\right\}$	$e^{-ax}(\sin x)^{2n}$ $n=0, 1, 2, \ldots$

2. Arbitrary Powers

	$Ng(y)$	$2Nf(x)$
406	$2^{\nu-1}\pi^{\frac{1}{2}}a^{\nu}(a^2+y^2)^{-\nu-\frac{1}{2}}\Gamma(\frac{1}{2}+\nu), \quad \nu>-\frac{1}{2}$ $N=2^{\nu-1}\pi^{\frac{1}{2}}a^{-\nu-1}\Gamma(\frac{1}{2}+\nu)$	$x^{\nu}K_{\nu}(ax)$

	$Ng(y)$	$2Nf(x)$
405	$\frac{1}{4}\pi\sec(\frac{1}{2}\pi\nu)(a^2+y^2)^{-\frac{1}{2}}$ $\cdot\{a^{-\nu}[y+(a^2+y^2)^{\frac{1}{2}}]^{\nu}$ $+a^{\nu}[y+(a^2+y^2)^{\frac{1}{2}}]^{-\nu}\}$ $N=\frac{1}{2}\pi a^{-1}\sec(\frac{1}{2}\pi\nu),\quad -1<\nu<1$	$K_\nu(ax)$

3. Exponential Functions

	$Ng(y)$	$2Nf(x)$
7	$\pi(2a)^{-1}e^{-ay}$ $N=\pi(2a)^{-1}$	$(a^2+x^2)^{-1}$
115	$\pi y^{-1}(1-e^{-ay})$ $N=\pi a$	$\log(1+a^2x^{-2})$
111	$\pi y^{-1}(e^{-by}-e^{-ay})$ $N=(a-b)\pi,\quad a>b$	$\log[(a^2+x^2)(b^2+x^2)^{-1}]$
9	$\frac{1}{2}\pi(a^2-b^2)^{-1}[b^{-1}e^{-by}-a^{-1}e^{-ay}]$ $N=\pi[2ab(a+b)]^{-1}$	$[(a^2+x^2)(b^2+x^2)]^{-1}$
73	$\frac{1}{2}(\pi/a)^{\frac{1}{2}}\exp(-y^2/4a)$ $N=\frac{1}{2}(\pi/a)^{\frac{1}{2}}$	$\exp(-ax^2)$
277	$(\frac{1}{2}a^{-2}+y^{-2})\exp(-\frac{1}{4}y^2/a^2)-y^{-2}$ $N=\frac{1}{4}a^{-2}$	$x\,\mathrm{Erfc}(ax)$
33	$\frac{1}{2}\pi(-1)^{m+n}(n!)^{-1}$ $\cdot(d^n/dz^n)[z^{m-\frac{1}{2}}\exp(-yz^{\frac{1}{2}})]$ $N=(-1)^{m+n}\pi(2m)!(2n-2m)!$ $\cdot 2^{-2n-1}[n!m!(n-m)!]^{-1}$ $\cdot z^{m-n-\frac{1}{2}},\quad z>0$ $n, m=0, 1, 2, \ldots,\quad n\geq m$	$x^{2m}(x^2+z)^{-n-1}$
447	$\frac{1}{2}\pi(ab)^{-1}\exp[-a(b^2+y^2)^{\frac{1}{2}}]$ $N=\frac{1}{2}\pi(ab)^{-1}e^{-ab}$	$(a^2+x^2)^{-\frac{1}{2}}K_1[b(a^2+x^2)^{\frac{1}{2}}]$
439	$\frac{1}{2}\pi(b^2+y^2)^{-\frac{1}{2}}\exp[-a(b^2+y^2)^{\frac{1}{2}}]$ $N=\frac{1}{2}\pi b^{-1}e^{-ab}$	$K_0[b(a^2+x^2)^{\frac{1}{2}}]$

	$Ng(y)$	$2Nf(x)$
99	$(\frac{1}{2}\pi)^{\frac{1}{2}}(a^2+y^2)^{-\frac{1}{2}}[a+(a^2+y^2)^{\frac{1}{2}}]^{\frac{1}{2}}$ $\cdot\exp[-b(a^2+y^2)^{\frac{1}{2}}]$ $N=(\pi/a)^{\frac{1}{2}}e^{-ab}$	$(b^2+x^2)^{-\frac{1}{2}}[(b^2+x^2)^{\frac{1}{2}}+b]^{\frac{1}{2}}$ $\cdot\exp[-a(b^2+x^2)^{\frac{1}{2}}]$
283	$2^{-\frac{1}{2}}a\exp(-a^2b)(a^4+y^2)^{-\frac{1}{2}}$ $\cdot[a^2+(a^4+y^2)^{\frac{1}{2}}]^{-\frac{1}{2}}\exp[-b(a^4+y^2)^{\frac{1}{2}}]$ $N=\frac{1}{2}a^{-2}\exp(-2a^2b)$	$\mathrm{Erfc}\{a[b+(b^2+x^2)^{\frac{1}{2}}]^{\frac{1}{2}}\}$
441	$\frac{1}{4}\pi a^{-\nu}(a^2+y^2)^{-\frac{1}{2}}\exp[-b(a^2+y^2)^{\frac{1}{2}}]$ $\cdot\{[(a^2+y^2)^{\frac{1}{2}}+y]^{\nu}+[(a^2+y^2)^{\frac{1}{2}}-y]^{\nu}\}$ $N=\frac{1}{2}\pi a^{-1}e^{-ab},\quad -1\leq\nu\leq 1$	$\cos[\nu\arctan(x/b)]$ $\cdot K_{\nu}[a(b^2+x^2)^{\frac{1}{2}}]$

4. Logarithmic Functions

	$Ng(y)$	$2Nf(x)$
304	$\frac{1}{2}\pi y^{-1}\log(1+y/a),\quad y<2a$ $\frac{1}{2}\pi y^{-1}\log[(y+a)(y-a)^{-1}],\quad y>2a$ $N=\frac{1}{2}\pi a^{-1}$	$[\mathrm{si}(ax)]^2$
305	$\frac{1}{2}\pi y^{-1}\log(1+y/a),\quad y<2a$ $\frac{1}{2}\pi y^{-1}\log(y^2/a^2-1),\quad y>2a$ $N=\frac{1}{2}\pi a^{-1}$	$[\mathrm{Ci}(ax)]^2$
61	$\frac{1}{2}\log[(a^2+y^2)(b^2+y^2)^{-1}]$ $N=\log(a/b),\quad a>b$	$x^{-1}(e^{-bx}-e^{-ax})$
223	$\frac{1}{4}\log\left[\frac{y^2+(a+b)^2}{y^2+(a-b)^2}\right]$ $N=\frac{1}{2}\log\left(\frac{a+b}{a-b}\right),\quad a>b$	$x^{-1}e^{-ax}\sinh(bx)$
190	$\log(1+e^{-\pi y})$ $N=\log 2$	$x^{-1}(x^{-1}-\mathrm{csch}x)$

5. Trigonometric Functions

	$Ng(y)$	$2Nf(x)$
1	$y^{-1}\sin(ay)$ $N=a$	$1,\quad x<a$ $0,\quad x>a$
172	$\pi y^{-1}e^{-ay}\sin(ay)$ $N=a\pi$	$\arctan(2a^2/x^2)$
2	$4y^{-2}\cos y\,\sin^2(\frac{1}{2}y)$ $N=1$	$x,\quad x<1$ $2-x,\quad 1<x<2$ $0,\quad x>2$
10	$\frac{1}{2}\pi a^{-3}\exp(-ay2^{-\frac{1}{2}})\sin(\frac{1}{4}\pi+2^{-\frac{1}{2}}ay)$ $N=2^{-\frac{3}{2}}\pi a^{-3}$	$(a^4+x^4)^{-1}$
8	$\pi b^{-1}\cos(ay)e^{-by}$ $N=\pi b^{-1}$	$[b^2+(a-x)^2]^{-1}$ $+[b^2+(a+x)^2]^{-1}$
12	$\frac{1}{2}\pi a^{-1}\csc(2\varphi)\exp(-ay\cos\varphi)$ $\cdot\sin(\varphi-ay\sin\varphi)$ $N=\frac{1}{4}\pi a^{-1}\sec\varphi,\quad -\frac{1}{2}\pi<\varphi<\frac{1}{2}\pi$	$x^2[x^4+2a^2x^2\cos(2\varphi)+a^4]^{-1}$
11	$\frac{1}{2}\pi a^{-3}\cos(2\varphi)\exp(-ay\cos\varphi)$ $\cdot\sin(\varphi+ay\sin\varphi)$ $N=\frac{1}{4}\pi a^{-3}\sec\varphi,\quad -\frac{1}{2}\pi<\varphi<\frac{1}{2}\pi$	$[x^4+2a^2x^2\cos(2\varphi)+a^4]^{-1}$
227	$\frac{1}{2}(\pi/b)^{\frac{1}{2}}\cos(\frac{1}{2}ay/b)\exp[\frac{1}{4}(a^2-y^2)/b]$ $N=\frac{1}{2}(\pi/b)^{\frac{1}{2}}\exp(\frac{1}{4}a^2/b)$	$\exp(-bx^2)\cosh(ax)$
312	$\frac{1}{4}y^{-1}\{\sin[2(ay)^{\frac{1}{2}}]-\cos[2(ay)^{\frac{1}{2}}]$ $+\exp[-2(ay)^{\frac{1}{2}}]\}$ $N=a$	$S(ax^{-1})$
84	$(\pi/a)^{\frac{1}{2}}\exp[-(2ay)^{\frac{1}{2}}]\cos[(2ay)^{\frac{1}{2}}]$ $N=(\pi/a)^{\frac{1}{2}}$	$x^{-\frac{3}{2}}e^{-a/x}$
376	$z^{-1}\sin(az)$ $z=(b^2+y^2)^{\frac{1}{2}},\quad N=b^{-1}\sin(ab)$ $ab\leq\tau_{0,1}$	$J_0[b(a^2-x^2)^{\frac{1}{2}}],\quad x<a$ $0,\quad x>a$

	$Ng(y)$	$2Nf(x)$
380	$2\cos(\frac{1}{2}ay)z^{-1}\sin(\frac{1}{2}az)$ $N=2b^{-1}\sin(\frac{1}{2}ab),\quad ab\leq 2\tau_{0,1}$ $z=(b^2+y^2)^{\frac{1}{2}}$	$J_0[b(ax-x^2)^{\frac{1}{2}}],\quad x<a$ $0,\quad x>a$
87	$(\pi/b)^{\frac{1}{2}}\exp(-2b^{\frac{1}{2}}u)\cos(2b^{\frac{1}{2}}v)$ $N=(\pi/b)^{\frac{1}{2}}\exp[-2(ab)^{\frac{1}{2}}]$ $\left.\begin{matrix}u\\v\end{matrix}\right\}=2^{-\frac{1}{2}}[(a^2+y^2)^{\frac{1}{2}}\pm a]^{\frac{1}{2}}$	$x^{-\frac{3}{2}}\exp[-ax-(b/x)]$

6. Inverse Trigonometric Functions

	$Ng(y)$	$2Nf(x)$
492	$\arccos(y/a),\quad y<a$ $0,\quad y>a$ $N=\frac{1}{2}\pi$	$x^{-1}\mathbf{H}_0(ax)$
291	$y^{-1}\arctan(y/a)$ $N=a^{-1}$	$-\mathrm{Ei}(-ax)$
296	$\pi\arctan(a/y)$ $N=\frac{1}{2}\pi^2$	$x^{-1}[e^{-ax}\mathrm{Ei}(ax)$ $-e^{ax}\mathrm{Ei}(-ax)]$
65	$a^{-\nu}\Gamma(\nu)(1+y^2/a^2)^{-\frac{1}{2}\nu}\cos(\nu\arctan y/a)$ $N=a^{-\nu}\Gamma(\nu),\quad \nu>0$	$x^{\nu-1}e^{-ax}$
64	$\Gamma(1+\nu)(a^2+y^2)^{-\frac{1}{2}-\frac{1}{2}\nu}e^{-ab}$ $\cdot\cos[by+(\nu+1)\arctan(y/a)]$ $N=a^{-\nu-1}\Gamma(1+\nu)e^{-ab},\quad \nu>-1$	$0,\quad x<b$ $(x-b)^{\nu}e^{-ax},\quad x>b$
125	$(a^2+y^2)^{-1}\{\frac{1}{6}a\pi^2+2y\arctan(y/a)$ $+a[\gamma+\frac{1}{2}\log(a^2+y^2)]^2$ $-a[\arctan(y/a)]^2\}$ $N=a^{-1}[\frac{1}{6}\pi^2+(\gamma+\log a)^2]$	$e^{-ax}(\log x)^2$
230	$\frac{1}{2}a\pi^{\frac{1}{2}}(b^2+y^2)^{\frac{3}{4}}\exp\{a^2b[4(b^2+y^2)]^{-1}\}$ $\cdot\cos[\frac{3}{2}\arctan(y/b)-\frac{1}{4}a^2y(b^2+y^2)^{-1}]$ $N=\frac{1}{2}ab^{\frac{3}{2}}\pi^{\frac{1}{2}}\exp(-\frac{1}{4}a^2/b)$	$\exp(-bx)\sinh(ax^{\frac{1}{2}})$

	$Ng(y)$	$2Nf(x)$
303	$\frac{1}{2}(b^2+y^2)^{-1}\{b\arctan[(a+y)/b]$ $-b\arctan[(y-a)/b]$ $-\frac{1}{2}y\log[b^2+(a+y)^2]$ $+\frac{1}{2}y\log[b^2+(a-y)^2]\}$ $N=b^{-1}\arctan(a/b)$	$e^{-bx}\,\mathrm{Si}(ax)$
293	$(a^2+y^2)^{-1}\{y\arctan[y(a+b)^{-1}]$ $+\frac{1}{2}a\log[(1+a/b)^2+y^2/b^2]\}$ $N=a^{-1}\log(1+a/b),\quad a>-b$	$-e^{-ax}\,\mathrm{Ei}(-bx)$
523	$-(\frac{1}{2}\pi)^{\frac{1}{2}}\nu^{-1}[y+(a+y^{\frac{1}{2}})^2]^{\nu}$ $\cdot\cos\{2\nu\arctan[y^{\frac{1}{2}}/(a+y^{\frac{1}{2}})]\}$ $N=-(\frac{1}{2}\pi)^{\frac{1}{2}}\nu^{-1}a^{2\nu},\quad \nu<0$	$x^{-\nu-1}\exp(\frac{1}{4}a^2x)\,D_{2\nu-1}(ax^{\frac{1}{2}})$

7. Hyperbolic Functions

	$Ng(y)$	$2Nf(x)$
240	$\frac{1}{2}\pi y^{-1}\tanh(\frac{1}{4}a^{-1}\pi y)$ $N=\frac{1}{8}a^{-1}\pi^2$	$\log[\mathrm{ctnh}(ax)]$
177	$\frac{1}{2}a^{-1}\pi\,\mathrm{sech}(\frac{1}{2}a^{-1}\pi y)$ $N=\frac{1}{2}a^{-1}\pi$	$\mathrm{sech}(ax)$
178	$\frac{1}{2}a^{-2}\pi y\,\mathrm{csch}(\frac{1}{2}a^{-1}\pi y)$ $N=a^{-1}$	$[\mathrm{sech}(ax)]^2$
127	$\frac{1}{2}[ay^{-2}-\pi y^{-1}\,\mathrm{csch}(a^{-1}\pi y)]$ $N=(12a)^{-1}\pi^2$	$\log(1+e^{-ax})$
128	$-\frac{1}{2}[ay^{-2}-\pi y^{-1}\,\mathrm{ctnh}(a^{-1}\pi y)]$ $N=(6a)^{-1}\pi^2$	$-\log(1-e^{-ax})$
239	$\frac{1}{2}\pi(1+y^2)^{-1}\,\mathrm{sech}(\frac{1}{2}\pi y)$ $N=\frac{1}{2}\pi$	$\cosh x\log(2\cosh x)$ $-x\sinh x$
179	$\frac{1}{4}a^{-3}\pi(a^2+y^2)\,\mathrm{sech}(\pi y/2a)$ $N=\frac{1}{4}a^{-1}\pi$	$[\mathrm{sech}(ax)]^3$

	$Ng(y)$	$2Nf(x)$
236	$(1-y^2)(1+y^2)^{-2}$ $+\frac{1}{2}\pi y(1+y^2)^{-1}\tanh(\frac{1}{2}\pi y)$ $N=1$	$-\log(1-e^{-2x})\cosh x$
205	$(\frac{1}{2}\pi)^{\frac{1}{2}}\{1+2\cosh[(2\pi/3)^{\frac{1}{2}}y]\}^{-1}$ $N=\frac{1}{3}(\frac{1}{2}\pi)^{\frac{1}{2}}$	$\{1+2\cosh[(2\pi/3)^{\frac{1}{2}}x]\}^{-1}$
243	$\pi y^{-1}\operatorname{sech}(\frac{1}{2}\pi y)\sinh[y\arcsin(a/b)]$ $N=\pi\arcsin(a/b),\quad a<b$	$\log[(b+a\operatorname{sech}x)(b-a\operatorname{sech}x)^{-1}]$
247	$\pi y^{-1}\sinh(by/a)\operatorname{sech}(\frac{1}{2}\pi y/a)$ $N=\pi a^{-1}b,\quad b\leq\frac{1}{2}\pi$	$\log\left[\frac{\cosh(ax)+\sin b}{\cosh(ax)-\sin b}\right]$
242	$2\pi y^{-1}\operatorname{csch}(\frac{1}{2}\pi y)\sinh^2(\frac{1}{2}y\arcsin a)$ $N=(\arcsin a)^2,\quad a\leq1$	$-\log(1-a^2\operatorname{sech}^2x)$
269	$\frac{1}{2}\pi y^{-1}\sin(ay/b)\operatorname{sech}(\frac{1}{2}\pi y/b)$ $N=\frac{1}{2}\pi a/b$	$\arctan[\sinh a\operatorname{sech}(bx)]$
202	$a^{-1}\pi\operatorname{csch}b\sin(by/a)\operatorname{csch}(\pi y/a)$ $N=a^{-1}b\operatorname{csch}b$	$[\cosh(ax)+\cosh b]^{-1}$
201	$a^{-1}\pi\csc b\sinh(by/a)\operatorname{csch}(\pi y/a)$ $N=a^{-1}b\csc b,\quad b<\pi$	$[\cosh(ax)+\cos b]^{-1}$
204	$\frac{1}{2}a^{-1}\pi\sec(\frac{1}{2}b)\cosh(a^{-1}by)\operatorname{sech}(a^{-1}\pi y)$ $N=\frac{1}{2}a^{-1}\pi\sec(\frac{1}{2}b),\quad b<\pi$	$\cosh(\frac{1}{2}ax)$ $+[\cosh(ax)+\cos b]^{-1}$
203	$\frac{1}{2}a^{-1}\pi\operatorname{sech}(\frac{1}{2}b)\cosh(by/a)\operatorname{sech}(\pi y/a)$ $N=\frac{1}{2}a^{-1}\pi\operatorname{sech}(\frac{1}{2}b)$	$\cosh(\frac{1}{2}ax)$ $\cdot[\cosh(ax)+\cosh b]^{-1}$
238	$\pi y^{-1}\operatorname{csch}(\pi y)[\cosh(\frac{1}{2}\pi y)-\cosh(ay)]$ $N=\frac{1}{8}\pi^2-\frac{1}{2}a^2,\quad a<\pi$	$\log(1+\cos a\operatorname{sech}x)$
248	$\pi y^{-1}\operatorname{csch}(\pi y/a)[\cosh(cy/a)-\cosh(by/a)]$ $N=\frac{1}{2}(c^2-b^2)/a,\quad c>b,\quad c,b\leq\pi$	$\log\left[\frac{\cosh(ax)+\cos b}{\cosh(ax)+\cos c}\right]$
185	$\frac{1}{2}b^{-1}\pi\sin(a\pi/b)$ $\cdot[\cos(a\pi/b)+\cosh(\pi y/b)]^{-1}$ $N=\frac{1}{2}b^{-1}\pi\tan(\frac{1}{2}a\pi/b),\quad a<b$	$\sinh(ax)\operatorname{csch}(bx)$

	$Ng(y)$	$2Nf(x)$
183	$\pi b^{-1}\cos(\frac{1}{2}\pi a/b)\cosh(\frac{1}{2}\pi y/b)$ $\cdot[\cos(a\pi/b)+\cosh(\pi y/b)]^{-1}$ $N=(\frac{1}{2}\pi/b)\sec(\frac{1}{2}a\pi/b),\quad a<b$	$\cosh(ax)\,\mathrm{sech}(bx)$
225	$\frac{1}{2}b^{-1}\pi\sin(2a\pi/b)$ $\cdot[\cosh(2\pi y/b)-\cos(2a\pi/b)]^{-1}$ $+\frac{1}{2}a(a^2+y^2)^{-1}$ $N=\frac{1}{2}[b^{-1}\pi\,\mathrm{ctn}(a\pi/b)+a^{-1}],\quad a<b$	$(e^{bx}-1)^{-1}\sinh(ax)$
357	$2^{-\frac{1}{2}}(a+\cosh y)^{-\frac{1}{2}}$ $N=2^{-\frac{1}{2}}(1+a)^{-\frac{1}{2}},\quad -1<a<1$	$\mathrm{sech}(\pi x)P_{-\frac{1}{2}+ix}(a)$
182	$\frac{1}{4}(\pi/a)^2[\mathrm{sech}(\frac{1}{2}\pi y/a)]^2$ $N=(\frac{1}{2}\pi/a)^2$	$x\,\mathrm{csch}(ax)$
67	$\frac{1}{2}y^{-2}-\frac{1}{2}(\pi/a)^2[\mathrm{csch}(\pi y/a)]^2$ $N=\frac{1}{6}(\pi/a)^2$	$x(e^{ax}-1)^{-1}$
191	$2a^3[\mathrm{sech}(ay)]^3$ $N=2a^3$	$(a^2+x^2)\,\mathrm{sech}(\frac{1}{2}\pi x/a)$
192	$\frac{3}{8}[a\,\mathrm{sech}(\frac{1}{2}ay)]^4$ $N=\frac{3}{8}a^4$	$x(a^2+x^2)\,\mathrm{csch}(\frac{1}{2}\pi x/a)$
271	$2^{-2a}\pi\Gamma(2a)[\mathrm{sech}(\frac{1}{2}y)]^{2a}$ $N=2^{-2a}\pi\Gamma(2a)$	$\lvert\Gamma(a+ix)\rvert^2$
362	$(\frac{1}{2}\pi)^{\frac{1}{2}}\Gamma(\mu)(1-a^2)^{\frac{1}{2}\mu-\frac{1}{4}}(a+\cosh y)^{-\mu}$ $N=(\frac{1}{2}\pi)^{\frac{1}{2}}\Gamma(\mu)(1-a)^{\frac{1}{2}\mu-\frac{1}{4}}(1+a)^{-\frac{1}{2}\mu-\frac{1}{4}}$ $-1<a<1,\quad \mu>0$	$\lvert\Gamma(\mu+ix)\rvert^2 P^{\frac{1}{2}-\mu}_{-\frac{1}{2}+ix}(a)$
435	$(y^2-b^2)^{-\frac{1}{2}}\sin[a(y^2-b^2)^{\frac{1}{2}}],\quad y>b$ $(b^2-y^2)^{-\frac{1}{2}}\sinh[a(b^2-y^2)^{\frac{1}{2}}],\quad y<b$ $N=b^{-1}\sinh(ab)$	$I_0[b(a^2-x^2)^{\frac{1}{2}}],\quad x<a$ $0,\quad x>a$
442	$2\cos(\frac{1}{2}ay)$ $\cdot\begin{cases}(b^2-y^2)^{-\frac{1}{2}}\sinh[\frac{1}{2}a(b^2-y^2)^{\frac{1}{2}}], & y<b\\(y^2-b^2)^{-\frac{1}{2}}\sin[\frac{1}{2}a(y^2-b^2)^{\frac{1}{2}}], & y>b\end{cases}$ $N=2b^{-1}\sinh(\frac{1}{2}ab)$	$I_0[b(ax-x^2)^{\frac{1}{2}}],\quad x<a$ $0,\quad x>a$
516	$\frac{1}{2}\pi\exp(-a\sinh y)$ $N=\frac{1}{2}\pi$	$S_{0,ix}(a)$

	$Ng(y)$	$2Nf(x)$
189	$\frac{1}{4}\log\left[\frac{1+\cosh(\pi y/b)}{\cosh(\pi y/b)+\cos(2a\pi/b)}\right]$ $N=-\frac{1}{2}\log[\cos(a\pi/b)],\quad a<\frac{1}{2}b$	$x^{-1}[\sinh(ax)]^2\operatorname{csch}(bx)$
200	$\frac{1}{2}\log\left[\frac{\cos(\pi b/c)+\cosh(\pi y/c)}{\cos(\pi a/c)+\cosh(\pi y/c)}\right]$ $N=\log\left[\frac{\cos(\frac{1}{2}\pi b/c)}{\cos(\frac{1}{2}\pi a/c)}\right],\quad c>a>b$	$x^{-1}\operatorname{csch}(cx)$ $\cdot[\cosh(ax)-\cosh(bx)]$
188	$\frac{1}{2}\log\left[\frac{\cosh(\frac{1}{2}\pi y/b)+\sin(\frac{1}{2}a\pi/b)}{\cosh(\frac{1}{2}\pi y/b)-\sin(\frac{1}{2}a\pi/b)}\right]$ $N=\frac{1}{2}\log\left[\frac{1+\sin(\frac{1}{2}a\pi/b)}{1-\sin(\frac{1}{2}a\pi/b)}\right],\quad a<b$	$x^{-1}\sinh(ax)\operatorname{sech}(bx)$
276	$\pi^{-1}(\frac{1}{2}\pi\cosh y)^{-\frac{1}{2}}$ $\cdot\log[(1+\cosh y)^{\frac{1}{2}}+(\cosh y)^{\frac{1}{2}}]$ $N=\pi^{-1}(\frac{1}{2}\pi)^{-\frac{1}{2}}\log(1+2^{\frac{1}{2}})$	$[\vert\Gamma(\frac{3}{4}+ix)\vert\cosh(\pi x)]^{-2}$
359	$\pi^{-1}(2\cosh y-2a)^{-\frac{1}{2}}$ $\cdot\log\left\{\frac{[(1+\cosh y)^{\frac{1}{2}}+(\cosh y-a)^{\frac{1}{2}}]^2}{1+a}\right\}$ $N=(2-2a)^{-\frac{1}{2}}\pi^{-1}\log\left\{\frac{[2^{\frac{1}{2}}+(1-a)^{\frac{1}{2}}]^2}{1+a}\right\}$ $-1<a<1$	$[\operatorname{sech}(\pi x)]^2P_{-\frac{1}{2}+ix}(a)$
364	$(\frac{1}{2}\pi)^{\frac{1}{2}}(a^2-1)^{-\frac{1}{2}\mu}\Gamma(\frac{1}{2}-\mu)$ $\cdot(a^2+\sinh^2 y)^{\frac{1}{2}\mu-\frac{1}{4}}\cos\{(\frac{1}{2}-\mu)$ $\cdot\arctan[(1/a)\sinh y]\}$ $N=(\frac{1}{2}\pi)^{\frac{1}{2}}(a^2-1)^{-\frac{1}{2}\mu}a^{\mu-\frac{1}{2}}\Gamma(\frac{1}{2}-\mu)$ $\mu<\frac{1}{2},\quad 0<a<1$	$\cosh(\frac{1}{2}\pi x)\,\vert\Gamma(\frac{1}{2}-\mu+ix)\vert^2$ $\cdot P^{\mu}_{-\frac{1}{2}+ix}(a)$
194	$2^{-\frac{1}{2}}\{\pi e^{-y}+2\sinh y\arctan(2^{-\frac{1}{2}}\operatorname{csch}y)$ $-\cosh y\log[(\cosh y+2^{-\frac{1}{2}})$ $\cdot(\cosh y-2^{-\frac{1}{2}})^{-1}]\}$ $N=2^{-\frac{1}{2}}[\pi-2\log(2^{\frac{1}{2}}+1)]$	$(1+x^2)^{-1}\operatorname{sech}(\frac{1}{4}\pi x)$

	$Ng(y)$	$2Nf(x)$
199	$\frac{1}{2}\pi \sin a\, e^{-y} - \frac{1}{2} \cos a \cosh y$ $\cdot \log\left[\frac{\cosh y + \sin a}{\cosh y - \sin a}\right]$ $+\sin a \sinh y \arctan(\cos a\, \operatorname{csch} y)$ $N = \frac{1}{2}\pi \sin a$ $-\cos a \log[\operatorname{ctn}(\frac{1}{4}\pi - a)], \quad a \leq \frac{1}{2}\pi$	$(1+x^2)^{-1} \sinh(ax) \operatorname{csch}(\frac{1}{2}\pi x)$
196	$\frac{1}{2}ye^{-y} - \frac{1}{2} + \cosh y \log(1+e^{-y})$ $N = \frac{1}{2}(2 \log 2 - 1)$	$x(1+x^2)^{-1} \operatorname{csch}(\pi x)$
195	$ye^{-y} + \cosh y \log(1+e^{-2y})$ $N = \log 2$	$(1+x^2)^{-1} \operatorname{sech}(\frac{1}{2}\pi x)$
224	$b^{-1}\pi \sin(a\pi/b) \cosh(\pi y/b)$ $\cdot[\cos(2\pi y/b) - \cos(2a\pi/b)]^{-1}$ $-\frac{1}{2}a(a^2+y^2)^{-1}$ $N = \frac{1}{2}b^{-1}\pi \csc(a\pi/b) - \frac{1}{2}a^{-1}, \quad a<b$	$(e^{bx}+1)^{-1} \sinh(ax)$
241	$2\pi y^{-1} \operatorname{csch}(\frac{1}{2}\pi y)$ $\cdot \sin^2\{\frac{1}{2}y \log[(1+a^{-2})^{\frac{1}{2}} + a^{-1}]\}$ $N = \{\log[(1+a^{-2})^{\frac{1}{2}} + a^{-1}]\}^2$	$\log(1+a^{-2} \operatorname{sech}^2 x)$
193	$2\cosh(\frac{1}{2}y) - e^{y} \arctan(e^{-\frac{1}{2}y})$ $-e^{-y} \arctan(e^{y/2})$ $N = 2 - \frac{1}{2}\pi$	$(1+x^2)^{-1} \operatorname{sech}(\pi x)$

7a. Orthogonal Polynomials

	$Ng(y)$	$2Nf(x)$
76	$(-1)^n 2^{-n-1} a^{-2n-1} \pi^{\frac{1}{2}}$ $\cdot \exp(-\frac{1}{4}y^2/a^2)\, He_{2n}(2^{-\frac{1}{2}}y/a)$ $N = (2a)^{-2n-1}\pi^{\frac{1}{2}}(2n)!/n!$ $n = 0, 1, 2, \ldots$	$x^{2n} \exp(-a^2x^2)$
401	$(-1)^n \frac{1}{2}\pi (2n)!(a^2+y^2)^{-n-\frac{1}{2}}$ $\cdot P_{2n}[y(a^2+y^2)^{-\frac{1}{2}}]$ $N = \pi(2a)^{-2n-1}[(2n)!/n!]^2$ $n = 0, 1, 2, \ldots$	$x^{2n} K_0(ax)$

8. Gamma Functions (Including Incomplete Gamma Functions) and Related Functions

	$Ng(y)$	$2Nf(x)$
138	$2^{-\alpha-1}\pi\Gamma(1+\alpha)[\Gamma(1+\frac{1}{2}\alpha+\frac{1}{2}y)$ $\cdot\Gamma(1+\frac{1}{2}\alpha-\frac{1}{2}y)]^{-1}$ $N=2^{-\alpha-1}\pi\Gamma(1+\alpha)[\Gamma(1+\frac{1}{2}\alpha)]^{-2}, \quad \alpha>-1$	$(\cos x)^{\alpha}, \quad x<\frac{1}{2}\pi$ $0, \quad x>\frac{1}{2}\pi$
137	$2^{-\alpha}\pi\cos(\frac{1}{2}y)\Gamma(1+\alpha)$ $\cdot[\Gamma(1+\frac{1}{2}\alpha+\frac{1}{2}y)\Gamma(1+\frac{1}{2}\alpha-\frac{1}{2}y)]^{-1}, \quad \alpha>-1$ $N=2^{-\alpha}\pi\Gamma(1+\alpha)[\Gamma(1+\frac{1}{2}\alpha)]^{-2}$	$(\sin x)^{\alpha}, \quad x<\pi$ $0, \quad x>\pi$
211	$\frac{1}{2}a^{-1}\pi^{\frac{1}{2}}[\Gamma(\frac{1}{2}\nu)\Gamma(\frac{1}{2}+\frac{1}{2}\nu)]^{-1}$ $\cdot\Gamma(\frac{1}{2}\nu+i\frac{1}{2}y/a)\Gamma(\frac{1}{2}\nu-i\frac{1}{2}y/a)$ $N=\frac{1}{2}a^{-1}\pi^{\frac{1}{2}}\Gamma(\frac{1}{2}\nu)[\Gamma(\frac{1}{2}+\frac{1}{2}\nu)]^{-1}, \quad \nu>0$	$[\operatorname{sech}(ax)]^{\nu}$
322	$\frac{1}{4}\pi\cos(\pi y)\Gamma(\frac{1}{4}+\frac{1}{2}y)\Gamma(\frac{1}{4}-\frac{1}{2}y)$ $\cdot[\Gamma(\frac{3}{4}+\frac{1}{2}y)\Gamma(\frac{3}{4}-\frac{1}{2}y)]^{-1}$ $N=\frac{1}{4}\pi[\Gamma(\frac{1}{4})]^{2}[\Gamma(\frac{3}{4})]^{-2}$	$K[\cos(\frac{1}{2}x)], \quad x<\pi$ $0, \quad x>\pi$
324	$\frac{1}{16}(a\pi)^{-1}\lvert\Gamma(\frac{1}{4}+i\frac{1}{4}y/a)\rvert^{4}$ $N=\frac{1}{16}(a\pi)^{-1}[\Gamma(\frac{1}{4})]^{4}$	$\operatorname{sech}(ax)\ K[\tanh(ax)]$
70	$\frac{1}{2}b^{-1}\{B[\nu, (a-iy)/b]+B[\nu, (a+iy)/b]\}$ $N=b^{-1}B(\nu, a/b), \quad \nu>0$	$e^{ax}(1-e^{-bx})^{\nu-1}$
341	$-\frac{1}{4}\pi^{-2}\sin(\nu\pi)\Gamma(-\frac{1}{2}\nu+\frac{1}{2}iy)$ $\cdot\Gamma(-\frac{1}{2}\nu-\frac{1}{2}iy)\Gamma(\frac{1}{2}\nu+\frac{1}{2}+\frac{1}{2}iy)$ $\cdot\Gamma(\frac{1}{2}\nu+\frac{1}{2}-\frac{1}{2}iy)$ $N=-\frac{1}{4}\pi^{-2}\sin(\nu\pi)[\Gamma(-\frac{1}{2}\nu)\Gamma(\frac{1}{2}+\frac{1}{2}\nu)]^{2}, \quad -1<\nu<0$	$\mathfrak{P}_{\nu}(\cosh x)$
325	$\frac{1}{8}\pi\Gamma(\frac{1}{4}+\frac{1}{4}iy)\Gamma(\frac{1}{4}-\frac{1}{4}iy)$ $\cdot[\Gamma(\frac{3}{4}+\frac{1}{4}iy)\Gamma(\frac{3}{4}-\frac{1}{4}iy)]^{-1}$ $N=\frac{1}{8}\pi[\Gamma(\frac{1}{4})]^{2}[\Gamma(\frac{3}{4})]^{-2}$	$\operatorname{sech}x\ K(\operatorname{sech}x)$
343	$2^{-\mu-2}\pi^{-\frac{1}{2}}\Gamma(\frac{1}{2}+\frac{1}{2}\nu-\frac{1}{2}\mu+\frac{1}{2}iy)$ $\cdot\Gamma(\frac{1}{2}+\frac{1}{2}\nu-\frac{1}{2}\mu-\frac{1}{2}iy)\Gamma(-\frac{1}{2}\nu-\frac{1}{2}\mu+\frac{1}{2}iy)$ $\cdot\Gamma(-\frac{1}{2}\nu-\frac{1}{2}\mu-\frac{1}{2}iy)$ $\cdot[\Gamma(-\nu-\mu)\Gamma(1+\nu-\mu)\Gamma(\frac{1}{2}-\mu)]^{-1}$ $N=2^{-\mu-2}\pi^{-\frac{1}{2}}[\Gamma(\frac{1}{2}+\frac{1}{2}\nu-\frac{1}{2}\mu)\Gamma(-\frac{1}{2}\nu-\frac{1}{2}\mu)]^{2}$ $\cdot[\Gamma(-\nu-\mu)\Gamma(1+\nu-\mu)\Gamma(\frac{1}{2}-\mu)]^{-1}, \quad \nu+\mu<0, \quad \mu-\nu<1$	$(\sinh x)^{\mu}\mathfrak{P}_{\nu}^{\mu}(\cosh x)$

	$Ng(y)$	$2Nf(x)$
350	$2^{\mu-2}\pi^{\frac{1}{2}}\Gamma(\frac{1}{2}-\mu)\Gamma[\frac{1}{2}(1+\nu+\mu+iy)]$ $\cdot\Gamma[\frac{1}{2}(1+\nu+\mu-iy)]$ $\cdot\{\Gamma[1+\frac{1}{2}(\nu-\mu+iy)]$ $\cdot\Gamma[1+\frac{1}{2}(\nu-\mu-iy)]\}^{-1}$ $N=2^{\mu-2}\pi^{\frac{1}{2}}\Gamma(\frac{1}{2}-\mu)$ $\cdot\{\Gamma[\frac{1}{2}(1+\nu+\mu)]\}^{2}\{\Gamma[1+\frac{1}{2}(\nu-\mu)]\}^{-2},$ $\mu+\nu+1>0,\quad \mu<\frac{1}{2}$	$(\sinh x)^{-\mu}$ $\cdot e^{-i\pi\mu}\mathfrak{Q}_{\nu}^{\mu}(\cosh x)$
351	$2^{\nu-2}[\Gamma(1+\nu)\Gamma(1+\nu-\mu)]^{-1}$ $\cdot\Gamma[\frac{1}{2}(1+\nu-\mu+iy)]\Gamma[\frac{1}{2}(1+\nu-\mu-iy)]$ $\cdot\Gamma[\frac{1}{2}(1+\nu+\mu+iy)]\Gamma[\frac{1}{2}(1+\nu+\mu-iy)]$ $N=2^{\nu-2}[\Gamma(1+\nu)\Gamma(1+\nu-\mu)]^{-1}$ $\cdot\{\Gamma[\frac{1}{2}(1+\nu-\mu)]\Gamma[\frac{1}{2}(1+\nu+\mu)]\}^{2},$ $\mu<\frac{1}{2},\quad \nu\pm\mu>-1$	$(\sinh x)^{-\nu-1}$ $\cdot e^{-i\pi\mu}\mathfrak{Q}_{\nu}^{\mu}(\mathrm{ctnh}\,x)$
226	$2^{-\nu-2}b^{-1}\Gamma(1+\nu)\dfrac{\Gamma[\frac{1}{2}b^{-1}(a-b\nu-iy)]}{\Gamma[\frac{1}{2}b^{-1}(a+b\nu-iy)+1]}$ $+\dfrac{\Gamma[\frac{1}{2}b^{-1}(a-b\nu+iy)]}{\Gamma[\frac{1}{2}b^{-1}(a+b\nu+iy)+1]},$ $\nu>-1,\quad b\nu<a$ $N=2^{-\nu-1}b^{-1}\Gamma(1+\nu)\Gamma(\frac{1}{2}ab^{-1}-\frac{1}{2}\nu)$ $\cdot[\Gamma(\frac{1}{2}ab^{-1}+\frac{1}{2}\nu+1)]^{-1}$	$e^{-ax}[\sinh(bx)]^{\nu}$
212	$2^{\nu}a^{-1}\pi\sin(\frac{1}{2}\nu\pi)\Gamma(1-\nu)$ $\cdot[\Gamma(1-\frac{1}{2}\nu+\frac{1}{2}iy/a)\Gamma(1-\frac{1}{2}\nu-\frac{1}{2}iy/a)]^{-1}$ $\cdot\cosh(\frac{1}{2}\pi y/a)[\cosh(\pi y/a)-\cos(\nu\pi)]^{-1},$ $0<\nu<1$ $N=\frac{1}{2}a^{-1}\Gamma(\frac{1}{2}\nu)\Gamma(\frac{1}{2}-\frac{1}{2}\nu)$	$(\mathrm{csch}\,ax)^{\nu}$
66	$\frac{1}{4}a^{-1}[\psi(i\frac{1}{2}y/a)+\psi(-i\frac{1}{2}y/a)$ $-\psi(\frac{1}{2}+i\frac{1}{2}y/a)-\psi(\frac{1}{2}-i\frac{1}{2}y/a)]$ $N=a^{-1}\log 2$	$(e^{ax}+1)^{-1}$
151	$y^{-1}\sin y[\gamma+\log 2+\frac{1}{2}\psi(1+\frac{1}{2}y/\pi)$ $+\frac{1}{2}\psi(1-\frac{1}{2}y/\pi)]$ $N=\log 2$	$\log[\csc(\pi x)],\quad x<1$ $0,\quad x>1$
150	$y^{-1}\sin y$ $\cdot[\gamma+\log 2+\frac{1}{2}\psi(1+y/\pi)+\frac{1}{2}\psi(1-y/\pi)]$ $N=\log 2$	$\log[\sec(\frac{1}{2}\pi x)],\quad x<1$ $0,\quad x>1$

	$Ng(y)$	$2Nf(x)$
237	$2^{\nu-2}[a\Gamma(\nu)]^{-1}\,\vert\Gamma(\tfrac{1}{2}\nu+\tfrac{1}{2}iy/a)\vert^2$ $\cdot\{\psi(\nu)-\log 2-\mathrm{Re}[\psi(\tfrac{1}{2}\nu+\tfrac{1}{2}iy/a)]\}$ $N=2^{\nu-3}a^{-1}B(\tfrac{1}{2}\nu,\tfrac{1}{2}\nu)$ $\cdot[\psi(\tfrac{1}{2}+\tfrac{1}{2}\nu)-\psi(\tfrac{1}{2}\nu)],\quad \nu>0$	$[\mathrm{sech}(ax)]^{\nu}$ $\cdot\log[\cosh(ax)]$
153	$2^{1-\nu}\Gamma(\nu)\cos(\tfrac{1}{2}y)$ $\cdot[\Gamma(\tfrac{1}{2}+\tfrac{1}{2}\nu+\tfrac{1}{2}y/\pi)\Gamma(\tfrac{1}{2}+\tfrac{1}{2}\nu-\tfrac{1}{2}y/\pi)]^{-1}$ $\cdot[\log 2+\tfrac{1}{2}\psi(\tfrac{1}{2}+\tfrac{1}{2}\nu+\tfrac{1}{2}y/\pi)$ $+\tfrac{1}{2}\psi(\tfrac{1}{2}+\tfrac{1}{2}\nu-\tfrac{1}{2}y/\pi)-\psi(\nu)]$ $N=2^{1-\nu}\Gamma(\nu)[\Gamma(\tfrac{1}{2}+\tfrac{1}{2}\nu)]^{-2}$ $\cdot[\log 2-\psi(\nu)+\psi(\tfrac{1}{2}+\tfrac{1}{2}\nu)],\quad \nu>0$	$[\sin(\pi x)]^{\nu-1}$ $\cdot\log[\csc(\pi x)],\quad x<1$ $0,\quad x>1$
152	$2^{1-\nu}\Gamma(\nu)[\Gamma(\tfrac{1}{2}+\tfrac{1}{2}\nu+y/\pi)]^{-1}$ $\cdot[\Gamma(\tfrac{1}{2}+\tfrac{1}{2}\nu-y/\pi)]^{-1}[\log 2-\psi(\nu)$ $+\tfrac{1}{2}\psi(\tfrac{1}{2}+\tfrac{1}{2}\nu+y/\pi)+\tfrac{1}{2}\psi(\tfrac{1}{2}+\tfrac{1}{2}\nu-y/\pi)]$ $N=2^{1-\nu}\Gamma(\nu)[\Gamma(\tfrac{1}{2}+\tfrac{1}{2}\nu)]^{-2}$ $\cdot[\log 2-\psi(\nu)+\psi(\tfrac{1}{2}+\tfrac{1}{2}\nu)],\quad \nu>0$	$[\cos(\tfrac{1}{2}\pi x)]^{\nu-1}$ $\cdot\log[\sec(\tfrac{1}{2}\pi x)],\quad x<1$ $0,\quad x>1$
184	$\frac{1}{4}b^{-1}\left\{\psi\left(\frac{3b-a+iy}{4b}\right)+\psi\left(\frac{3b-a-iy}{4b}\right)-\psi\left(\frac{3b+a+iy}{4b}\right)-\psi\left(\frac{3b+a-iy}{4b}\right)+2\pi\sin\left(\frac{a\pi}{b}\right)\left[\cos\left(\frac{a\pi}{b}\right)+\cosh\left(\frac{\pi y}{b}\right)\right]^{-1}\right\}$ $N=\tfrac{1}{2}b^{-1}[\pi\tan(\tfrac{1}{2}a\pi/b)$ $+\psi(\tfrac{3}{4}-\tfrac{1}{4}a/b)-\psi(\tfrac{3}{4}+\tfrac{1}{4}a/b)],\quad a<b$	$\sinh(ax)\,\mathrm{sech}(bx)$
187	$-\Gamma(s-1)\Gamma(s)\{2^{-s}(s-1)[\zeta(s,\tfrac{1}{2}+\tfrac{1}{2}iy)$ $+\zeta(s,\tfrac{1}{2}-\tfrac{1}{2}iy)]-y^{1-s}\sin(\tfrac{1}{2}\pi s)\}$ $N=2\Gamma(s)\zeta(s)(2^{-s}-1),\quad -1<s<1$	$x^{s-1}(x^{-1}-\mathrm{csch}\,x)$
294	$\pi y^{-1}\,\mathrm{Erf}(\tfrac{1}{2}ya^{-\frac{1}{2}})$ $N=(\pi/a)^{\frac{1}{2}}$	$-\mathrm{Ei}(-ax^2)$
281	$\tfrac{1}{2}\pi\,\mathrm{Erfc}(\tfrac{1}{2}y/a)$ $N=\tfrac{1}{2}\pi$	$-ix^{-1}\exp(-a^2x^2)\,\mathrm{Erf}(iax)$

	$Ng(y)$	$2Nf(x)$
295	$\frac{1}{2}\pi(\pi/a)^{\frac{1}{2}}\exp(\frac{1}{4}y^2/a)\ \mathrm{Erfc}(\frac{1}{2}ya^{-\frac{1}{2}}+b^{\frac{1}{2}})$ $N=\frac{1}{2}\pi(\pi/a)^{\frac{1}{2}}\ \mathrm{Erfc}(b^{\frac{1}{2}})$	$-\exp(ax^2)\ \mathrm{Ei}[-(ax^2+b)]$
229	$(\frac{1}{2}\pi)^{-\frac{1}{2}}a[\exp(-\frac{1}{8}y^2/a^2)$ $-2^{-\frac{3}{2}}a^{-1}y\exp(-\frac{1}{16}y^2/a^2)\ \mathrm{Erfc}(2^{-\frac{3}{2}}a^{-1}y)]$ $N=(\frac{1}{2}\pi)^{-\frac{1}{2}}a$	$x^{-9/4}\exp(-a^2x^2)\sinh(a^2x^2)$
297	$\frac{1}{2}\pi(\pi/a)^{\frac{1}{2}}[\exp(\frac{1}{4}y^2/a)\ \mathrm{Erfc}(\frac{1}{2}ya^{-\frac{1}{2}})$ $+i\exp(-\frac{1}{4}y^2/a)\ \mathrm{Erf}(i\frac{1}{2}ya^{-\frac{1}{2}})]$ $N=\frac{1}{2}\pi(\pi/a)^{\frac{1}{2}}$	$\exp(-ax^2)\ \mathrm{Ei}(ax^2)$ $-\exp(ax^2)\ \mathrm{Ei}(-ax^2)$
78	$\frac{1}{4}b^{-1}\pi\exp(a^2b^2)[\exp(-by)\ \mathrm{Erfc}(ab-\frac{1}{2}y/a)$ $+\exp(by)\ \mathrm{Erfc}(ab+\frac{1}{2}y/a)]$ $N=\frac{1}{2}b^{-1}\pi\exp(a^2b^2)\ \mathrm{Erfc}(ab)$	$(b^2+x^2)^{-1}\exp(-a^2x^2)$
79	$\frac{1}{4}b^{-1}\pi^{\frac{1}{2}}\left\{\exp\left[\left(\frac{a+iy}{2b}\right)^2\right]\mathrm{Erfc}\left(\frac{a+iy}{2b}\right)\right.$ $\left.+\exp\left[\left(\frac{a-iy}{2b}\right)^2\right]\mathrm{Erfc}\left(\frac{a-iy}{2b}\right)\right\}$ $N=\frac{1}{2}b^{-1}\pi^{\frac{1}{2}}\exp(\frac{1}{4}a^2/b^2)\ \mathrm{Erfc}(\frac{1}{2}a/b)$	$\exp(-ax-b^2x^2)$
32	$(2a)^{-\frac{1}{2}}\pi\ \mathrm{Erfc}[(ay)^{\frac{1}{2}}]$ $N=(2a)^{-\frac{1}{2}}\pi$	$(a^2+x^2)^{-\frac{1}{2}}$ $\cdot[a+(a^2+x^2)^{\frac{1}{2}}]^{-\frac{1}{2}}$
228	$(\frac{1}{2}\pi)^{\frac{1}{2}}\{\exp(-\frac{1}{8}y^2)-\frac{1}{2}(\frac{1}{2}\pi)^{\frac{1}{2}}y\ \mathrm{Erfc}(2^{-\frac{3}{2}}y)\}$ $N=(\frac{1}{2}\pi)^{\frac{1}{2}}$	$x^{-2}\exp(-x^2)\sinh x^2$
28	$(2a)^{-\frac{1}{2}}\pi e^{ay}\ \mathrm{Erfc}[(ay)^{\frac{1}{2}}]$ $N=\pi(2a)^{-\frac{1}{2}}$	$x^{-\frac{1}{2}}[a+x+(2ax)^{\frac{1}{2}}]^{-1}$
103	$(2a)^{-\frac{1}{2}}\pi e^{ab}\mathrm{Erfc}\{a^{\frac{1}{2}}[(b^2+y^2)^{\frac{1}{2}}+b]^{\frac{1}{2}}\}$ $N=(2a)^{-\frac{1}{2}}\pi e^{ab}\ \mathrm{Erfc}[(2ab)^{\frac{1}{2}}]$	$(a^2+x^2)^{-\frac{1}{2}}[a+(a^2+x^2)^{\frac{1}{2}}]^{-\frac{1}{2}}$ $\cdot\exp[-b(a^2+x^2)^{\frac{1}{2}}]$
529	$\frac{1}{2}(\pi\ \mathrm{sech}y)^{\frac{1}{2}}\exp(\frac{1}{2}a^2\ \mathrm{sech}y)$ $\cdot\mathrm{Erfc}[a(\frac{1}{2}+\frac{1}{2}\ \mathrm{sech}y)^{\frac{1}{2}}]$ $N=\frac{1}{2}\pi^{\frac{1}{2}}\ \mathrm{Erfc}(a)$	$\mathrm{sech}(\pi x)$ $\cdot D_{-\frac{1}{2}+ix}(a)D_{-\frac{1}{2}-ix}(a)$

	$Ng(y)$	$2Nf(x)$
109	$-\frac{1}{2}a^{-1}\pi\{e^{-ay}[\gamma-\log(2a/y)]$ $-e^{ay}\,\mathrm{Ei}(-2ay)\}$ $N=a^{-1}\pi\log(2a),\quad a\geq 1$	$(a^2+x^2)^{-1}\log(a^2+x^2)$
292	$y^{-1}[\sin(ay)\,\mathrm{Ei}(-ab)$ $-\arctan(y/b)-\frac{1}{2}i\,\mathrm{Ei}(-ab-iay)$ $+\frac{1}{2}i\,\mathrm{Ei}(-ab+iay)]$ $N=b^{-1}-a\,\mathrm{Ei}(-ab)$	$\mathrm{Ei}(-bx),\quad x<a$ $0,\quad x>a$
279	$\frac{1}{2}[\mathrm{Ei}(-\frac{1}{4}y^2/a^2)-\mathrm{Ei}(-\frac{1}{4}y^2/b^2)]$ $N=\log(b/a),\quad a<b$	$x^{-1}[\mathrm{Erfc}(ax)-\mathrm{Erfc}(bx)]$
290	$-\frac{1}{2}\pi^{-\frac{1}{2}}a^{-1}\exp(\frac{1}{4}y^2/a^2)\,\mathrm{Ei}[-(b^2+\frac{1}{4}y^2/a^2)]$ $N=-\frac{1}{2}\pi^{-\frac{1}{2}}a^{-1}\,\mathrm{Ei}(-b^2)$	$\exp(a^2x^2)\,\mathrm{Erfc}(ax+b)$
450	$-\frac{1}{4}\pi a^{-1}[e^{-ay}\,\mathrm{Ei}(-z_2)+e^{ay}\,\mathrm{Ei}(-z_1)]$ $N=-\frac{1}{2}\pi a^{-1}\,\mathrm{Ei}(-ab)$ $z_{1,2}=a[(b^2+y^2)^{\frac{1}{2}}\pm y]$	$(a^2+x^2)^{-1}$ $\cdot K_0[b(a^2+x^2)^{\frac{1}{2}}]$
452	$\pi(b^2+y^2)^{-\frac{1}{2}}\{\exp[-a(b^2+y^2)^{\frac{1}{2}}]$ $\cdot[\log(ab)-\frac{1}{2}\log(b^2+y^2)]$ $-\exp[a(b^2+y^2)^{\frac{1}{2}}]\,\mathrm{Ei}[-2a(b^2+y^2)^{\frac{1}{2}}]\}$ $N=\pi b^{-1}[e^{-ab}\log a-e^{ab}\,\mathrm{Ei}(-2ab)]$	$\log(a^2+x^2)\,K_0[b(a^2+x^2)^{\frac{1}{2}}]$
106	$y^{-1}\,\mathrm{Si}(y)$ $N=1$	$-\log x,\quad x<1$ $0,\quad x>1$
175	$\frac{1}{2}\pi[\mathrm{Ci}(y)-Ji_0(y)]$ $N=\frac{1}{2}\pi\log 2$	$x^{-1}\arcsin x,\quad x<1$ $0,\quad x>1$
114	$\frac{1}{2}a^{-1}\pi[\sin(ay)\,\mathrm{Ci}(ay)$ $-\cos(ay)\,\mathrm{si}(ay)]$ $N=\frac{1}{4}a^{-1}\pi^2$	$(x^2-a^2)^{-1}\log(x/a)$
116	$-y^{-1}\{\sin(ay)[\mathrm{Ci}(ay)-\gamma-\log y]$ $-\cos(ay)\,\mathrm{Si}(ay)\}$ $N=a(1-\log a),\quad a\leq 1$	$-\log(a-x),\quad x<a$ $0,\quad x>a$
118	$y^{-1}\{\cos(ay)\,\mathrm{Si}(2ay)+\sin(ay)$ $\cdot[\gamma+\log(\frac{1}{2}y/a)-\mathrm{Ci}(2ay)]\}$ $N=2a(1-\log a),\quad a\leq 1$	$-\log(a^2-x^2),\quad x<a$ $0,\quad x>a$

	$Ng(y)$	$2Nf(x)$
4	$\cos(ay)[\mathrm{Ci}(ay+by)-\mathrm{Ci}(ay)]$ $+\sin(ay)[\mathrm{si}(ay+by)-\mathrm{si}(ay)]$ $N=\log(1+b/a)$	$(a+x)^{-1}, \quad x<b$ $0, \quad x>b$
6	$a^{-1}[\cos(ay)\ \mathrm{Ci}(ay+by)$ $+\sin(ay)\ \mathrm{si}(ay+by)-\mathrm{Ci}(by)]$ $N=a^{-1}\log(1+a/b)$	$0, \quad x<b$ $[x(a+x)]^{-1}, \quad x>b$
19	$(2a)^{-1}\{\sin(ay)[\mathrm{si}(by-ay)+\mathrm{si}(by+ay)]$ $-\cos(ay)[\mathrm{Ci}(by-ay)-\mathrm{Ci}(by+ay)]\}$ $N=(2a)^{-1}\log[(b+a)(b-a)^{-1}], \quad b>a$	$0, \quad x<b$ $(x^2-a^2)^{-1}, \quad x>b$
17	$\cos(ay)[\mathrm{Ci}(ay)-\mathrm{Ci}(ay-by)]$ $+\sin(ay)[\mathrm{si}(ay)-\mathrm{si}(ay-by)]$ $N=-\log[1-(b/a)], \quad b<a$	$(a-x)^{-1}, \quad x<b$ $0, \quad x>b$
18	$(2a)^{-1}\{\cos(ay)[\mathrm{Ci}(ay+by)-\mathrm{Ci}(ay-by)]$ $+\sin(ay)[\mathrm{si}(ay+by)-\mathrm{si}(ay-by)]\}$ $N=(2a)^{-1}\log[(a+b)(a-b)^{-1}], \quad b<a$	$(a^2-x^2)^{-1}, \quad x<b$ $0, \quad x>b$
117	$-y^{-1}\{\sin(by)\log(a-b)+\sin(ay)[\mathrm{Ci}(ay)$ $-\mathrm{Ci}(ay-by)]-\cos(ay)[\mathrm{Si}(ay)$ $-\mathrm{Si}(ay-by)]\}$ $N=b-a\log a+(a-b)\log(a-b), \quad b<a\leq 1$	$-\log(a-x), \quad x<b$ $0, \quad x>b$
107	$y^{-1}\{\sin(by)\log(a+b)-\cos(ay)[\mathrm{si}(ay+by)$ $-\mathrm{si}(ay)]+\sin(ay)[\mathrm{Ci}(ay+by)$ $-\mathrm{Ci}(ay)]\}$ $N=(a+b)\log(a+b)-a\log a-b, \quad a\geq 1$	$\log(a+x), \quad x<b$ $0, \quad x>b$
120	$-y^{-1}\{\sin(by)\log(a^2-b^2)-\cos(ay)$ $\cdot[\mathrm{si}(ay+by)-\mathrm{si}(ay-by)]+\sin(ay)$ $\cdot[\mathrm{Ci}(ay+by)-\mathrm{Ci}(ay-by)]\}$ $N=2b+(a-b)\log(a-b)-(a+b)$ $\cdot\log(a+b), \quad b<a\leq 1$	$-\log(a^2-x^2), \quad x<b$ $0, \quad x>b$
311	$\frac{1}{2}(2a\pi)^{-\frac{1}{2}}[\sin(\frac{1}{4}y^2/a)\ \mathrm{Ci}(\frac{1}{4}y^2/a)$ $-\cos(\frac{1}{4}y^2/a)\ \mathrm{si}(\frac{1}{4}y^2/a)]$ $N=\frac{1}{4}(\frac{1}{2}\pi/a)^{\frac{1}{2}}$	$[\frac{1}{2}-C(ax^2)]\cos(ax^2)$ $+[\frac{1}{2}-S(ax^2)]\sin(ax^2)$
432	$\frac{1}{2}y^{-1}[\mathrm{Ci}(\frac{1}{4}a^2/y)\sin(\frac{1}{4}a^2/y)$ $-\mathrm{si}(\frac{1}{4}a^2/y)\cos(\frac{1}{4}a^2/y)]$ $N=2a^{-2}$	$K_0(ax^{\frac{1}{2}})$

	$Ng(y)$	$2Nf(x)$
385	$-4z^{-1}\{\sin(\frac{1}{2}az)[\mathrm{Ci}(az)-\frac{1}{2}\mathrm{Ci}(\frac{1}{2}az+\frac{1}{2}ay)$ $-\frac{1}{2}\mathrm{Ci}(\frac{1}{2}az-\frac{1}{2}ay)+\log(\frac{1}{2}ab)-\log z]$ $-\cos(\frac{1}{2}az)[\mathrm{Si}(az)-\frac{1}{2}\mathrm{Si}(\frac{1}{2}az+\frac{1}{2}ay)$ $-\frac{1}{2}\mathrm{Si}(\frac{1}{2}az-\frac{1}{2}ay)\}\cos(\frac{1}{2}ay)$ $N=-4b^{-1}\{\sin(\frac{1}{2}ab)[\mathrm{Ci}(ab)-\mathrm{Ci}(\frac{1}{2}ab)$ $+\log(\frac{1}{2}a)]-\cos(\frac{1}{2}ab)[\mathrm{Si}(ab)$ $-\mathrm{Si}(\frac{1}{2}ab)]\}, \quad a\leq 2, \quad ab\leq 2\tau_{0,1}$ $z=(b^2+y^2)^{\frac{1}{2}}$	$-\log(ax-x^2)J_0[b(ax-x^2)^{\frac{1}{2}}], \quad x<a$ $0, \quad x>a$
379	$-2z^{-1}\{\sin(az)[\mathrm{Ci}(2az)-\frac{1}{2}\mathrm{Ci}(az+ay)$ $-\frac{1}{2}\mathrm{Ci}(az-ay)+\log(ab)-\log z]$ $-\cos(az)[\mathrm{Si}(2az)-\frac{1}{2}\mathrm{Si}(az+ay)$ $-\frac{1}{2}\mathrm{Si}(az-ay)]\}$ $N=-2b^{-1}\{\sin(ab)[\mathrm{Ci}(2ab)-\mathrm{Ci}(ab)$ $+\log a]-\cos(ab)[\mathrm{Si}(2ab)$ $-\mathrm{Si}(ab)]\}, \quad a\leq 1, \quad ab\leq\tau_{0,1},$ $z=(b^2+y^2)^{\frac{1}{2}}$	$-\log(a^2-x^2)$ $\cdot J_0[b(a^2-x^2)^{\frac{1}{2}}], \quad x<a$ $0, \quad x>a$
386	$-(\pi z)^{-1}\{\sin(az)[\mathrm{Ci}(az+ay)+\mathrm{Ci}(az-ay)]$ $-\cos(az)[\mathrm{Si}(az+ay)+\mathrm{Si}(az-ay)]\}$ $N=-2(b\pi)^{-1}[\sin(ab)\,\mathrm{Ci}(ab)$ $-\cos(ab)\,\mathrm{Si}(ab)], \quad ab\leq\zeta_{0,1},$ $z=(b^2+y^2)^{\frac{1}{2}}$	$-Y_0[b(a^2-x^2)^{\frac{1}{2}}], \quad x<a$ $0, \quad x>a$
383	$-2(\pi z)^{-1}\cos(\frac{1}{2}ay)\{\sin(\frac{1}{2}az)[\mathrm{Ci}(\frac{1}{2}az+\frac{1}{2}ay)$ $+\mathrm{Ci}(\frac{1}{2}az-\frac{1}{2}ay)]-\cos(\frac{1}{2}az)$ $\cdot[\mathrm{Si}(\frac{1}{2}az+\frac{1}{2}ay)+\mathrm{Si}(\frac{1}{2}az-\frac{1}{2}ay)]\}$ $N=-4(b\pi)^{-1}[\sin(\frac{1}{2}ab)\,\mathrm{Ci}(\frac{1}{2}ab)-\cos(\frac{1}{2}ab)$ $\cdot\mathrm{Si}(\frac{1}{2}ab)], \quad ab\leq 2\zeta_{0,1}, \quad z=(b^2+y^2)^{\frac{1}{2}}$	$-Y_0[b(ax-x^2)^{\frac{1}{2}}], \quad x<a$ $0, \quad x>a$
451	$\frac{1}{2}a\alpha^{-1}\cos(ay)\{\sin\alpha[\mathrm{Ci}(z_1)+\mathrm{Ci}(z_2)]$ $-\cos\alpha[\mathrm{si}(z_1)+\mathrm{si}(z_2)]\}+\frac{1}{2}a\alpha^{-1}\sin(ay)$ $\cdot\{\cos\alpha[\mathrm{Ci}(z_1)-\mathrm{Ci}(z_2)]+\sin\alpha[\mathrm{si}(z_1)$ $-\mathrm{si}(z_2)]\}$ $\alpha=a(b^2+y^2)^{\frac{1}{2}}, \quad z_{\frac{1}{2}}=a[(b^2+y^2)^{\frac{1}{2}}\pm y]$ $N=b^{-1}[\sin(ab)\,\mathrm{Ci}(ab)-\cos(ab)\,\mathrm{si}(ab)]$	$K_0[b(2ax+x^2)^{\frac{1}{2}}]$
456	$\frac{1}{2}a\alpha^{-1}\{\sin\alpha[\mathrm{Ci}(z_1)+\mathrm{Ci}(z_2)]$ $-\cos\alpha[\mathrm{si}(z_1)+\mathrm{si}(z_2)]\}$ $N=b^{-1}[\sin(ab)\,\mathrm{Ci}(ab)-\cos(ab)\,\mathrm{si}(ab)]$ $\alpha=a(b^2+y^2)^{\frac{1}{2}}, \quad z_{\frac{1}{2}}=a[(b^2+y^2)^{\frac{1}{2}}\pm y]$	$0, \quad x<a$ $K_0[b(x^2-a^2)^{\frac{1}{2}}], \quad x>a$

	$Ng(y)$	$2Nf(x)$
453	$-\frac{1}{2}a\alpha^{-1}\{\sin\alpha[\mathrm{Ci}(z_1)+\mathrm{Ci}(z_2)]$ $-\cos\alpha[\mathrm{Si}(z_1)-\mathrm{Si}(z_2)]\}$ $N=\frac{1}{2}b^{-1}\{i\cosh(ab)[\mathrm{Si}(-iab)$ $-\mathrm{Si}(iab)]-\sinh(ab)[\mathrm{Ci}(iab)$ $+\mathrm{Ci}(-iab)]\}$ $\alpha=a(y^2-b^2)^{\frac{1}{2}},\quad z_{\substack{1\\2}}=a[y\pm(y^2-b^2)^{\frac{1}{2}}]$ $\arg(y^2-b^2)=\begin{cases}0, & y>b\\ \frac{1}{2}\pi, & y<b\end{cases}$	$K_0[b(a^2-x^2)^{\frac{1}{2}}],\quad x<a$ $0,\quad x>a$
517	$\sin(a\cosh y)\ \mathrm{Ci}(a\cosh y)$ $-\cos(a\cosh y)\ \mathrm{si}(a\cosh y)$ $N=\sin a\ \mathrm{Ci}(a)-\cos a\ \mathrm{si}(a)$	$\mathrm{sech}(\frac{1}{2}\pi x)\, S_{0,ix}(a)$
518	$-\cos(a\cosh y)\ \mathrm{Ci}(a\cosh y)$ $-\sin(a\cosh y)\ \mathrm{si}(a\cosh y)$ $N=-\cos a\ \mathrm{Ci}(a)-\sin a\ \mathrm{si}(a)$	$x\ \mathrm{csch}(\pi x)\, S_{-1,ix}(a)$
3	$(2\pi/y)^{\frac{1}{2}}C(y)$ $N=2$	$x^{-\frac{1}{2}},\quad x<1$ $0,\quad x>1$
20	$b^{-\frac{1}{2}}\pi[1-C(by)-S(by)]$ $N=b^{-\frac{1}{2}}\pi$	$0,\quad x<b$ $x^{-1}(x-b)^{-\frac{1}{2}},\quad x>b$
16	$(2\pi/y)^{\frac{1}{2}}[\cos(ay)C(ay)+\sin(ay)S(ay)]$ $N=2a^{\frac{1}{2}}$	$(a-x)^{-\frac{1}{2}},\quad x<a$ $0,\quad x>a$
13	$\pi a^{-\frac{1}{2}}\{\cos(ay)[1-C(ay)-S(ay)]$ $+\sin(ay)[C(ay)-S(ay)]\}$ $N=\pi a^{-\frac{1}{2}}$	$x^{-\frac{1}{2}}(a+x)^{-1}$
22	$(2b)^{-\frac{1}{2}}\pi\{\cos(by)[1-C(2by)-S(2by)]$ $+\sin(by)[C(2by)-S(2by)]\}$ $N=(2b)^{-\frac{1}{2}}\pi$	$0,\quad x<b$ $(x-b)^{-\frac{1}{2}}(x+b)^{-1},\quad x>b$
23	$(a+b)^{-\frac{1}{2}}\pi\{\cos(ay)[1-C(ay+by)$ $-S(ay+by)]+\sin(ay)[C(ay+by)$ $-S(ay+by)]\}$ $N=(a+b)^{-\frac{1}{2}}\pi$	$0,\quad x<b$ $(x-b)^{-\frac{1}{2}}(a+x)^{-1},\quad x>b$
308	$(2a/\pi)^{-\frac{1}{2}}\{\cos(\frac{1}{4}y^2/a)[\frac{1}{2}-S(\frac{1}{4}y^2/a)]$ $-\sin(\frac{1}{4}y^2/a)[\frac{1}{2}-C(\frac{1}{4}y^2/a)]\}$ $N=\frac{1}{2}(2a/\pi)^{-\frac{1}{2}}$	$-\sin(ax^2)\ \mathrm{si}(ax^2)$ $-\cos(ax^2)\ \mathrm{Ci}(ax^2)$

	$Ng(y)$	$2Nf(x)$
307	$\pi(2a/\pi)^{-\frac{1}{2}}\{\sin(\frac{1}{4}y^2/a)[\frac{1}{2}-S(\frac{1}{4}y^2/a)]$ $+\cos(\frac{1}{4}y^2/a)[\frac{1}{2}-C(\frac{1}{4}y^2/a)]\}$ $N=(\frac{1}{2}\pi)^{\frac{3}{2}}a^{-\frac{1}{2}}$	$\sin(ax^2)\ \mathrm{Ci}(ax^2)$ $-\cos(ax^2)\ \mathrm{si}(ax^2)$
313	$\frac{1}{2}\pi\{\frac{1}{2}+[C(\frac{1}{4}y^2/a)]^2+[S(\frac{1}{4}y^2/a)]^2$ $-C(\frac{1}{4}y^2/a)-S(\frac{1}{4}y^2/a)\}$ $N=\frac{1}{4}\pi$	$x^{-1}\{\cos(ax^2)[C(ax^2)-S(ax^2)]$ $+\sin(ax^2)[C(ax^2)+S(ax^2)-1]\}$
90	$ay^{-1}(2y/\pi)^{-\frac{1}{2}}\{\cos(\frac{1}{4}a^2/y)[\frac{1}{2}-C(\frac{1}{4}a^2/y)]$ $+\sin(\frac{1}{4}a^2/y)[\frac{1}{2}-S(\frac{1}{4}a^2/y)]\}$ $N=2a^{-2}$	$\exp(-ax^{\frac{1}{2}})$
91	$(2\pi/y)^{\frac{1}{2}}\{\cos(\frac{1}{4}a^2/y)[\frac{1}{2}-S(\frac{1}{4}a^2/y)]$ $-\sin(\frac{1}{4}a^2/y)[\frac{1}{2}-S(\frac{1}{4}a^2/y)]\}$ $N=2a^{-1}$	$x^{-\frac{1}{2}}\exp(-ax^{\frac{1}{2}})$
14	$2a^{-\frac{1}{2}}-(2\pi y)^{\frac{1}{2}}\{\cos(ay)[1-S(ay)^{\frac{1}{2}}]$ $-\sin(ay)[1-2C(ay)^{\frac{1}{2}}]\}$ $N=2a^{-\frac{1}{2}}$	$(a+x)^{-\frac{3}{2}}$
38	$-\frac{1}{2}iy^{-\nu-1}\{\exp[-i(\frac{1}{2}\pi\nu-by)]\gamma(\nu+1,iby)$ $-\exp[i(\frac{1}{2}\pi\nu-by)]\gamma(\nu+1,-iby)\}$ $N=b^{\nu+1}(\nu+1)^{-1},\quad \nu>-1$	$(b-x)^{\nu},\quad x<b$ $0,\quad x>b$
317	$\frac{1}{2}\pi^{\frac{1}{2}}(\frac{1}{4}y^2)^{\nu-\frac{1}{2}}\gamma(\frac{1}{2}-\nu,\frac{1}{4}y^2/a)$ $N=\pi^{\frac{1}{2}}a^{\nu-\frac{1}{2}}(1-2\nu)^{-1},\quad \nu<\frac{1}{2}$	$x^{-2\nu}\Gamma(\nu,ax^2)$
316	$\frac{1}{2}\pi^{\frac{1}{2}}(\frac{1}{4}y^2)^{\nu-\frac{1}{2}}\Gamma(\frac{1}{2}-\nu,\frac{1}{4}y^2/a)$ $N=\pi^{\frac{1}{2}}a^{\nu-\frac{1}{2}}(2\nu-1)^{-1},\quad \nu>\frac{1}{2}$	$x^{-2\nu}\gamma(\nu,ax^2)$
318	$\frac{1}{2}(\pi/a)^{-\frac{1}{2}}\Gamma(\frac{1}{2}+\nu)[\Gamma(1-\nu)]^{-1}$ $\cdot\exp(\frac{1}{4}y^2/a)\Gamma(\frac{1}{2}-\nu,\frac{1}{4}y^2/a)$ $N=\frac{1}{2}\pi(\pi/a)^{\frac{1}{2}}[\Gamma(1-\nu)\ \cos(\pi\nu)]^{-1},$ $-1<\nu<0$	$\exp(ax^2)\Gamma(\nu,ax^2)$
521	$\pi^{\frac{1}{2}}(2a)^{\nu}[\Gamma(-\frac{1}{2}\nu)]^{-1}y^{-\nu-1}$ $\cdot\exp(\frac{1}{2}y^2/a^2)\Gamma(\frac{1}{2}+\frac{1}{2}\nu,\frac{1}{2}y^2/a^2)$ $N=\pi^{\frac{1}{2}}2^{\frac{1}{2}\nu-\frac{1}{2}}\Gamma(-\frac{1}{2}-\frac{1}{2}\nu)a^{-1}$ $\cdot[\Gamma(-\frac{1}{2}\nu)\Gamma(\frac{1}{2}-\frac{1}{2}\nu)]^{-1},\quad \nu<-1$	$\exp(\frac{1}{4}a^2x^2)D_{\nu}(ax)$

9. Elliptic Integrals and Legendre Functions

	$Ng(y)$	$2Nf(x)$
414	$\pi(4a^2+y^2)^{-\frac{1}{2}}K[y(4a^2+y^2)^{-\frac{1}{2}}]$ $N=\frac{1}{4}\pi^2a^{-1}$	$[K_0(ax)]^2$
415	$\pi[(a+b)^2+y^2]^{-\frac{1}{2}}$ $\cdot K\{[y^2+(a-b)^2]^{\frac{1}{2}}[y^2+(a+b)^2]^{-\frac{1}{2}}]\}$ $N=\pi(a+b)^{-1}K[(a-b)^{\frac{1}{2}}(a+b)^{-\frac{1}{2}}]$	$K_0(ax)K_0(bx)$
409	$[(a+b)^2+y^2]^{-\frac{1}{2}}$ $\cdot K\{2(ab)^{\frac{1}{2}}[(a+b)^2+y^2]^{-\frac{1}{2}}\}$ $N=(a+b)^{-1}K[2(ab)^{\frac{1}{2}}(a+b)^{-1}], \quad a>b$	$I_0(bx)K_0(ax)$
403	$\frac{1}{4}z(2\pi z)^{\frac{1}{2}}\{2E[(\frac{1}{2}-\frac{1}{2}yz)^{\frac{1}{2}}+2E[(\frac{1}{2}+\frac{1}{2}yz)^{\frac{1}{2}}]$ $-K[(\frac{1}{2}-\frac{1}{2}yz)^{\frac{1}{2}}]-K[(\frac{1}{2}+\frac{1}{2}yz)^{\frac{1}{2}}]\}$ $N=(2a)^{-\frac{1}{2}}a^{-1}[\Gamma(\frac{3}{4})]^2, \quad z=(a^2+y^2)^{-\frac{1}{2}}$	$x^{\frac{1}{2}}K_0(ax)$
404	$(\frac{1}{2}\pi z)^{\frac{1}{2}}\{K[(\frac{1}{2}+\frac{1}{2}yz)^{\frac{1}{2}}]+K[(\frac{1}{2}-\frac{1}{2}yz)^{\frac{1}{2}}]\}$ $N=\frac{1}{2}(2a)^{-\frac{1}{2}}[\Gamma(\frac{1}{4})]^2, \quad z=(a^2+y)^{-\frac{1}{2}}$	$x^{-\frac{1}{2}}K_0(ax)$
273	$2b^{-1}\pi^{-2}\operatorname{sech}(\frac{1}{4}y/b)K[\tanh(\frac{1}{4}y/b)]$ $N=(b\pi)^{-1}$	$\lvert\Gamma(\frac{1}{4}+ibx)\rvert^4$
366	$\pi^{-1}\operatorname{sech}(\frac{1}{2}y)K[(1-a^2)^{\frac{1}{2}}\operatorname{sech}(\frac{1}{2}y)]$ $N=\pi^{-1}K[(1-a^2)^{\frac{1}{2}}], \quad a\leq 1$	$\operatorname{sech}(\pi x)[P_{-\frac{1}{2}+ix}(a)]^2$
368	$\pi^{-1}\operatorname{sech}(\frac{1}{2}y)K\{[1-(1-a^2)\operatorname{sech}(\frac{1}{2}y)^2]^{\frac{1}{2}}\}$ $N=\pi^{-1}K(a), \quad a<1$	$[\operatorname{sech}(\pi x)]^2$ $\cdot P_{-\frac{1}{2}+ix}(a)P_{-\frac{1}{2}+ix}(-a)$
369	$\pi^{-1}[a^2+\sinh^2(\frac{1}{2}y)]^{-\frac{1}{2}}$ $\cdot K\{a(1-a^{-2})^{\frac{1}{2}}[a^2+\sinh^2(\frac{1}{2}y)]^{-\frac{1}{2}}\}$ $N=(a\pi)^{-1}K[(1-a^{-2})^{-\frac{1}{2}}], \quad a>1$	$\operatorname{sech}(\pi x)$ $\cdot[\mathfrak{P}_{-\frac{1}{2}+ix}(a)]^2$
360	$2(2\pi)^{\frac{1}{2}}[(1-a^2)^{\frac{1}{2}}+\cosh y]^{-\frac{1}{2}}$ $\cdot K\{[\frac{1}{2}+\frac{1}{2}(1-a^2)^{-\frac{1}{2}}\cosh y]^{-\frac{1}{2}}\}$ $N=2(2\pi)^{\frac{1}{2}}[1+(1-a^2)^{\frac{1}{2}}]^{-\frac{1}{2}}$ $\cdot K\{[\frac{1}{2}+\frac{1}{2}(1-a^2)^{-\frac{1}{2}}]^{-\frac{1}{2}}\}, \quad a<1$	$\lvert\Gamma(\frac{1}{4}+\frac{1}{2}ix)\rvert^2P_{-\frac{1}{2}+ix}(a)$
493	$(2\pi)^{\frac{1}{2}}(1-y^2)^{\frac{1}{2}\nu+\frac{1}{4}}P_{\nu-\frac{1}{2}}^{-\nu-\frac{1}{2}}(y), \quad y<1$ $0, \quad y>1$ $N=2^{-\nu-1}\pi[\Gamma(1+\nu)]^{-1}, \quad \nu\geq\frac{1}{2}$	$x^{-\nu-1}\mathbf{H}_\nu(x)$

	$Ng(y)$	$2Nf(x)$
510	$\frac{1}{4}(\frac{1}{2}\pi/a)^{\frac{1}{2}}2^{-\mu}\Gamma(-\frac{1}{2}\nu-\frac{1}{2}\mu)\Gamma(\frac{1}{2}\nu-\frac{1}{2}\mu)$ $\cdot\begin{cases}(y^2-a^2)^{\frac{1}{2}\mu+\frac{1}{4}}\mathfrak{P}^{\mu+\frac{1}{2}}_{\nu-\frac{1}{2}}(y/a), & y>a\\(a^2-y^2)^{\frac{1}{2}\mu+\frac{1}{4}}P^{\mu+\frac{1}{2}}_{\nu-\frac{1}{2}}(y/a), & y<a\end{cases}$ $N=\frac{1}{4}\pi a^{\mu}\Gamma(-\frac{1}{2}\mu-\frac{1}{2}\nu)\Gamma(-\frac{1}{2}\mu+\frac{1}{2}\nu)$ $\cdot[\Gamma(\frac{1}{2}-\frac{1}{2}\mu+\frac{1}{2}\nu)\Gamma(\frac{1}{2}-\frac{1}{2}\mu-\frac{1}{2}\nu)]^{-1},$ $\mu\pm\nu<0$	$x^{-\mu-1}S_{\mu,\nu}(ax)$
416	$\frac{1}{4}\pi^2\sec(\pi\nu)(ab)^{-\frac{1}{2}}$ $\cdot\mathfrak{P}_{\nu-\frac{1}{2}}[(2ab)^{-1}(y^2+a^2+b^2)]$ $N=\frac{1}{4}\pi^2(ab)^{-\frac{1}{2}}\sec(\pi\nu)\mathfrak{P}_{\nu-\frac{1}{2}}(\frac{1}{2}a/b+\frac{1}{2}b/a),$ $-\frac{1}{2}<\nu<\frac{1}{2}$	$K_{\nu}(ax)K_{\nu}(bx)$
410	$\frac{1}{2}(ab)^{-\frac{1}{2}}\mathfrak{Q}_{\nu-\frac{1}{2}}[(2ab)^{-1}(a^2+b^2+y^2)]$ $N=\frac{1}{2}(ab)^{-\frac{1}{2}}\mathfrak{Q}_{\nu-\frac{1}{2}}(\frac{1}{2}a/b+\frac{1}{2}b/a),\quad a>b,$ $\nu>-\frac{1}{2}$	$I_{\nu}(bx)K_{\nu}(ax)$
494	$(\frac{1}{2}\pi)^{-\frac{1}{2}}\cos(\pi\nu)a^{\nu}\Gamma(1+2\nu)y^{-\nu-\frac{1}{2}}$ $\cdot\begin{cases}(a^2-y^2)^{-\frac{1}{2}\nu-\frac{1}{4}}\mathfrak{P}^{-\nu-\frac{1}{2}}_{-\nu-\frac{1}{2}}(a/y), & y<a\\(y^2-a^2)^{-\frac{1}{2}\nu-\frac{1}{4}}P^{-\nu-\frac{1}{2}}_{-\nu-\frac{1}{2}}(a/y), & y>a\end{cases}$ $N=-2^{\nu}\pi^{\frac{1}{2}}\csc(\pi\nu)a^{-\nu-1}[\Gamma(\frac{1}{2}-\nu)]^{-1},$ $-\frac{1}{2}<\nu<0$	$x^{\nu}[\mathbf{H}_{\nu}(ax)-Y_{\nu}(ax)]$
373	$2^{3\nu-\frac{1}{2}}\pi^{-\frac{1}{2}}a^{\nu-\frac{1}{2}}y^{-\nu-\frac{1}{2}}[\Gamma(\frac{1}{2}-\nu)]^{-1}(4a^2-y^2)^{-\nu}$ $\cdot[\pi\mathfrak{P}^{\nu}_{\nu-\frac{1}{2}}(a/y+\frac{1}{4}y/a)$ $-2e^{-i\pi\nu}\mathfrak{Q}^{\nu}_{\nu-\frac{1}{2}}(a/y+\frac{1}{4}y/a)],\quad y<2a$ $-2^{3\nu+\frac{1}{2}}\pi^{-\frac{1}{2}}\sin(\pi\nu)a^{\nu-\frac{1}{2}}y^{-\nu-\frac{1}{2}}[\Gamma(\frac{1}{2}-\nu)]^{-1}$ $\cdot(y^2-4a^2)^{-\nu}e^{-i\pi\nu}\mathfrak{Q}^{\nu}_{\nu-\frac{1}{2}}(a/y+\frac{1}{4}y/a),$ $y>2a$ $N=\frac{1}{2}[\pi\Gamma(\frac{1}{4}-\nu)]^{-1}\Gamma(-\nu)\Gamma(\frac{1}{2}+2\nu),$ $-\frac{1}{4}<\nu<0$	$[x^{\nu}J_{\nu}(ax)]^2$
412	$2^{-3\nu-1}\pi^{\frac{1}{2}}\Gamma(\frac{1}{2}-\nu)(y/a)^{\nu}$ $\cdot(y^2+4a^2)^{\nu-\frac{1}{2}}P^{-\nu}_{\nu-\frac{1}{2}}[(y^2-4a^2)(y^2+4a^2)^{-1}]$ $N=\frac{1}{4}a^{2\nu-1}\Gamma(\nu)\Gamma(\frac{1}{2}-\nu)[\Gamma(\frac{1}{2}+2\nu)]^{-1},$ $0<\nu<\frac{1}{2}$	$x^{-2\nu}I_{\nu}(ax)K_{\nu}(ax)$
407	$\frac{1}{4}\pi\sec[\frac{1}{2}\pi(\mu-\nu)]\Gamma(1+\mu+\nu)$ $\cdot(a^2+y^2)^{-\frac{1}{2}\mu-\frac{1}{2}}\{P^{-\nu}_{\mu}[y(a^2+y^2)^{-\frac{1}{2}}]$ $+P^{-\nu}_{\mu}[-y(a^2+y^2)^{-\frac{1}{2}}]\}$ $N=a^{-\mu-1}2^{\mu-1}\Gamma(\frac{1}{2}+\frac{1}{2}\mu+\frac{1}{2}\nu)\Gamma(\frac{1}{2}+\frac{1}{2}\mu-\frac{1}{2}\nu),$ $\mu\pm\nu>-1$	$x^{\mu}K_{\nu}(ax)$

	$Ng(y)$	$2Nf(x)$
402	$(-1)^n(2n+1)!(a^2+y^2)^{-n-1}Q_{2n}[y(a^2+y^2)^{-\frac{1}{2}}]$ $N=a^{-2n-2}2^{2n}(n!)^2, \quad n=0, 1, 2, \ldots$	$x^{2n+1}K_0(ax)$
418	$\Gamma(\frac{1}{4}+\nu)[\Gamma(\frac{1}{4}-\nu)]^{-1}e^{i2\pi\nu}(\frac{1}{2}\pi/y)^{\frac{1}{2}}$ $\cdot\{\mathfrak{Q}_{-\frac{3}{4}}^{-\nu}[(1+4a^2y^{-2})^{\frac{1}{2}}]\}^2$ $N=\frac{1}{4}(2a\pi)^{-\frac{1}{2}}\Gamma^2(\frac{1}{4})\Gamma(\frac{1}{4}+\nu)\Gamma(\frac{1}{4}-\nu),$ $-\frac{1}{4}<\nu<\frac{1}{4}$	$x^{-\frac{1}{2}}[K_\nu(ax)]^2$
411	$\Gamma(\frac{1}{4}+\nu)[\Gamma(\frac{1}{4}-\nu)]^{-1}(\frac{1}{2}\pi/y)^{\frac{1}{2}}e^{i\pi\nu}$ $\cdot\mathfrak{Q}_{-\frac{3}{4}}^{-\nu}[(1+4a^2/y^2)^{\frac{1}{2}}]\mathfrak{P}_{-\frac{3}{4}}^{-\nu}[(1+4a^2/y^2)^{\frac{1}{2}}]$ $N=\frac{1}{2}(\frac{1}{2}\pi)^{\frac{1}{2}}\Gamma(\frac{1}{4})\Gamma(\frac{1}{4}+\nu)[\Gamma(\frac{3}{4})\Gamma(\frac{3}{4}+\nu)]^{-1},$ $\nu>-\frac{1}{4}$	$x^{-\frac{1}{2}}I_\nu(ax)K_\nu(ax)$
417	$\Gamma(\frac{3}{4}+\nu)[\Gamma(-\frac{1}{4}-\nu)]^{-1}e^{i2\pi\nu}(4a^2+y^2)^{-\frac{1}{2}}$ $\cdot(\frac{1}{2}\pi y)^{\frac{1}{2}}\mathfrak{Q}^{\nu}_{-\frac{1}{4}}[(1+4a^2/y^2)^{\frac{1}{2}}]$ $\cdot\mathfrak{Q}_{-\frac{5}{4}}^{-\nu}[(1+4a^2/y^2)^{\frac{1}{2}}]$ $N=a^{-1}(2\pi a)^{-\frac{1}{2}}\Gamma^2(\frac{3}{4})\Gamma(\frac{3}{4}+\nu)\Gamma(\frac{3}{4}-\nu),$ $-\frac{3}{4}<\nu<\frac{3}{4}$	$x^{\frac{1}{2}}[K_\nu(ax)]^2$
272	$[b\pi\sin(2a\pi)]^{-1}P_{2a-1}[\cosh(\frac{1}{2}y/b)]$ $N=[b\pi\sin(2a\pi)]^{-1}, \quad 0<a<\frac{1}{2}$	$\lvert\Gamma(a+ibx)\Gamma(\frac{1}{2}-a+ibx)\rvert^2$
274	$2^{\frac{1}{2}-b-c}a^{-1}\pi^{\frac{3}{2}}\Gamma(2b)\Gamma(2c)\Gamma(b+c)$ $\cdot[\sinh(\frac{1}{2}y/a)]^{\frac{1}{2}-b-c}\mathfrak{P}_{b-c-\frac{1}{2}}^{\frac{1}{2}-b-c}[\cosh(\frac{1}{2}y/a)]$ $N=2^{1-2b-2c}\pi^{\frac{3}{2}}a^{-1}\Gamma(2b)\Gamma(2c)\Gamma(b+c)$ $\cdot[\Gamma(\frac{1}{2}+b+c)]^{-1}$	$\lvert\Gamma(b+iax)\Gamma(c+iax)\rvert^2$
361	$2^{-\mu-\frac{1}{2}}(a+1)^{\frac{1}{2}\mu}(a+\cosh y)^{-\frac{1}{2}-\frac{1}{2}\mu}$ $\cdot\mathfrak{P}_\mu^\mu[(1+\cosh y)^{\frac{1}{2}}(a+\cosh y)^{-\frac{1}{2}}]$ $N=2^{-\mu-\frac{1}{2}}(1+a)^{-\frac{1}{2}}\mathfrak{P}_\mu^\mu[(\frac{1}{2}+\frac{1}{2}a)^{-\frac{1}{2}}],$ $-1<a<1, \quad \mu<1$	$\operatorname{sech}x\, P^\mu_{-\frac{1}{2}+ix}(a)$
370	$(a^2-1)^{-\frac{1}{2}}\mathfrak{Q}_{-\mu-\frac{1}{2}}[1+2(a^2-1)^{-1}\cosh^2(\frac{1}{2}y)]$ $N=(a^2-1)^{-\frac{1}{2}}\mathfrak{Q}_{-\mu-\frac{1}{2}}[1+2(a^2-1)^{-1}],$ $a>1, \quad \mu<\frac{1}{2}$	$\lvert\Gamma(\frac{1}{2}-\mu+ix)\rvert^2$ $\cdot[\mathfrak{P}^\mu_{-\frac{1}{2}+ix}(a)]^2$
367	$(1-a^2)^{-\frac{1}{2}}\mathfrak{Q}_{-\mu-\frac{1}{2}}[2(1-a^2)^{-1}\cosh^2(\frac{1}{2}y)-1]$ $N=(1-a^2)^{-\frac{1}{2}}\mathfrak{Q}_{-\mu-\frac{1}{2}}[(1+a^2)(1-a^2)^{-1}],$ $a\leq 1, \quad \mu<\frac{1}{2}$	$\lvert\Gamma(\frac{1}{2}-\mu+ix)\rvert^2$ $\cdot[P^\mu_{-\frac{1}{2}+ix}(a)]^2$

	$Ng(y)$	$2Nf(x)$
363	$\pi^{\frac{1}{2}}2^{\mu+1}(1-a^2)^{-\frac{1}{4}}\mathfrak{Q}_{-\mu-\frac{1}{2}}[(1-a^2)^{-\frac{1}{2}}\cosh y]$ $N=\pi^{\frac{1}{2}}2^{\mu+1}(1-a^2)^{-\frac{1}{4}}\mathfrak{Q}_{-\mu-\frac{1}{2}}[(1-a^2)^{-\frac{1}{2}}]$, $a<1, \quad \mu<\frac{1}{2}$	$\lvert\Gamma(\frac{1}{4}-\frac{1}{2}\mu+\frac{1}{2}ix)\rvert^2$ $\cdot P^{\mu}_{-\frac{1}{2}+ix}(a)$
275	$\frac{1}{2}(2\pi)^{\frac{1}{2}}2^{b-a}c^{-1}[\Gamma(b-a)]^{-1}\sinh(\frac{1}{2}y/c)$ $\cdot e^{-i\pi(a-b+\frac{1}{2})}\mathfrak{Q}^{a-b+\frac{1}{2}}_{a+b-\frac{3}{2}}[\cosh(\frac{1}{2}y/c)]$ $N=2^{2b-2a-2}c^{-1}\Gamma(2a)[\Gamma(2b-1)]^{-1}$ $\cdot B(b-a-\frac{1}{2},\frac{1}{2}), \quad b-a>\frac{1}{2}$	$\lvert\Gamma(a+icx)\Gamma(b+icx)\rvert^{-2}$
365	$2^{-\mu+1}(1-a^2)^{\frac{1}{2}\mu-\frac{1}{4}}\Gamma(2\mu)(\cosh y-a)^{-\frac{1}{2}\mu-\frac{1}{4}}$ $\cdot e^{-i\pi(\frac{1}{2}-\mu)}\mathfrak{Q}^{\frac{1}{2}-\mu}_{\mu-\frac{1}{2}}[\cosh(\frac{1}{2}y)(\frac{1}{2}\cosh y-\frac{1}{2}a)^{-\frac{1}{2}}]$ $N=2^{1-\mu}(1+a)^{\frac{1}{2}\mu-\frac{1}{4}}(1-a)^{-\frac{1}{2}}\Gamma(2\mu)$ $\cdot e^{-i\pi(\frac{1}{2}-\mu)}\mathfrak{Q}^{\frac{1}{2}-\mu}_{\mu-\frac{1}{2}}[(\frac{1}{2}-\frac{1}{2}a)^{-\frac{1}{2}}]$, $-1<a<1, \quad \mu>0$	$\operatorname{sech}(\pi x)\lvert\Gamma(\mu+ix)\rvert^2$ $\cdot P^{\frac{1}{2}-\mu}_{-\frac{1}{2}+\frac{1}{2}ix}(a)$
135	$2^{-\frac{1}{2}}\pi P_{-\frac{1}{2}+y}(\cos\delta)$ $N=2^{\frac{1}{2}}K(\sin\frac{1}{2}\delta), \quad \delta<\pi$	$(\cos x-\cos\delta)^{-\frac{1}{2}}, \quad x<\delta$ $0, \quad x>\delta$
136	$(\frac{1}{2}\pi)^{\frac{1}{2}}(\sin\delta)^{\nu}\Gamma(\nu+\frac{1}{2})P^{-\nu}_{y-\frac{1}{2}}(\cos\delta)$ $N=(\frac{1}{2}\pi)^{\frac{1}{2}}(\sin\delta)^{\nu}\Gamma(\nu+\frac{1}{2})P^{-\nu}_{-\frac{1}{2}}(\cos\delta)$, $\delta<\pi, \quad \nu>-\frac{1}{2}$	$(\cos x-\cos\delta)^{\nu-\frac{1}{2}}, \quad x<\delta$ $0, \quad x>\delta$
160	$\pi 2^{-\frac{1}{2}}\{P_{-\frac{1}{2}+y}(\cos\delta)[-\gamma-\log 4$ $-\psi(\frac{1}{2}+y)+\log(\sin\delta)]-Q_{-\frac{1}{2}-y}(\cos\delta)\}$ $N=\pi 2^{-\frac{1}{2}}[2\pi^{-1}K(\sin\frac{1}{2}\delta)\log(\sin\delta)$ $-K(\cos\frac{1}{2}\delta)], \quad \delta<\pi$	$(\cos x-\cos\delta)^{-\frac{1}{2}}$ $\cdot\log(\cos x-\cos\delta), \quad x<\delta$ $0, \quad x>\delta$
323	$\frac{1}{2}a^{-1}\cos(\frac{1}{2}\pi y)\mathfrak{Q}_{-\frac{1}{2}+\frac{1}{2}y}(p)\mathfrak{Q}_{-\frac{1}{2}-\frac{1}{2}y}(p)$ $N=2^{-\frac{1}{2}}a^{-1}(p+1)^{-\frac{1}{2}}K[(\frac{1}{2}p+\frac{1}{2})^{-\frac{1}{2}}]$ $p=(1+a^{-2})^{\frac{1}{2}}$	$(1+a^2\cos^2 x)^{-\frac{1}{2}}$ $\cdot K[a\cos x(1+a^2\cos^2 x)^{-\frac{1}{2}}], \quad x<\frac{1}{2}\pi$ $0, \quad x>\frac{1}{2}\pi$
340	$\frac{1}{2}\pi\mathfrak{P}^{\frac{1}{2}y}_{\nu}[(1+a^{-2})^{\frac{1}{2}}]\mathfrak{P}^{-\frac{1}{2}y}_{\nu}[(1+a^{-2})^{\frac{1}{2}}]$ $N=\frac{1}{2}\pi\{\mathfrak{P}_{\nu}[(1+a^{-2})^{\frac{1}{2}}]\}^2$	$\mathfrak{P}_{\nu}(1+2a^{-2}\cos^2 x), \quad x<\frac{1}{2}\pi$ $0, \quad x>\frac{1}{2}\pi$
206	$2^{-\frac{1}{2}}\pi P_{-\frac{1}{2}+iy}(\cos b)\operatorname{sech}(\pi y)$ $N=2^{\frac{1}{2}}K(\sin\frac{1}{2}b), \quad b<\pi$	$(\cosh x+\cos b)^{-\frac{1}{2}}$
249	$2^{-\frac{1}{2}}\pi^2 P_{-\frac{1}{2}+iy}(a)[\operatorname{sech}(\pi y)]^2$ $N=2^{\frac{1}{2}}\pi K[(\frac{1}{2}-\frac{1}{2}a)^{\frac{1}{2}}], \quad a<1$	$(\cosh x-a)^{-\frac{1}{2}}$ $\cdot\log\left[\frac{(\cosh x+1)^{\frac{1}{2}}+(\cosh x-a)^{\frac{1}{2}}}{(\cosh x+1)^{\frac{1}{2}}-(\cosh x-a)^{\frac{1}{2}}}\right]$

	$Ng(y)$	$2Nf(x)$
329	$\frac{1}{4}(\frac{1}{2}\pi/a)^{\frac{1}{2}}\Gamma(\frac{1}{4}+\frac{1}{2}iy)\Gamma(\frac{1}{4}-\frac{1}{2}iy)$ $\cdot P_{-\frac{1}{2}+iy}[(1-a^{-2})^{\frac{1}{2}}]$ $N=\frac{1}{2}(2\pi a)^{-\frac{1}{2}}[\Gamma(\frac{1}{4})]^2 K\{[\frac{1}{2}-\frac{1}{2}(1-a^{-2})^{\frac{1}{2}}]^{\frac{1}{2}}\}$	$(1+a\cosh x)^{-\frac{1}{2}}$ $\cdot K\{[\frac{1}{2}+\frac{1}{2}a\cosh x]^{-\frac{1}{2}}\}$
214	$(\frac{1}{2}\pi)^{\frac{1}{2}}[\Gamma(\nu)]^{-1}(\sin b)^{\frac{1}{2}-\nu}\Gamma(\nu+iy)\Gamma(\nu-iy)$ $\cdot P^{\frac{1}{2}-\nu}_{-\frac{1}{2}+iy}(\cos b)$ $N=(\frac{1}{2}\pi)^{\frac{1}{2}}\Gamma(\nu)(\sin b)^{\frac{1}{2}-\nu}P^{\frac{1}{2}-\nu}_{-\frac{1}{2}}(\cos b),$ $b<\pi,\quad \nu>0$	$(\cos b+\cosh x)^{-\nu}$
349	$2^{\nu-\frac{3}{2}}(\pi/a)^{\frac{1}{2}}\Gamma(\frac{1}{2}+\frac{1}{2}\nu+\frac{1}{2}iy)\Gamma(\frac{1}{2}+\frac{1}{2}\nu-\frac{1}{2}iy)$ $\cdot P^{-\nu-\frac{1}{2}}_{-\frac{1}{2}+iy}[(1-a^{-2})^{\frac{1}{2}}]$ $N=2^{\nu-\frac{3}{2}}(\pi/a)^{\frac{1}{2}}[\Gamma(\frac{1}{2}+\frac{1}{2}\nu)]^2$ $\cdot P^{-\nu-\frac{1}{2}}_{-\frac{1}{2}}[(1-a^{-2})],\qquad a\geq 1,\quad \nu>-1$	$\mathfrak{Q}_\nu(a\cosh x)$
332	$2^{-7/2}(\pi/a)^{\frac{1}{2}}\operatorname{sech}(\pi y)\Gamma(\frac{1}{4}+\frac{1}{2}iy)\Gamma(\frac{1}{4}-\frac{1}{2}iy)$ $\cdot[P_{-\frac{1}{2}+iy}(s)+P_{-\frac{1}{2}+iy}(-s)]$ $N=2^{-\frac{5}{2}}(a\pi)^{-\frac{1}{2}}[\Gamma(\frac{1}{4})]^2\{K[2^{-\frac{1}{2}}(1-s)]$ $+K[2^{-\frac{1}{2}}(1+s)]\}$ $s=(1-a^{-2})^{\frac{1}{2}},\quad a>1$	$(1+a\cosh x)^{-\frac{1}{2}}$ $\cdot K\left[\left(\frac{a\cosh x-1}{a\cosh x+1}\right)^{\frac{1}{2}}\right]$
270	$(\frac{1}{2}\pi/c)^{\frac{1}{2}}[\Gamma(v)]^{-1}\cosh(\frac{1}{2}\pi y)$ $\cdot\lvert\Gamma(\nu+iy)\rvert^2(1-c^2)^{\frac{1}{4}-\frac{1}{2}\nu}P^{\frac{1}{2}-\nu}_{-\frac{1}{2}+iy}(c^{-1})$ $N=(\frac{1}{2}\pi c)^{\frac{1}{2}}\Gamma(\nu)(1-c^2)^{\frac{1}{4}-\frac{1}{2}\nu}P^{\frac{1}{2}-\nu}_{-\frac{1}{2}}(c^{-1}),\quad \nu\leq 1$	$(1+c^2\sinh^2 x)^{-\frac{1}{2}\nu}$ $\cdot\cosh[\nu\arctan(c\sinh x)]$
342	$-2^{\nu-1}(2a\pi)^{-\frac{1}{2}}\sin(\pi\nu)[\cosh(\pi y)-\cos(\pi\nu)]^{-1}$ $\cdot\Gamma(\frac{1}{2}+\frac{1}{2}\nu+\frac{1}{2}iy)\Gamma(\frac{1}{2}+\frac{1}{2}\nu-\frac{1}{2}iy)$ $\cdot\{P^{-\nu-\frac{1}{2}}_{-\frac{1}{2}+iy}[(1-a^{-2})^{\frac{1}{2}}]$ $+P^{-\nu-\frac{1}{2}}_{-\frac{1}{2}+iy}[-(1-a^{-2})^{\frac{1}{2}}]\}$ $N=-2^{\nu-1}(2a\pi)^{-\frac{1}{2}}\cot(\frac{1}{2}\pi\nu)[\Gamma(\frac{1}{2}+\frac{1}{2}\nu)]^2$ $\cdot\{P^{-\nu-\frac{1}{2}}_{-\frac{1}{2}}[(1-a^{-2})^{\frac{1}{2}}]$ $+P^{-\nu-\frac{1}{2}}_{-\frac{1}{2}}[-(1-a^{-2})^{\frac{1}{2}}]\},$ $a\geq 1,\quad -1<\nu<0$	$\mathfrak{P}_\nu(a\cosh x)$
209	$2^{-\frac{1}{2}}\pi\mathfrak{P}_{-\frac{1}{2}+iy}(\cosh a)$ $N=2^{\frac{1}{2}}\operatorname{sech}(\frac{1}{2}a)K[\tanh(\frac{1}{2}a)]$	$(\cosh a-\cosh x)^{-\frac{1}{2}},\quad x<a$ $0,\qquad x>a$
207	$2^{-\frac{1}{2}}\pi\mathfrak{P}_{-\frac{1}{2}+iy}(\cosh b)\operatorname{sech}(\pi y)$ $N=2^{\frac{1}{2}}\operatorname{sech}(\frac{1}{2}b)K[\tanh(\frac{1}{2}b)]$	$(\cosh x+\cosh b)^{-\frac{1}{2}}$
216	$(\frac{1}{2}\pi)^{\frac{1}{2}}\Gamma(1-\nu)(\sinh a)^{\frac{1}{2}-\nu}\mathfrak{P}^{\nu-\frac{1}{2}}_{-\frac{1}{2}+iy}(\cosh a)$ $N=(\sinh a)^{-\nu}\mathfrak{Q}_{-\nu}(\operatorname{ctnh} a),\quad \nu<1$	$(\cosh a-\cosh x)^{-\nu},\quad x<a$ $0,\qquad x>a$

	$Ng(y)$	$2Nf(x)$
213	$(\frac{1}{2}\pi)^{\frac{1}{2}}[\Gamma(\nu)]^{-1}(\sinh a)^{\frac{1}{2}-\nu}$ $\cdot\Gamma(\nu+iy)\Gamma(\nu-iy)\mathfrak{P}^{\frac{1}{2}-\nu}_{-\frac{1}{2}+iy}(\cosh a)$ $N=(\sinh a)^{-\nu}\mathfrak{Q}_{\nu-1}(\operatorname{ctnh} a),\quad \nu>0$	$(\cosh a+\cosh x)^{-\nu}$
208	$2^{-\frac{1}{2}}[Q_{-\frac{1}{2}+iy}(\cos b)+Q_{-\frac{1}{2}-iy}(\cos b)]$ $N=2^{\frac{1}{2}}K(\cos\frac{1}{2}b),\quad b<\pi$	$(\cosh x-\cos b)^{-\frac{1}{2}}$
210	$2^{-\frac{1}{2}}[\mathfrak{Q}_{\frac{1}{2}+iy}(\cosh a)+\mathfrak{Q}_{\frac{1}{2}-iy}(\cosh a)]$ $N=2^{\frac{1}{2}}\operatorname{sech}(\frac{1}{2}a)K[\operatorname{sech}(\frac{1}{2}a)]$	$0,\quad x<a$ $(\cosh x-\cosh a)^{-\frac{1}{2}},\quad x>a$
215	$(2\pi)^{-\frac{1}{2}}\Gamma(1-\nu)(\sinh a)^{\frac{1}{2}-\nu}e^{-i\pi(\nu-\frac{1}{2})}$ $\cdot[\mathfrak{Q}^{\nu-\frac{1}{2}}_{-\frac{1}{2}-iy}(\cosh a)+\mathfrak{Q}^{\nu-\frac{1}{2}}_{-\frac{1}{2}+iy}(\cosh a)]$ $N=(\frac{1}{2}\pi)^{-\frac{1}{2}}\Gamma(1-\nu)(\sinh a)^{\frac{1}{2}-\nu}$ $\cdot e^{-i\pi(\nu-\frac{1}{2})}\mathfrak{Q}^{\nu-\frac{1}{2}}_{-\frac{1}{2}}(\cosh a),\quad \nu<1$	$(\cosh x-\cosh a)^{-\nu},\quad x>a$ $0,\quad x<a$
348	$(a\pi)^{-1}\operatorname{Re}[\mathfrak{Q}^{-\nu-\frac{1}{2}}_{-\frac{1}{2}+\frac{1}{2}iy}(p)\mathfrak{Q}^{\nu+\frac{1}{2}}_{-\frac{1}{2}+\frac{1}{2}iy}(p)]$ $N=-\frac{1}{2}\pi\csc(\pi\nu)[\mathfrak{P}_\nu(ap)]^2,\quad -1<\nu<0$ $p=(1+a^{-2})^{\frac{1}{2}}$	$0,\quad \sinh x<a^{-1}$ $\mathfrak{P}_\nu(2a^2\sinh^2 x-1),\quad \sinh x>a^{-1}$
244	$2^{-\frac{1}{2}}\pi\operatorname{sech}(\pi y)\{P_{-\frac{1}{2}+iy}(\cos\delta)[-\gamma-\log 4$ $+\log(\sin\delta)-\frac{1}{2}\psi(\frac{1}{2}+iy)-\frac{1}{2}\psi(\frac{1}{2}-iy)]$ $+\frac{1}{2}Q_{-\frac{1}{2}+iy}(\cos\delta)+\frac{1}{2}Q_{-\frac{1}{2}-iy}(\cos\delta)\}$ $N=2^{-\frac{1}{2}}\pi[\log(\sin\delta)2\pi^{-1}K(\sin\frac{1}{2}\delta)+K(\cos\frac{1}{2}\delta)],$ $\delta<\pi$	$(\cosh x+\cos\delta)^{-\frac{1}{2}}$ $\cdot\log(\cosh x+\cos\delta)$
245	$2^{-\frac{1}{2}}\pi\operatorname{sech}(\pi y)\{\mathfrak{P}_{-\frac{1}{2}+iy}(\cosh a)[\log(\sinh a)$ $-\gamma-\log 4-\frac{1}{2}\psi(\frac{1}{2}+iy)-\frac{1}{2}\psi(\frac{1}{2}-iy)]$ $+\frac{1}{2}\mathfrak{Q}_{-\frac{1}{2}+iy}(\cosh a)+\frac{1}{2}\mathfrak{Q}_{-\frac{1}{2}-iy}(\cosh a)\}$ $N=2^{-\frac{1}{2}}\pi\{\operatorname{sech}(\frac{1}{2}a)K[\operatorname{sech}(\frac{1}{2}a)]$ $+(2/\pi)\log(\sinh a)\operatorname{sech}(\frac{1}{2}a)$ $\cdot K[\tanh(\frac{1}{2}a)]\}$	$(\cosh x+\cosh a)^{-\frac{1}{2}}$ $\cdot\log(\cosh x+\cosh a)$
246	$-2^{-\frac{1}{2}}\pi\operatorname{sech}(\pi y)\{\mathfrak{P}_{-\frac{1}{2}+iy}(\cosh a)[\log(\sinh a)$ $-\gamma-\log 4-\frac{1}{2}\psi(\frac{1}{2}+iy)-\frac{1}{2}\psi(\frac{1}{2}-iy)]$ $-\frac{1}{2}\mathfrak{Q}_{-\frac{1}{2}+iy}(\cosh a)-\frac{1}{2}\mathfrak{Q}_{-\frac{1}{2}-iy}(\cosh a)\}$ $N=2^{-\frac{1}{2}}\pi\{2\pi^{-1}\log(\sinh a)\operatorname{sech}(\frac{1}{2}a)$ $\cdot K[\tanh(\frac{1}{2}a)]-\operatorname{sech}(\frac{1}{2}a)K[\operatorname{sech}(\frac{1}{2}a)]\},$ $\cosh a\leq 2$	$-(\cosh a-\cosh x)^{-\frac{1}{2}}$ $\cdot\log(\cosh a-\cosh x),\quad x<a$ $0,\quad x>a$
326	$\frac{1}{4}\pi^2\operatorname{sech}(\frac{1}{2}\pi y)\{P_{-\frac{1}{2}+\frac{1}{2}iy}[(1-a^2)^{\frac{1}{2}}]\}^2$ $N=\{K[2^{-\frac{1}{2}}(1-(1-a^2)^{\frac{1}{2}})^{\frac{1}{2}}]\}^2,\quad a\leq 1$	$\operatorname{sech} x\, K(a\operatorname{sech} x)$

	$Ng(y)$	$2Nf(x)$
356	$\frac{1}{4}a^{-1}\pi\,\vert\Gamma(1+\nu+\frac{1}{2}iy)\vert^2\{P^{-\nu-\frac{1}{2}}_{-\frac{1}{2}+\frac{1}{2}iy}[(1-a^{-2})^{\frac{1}{2}}]\}^2$ $N=\frac{1}{4}a^{-1}\pi[\Gamma(1+\nu)]^2$ $\cdot\{P^{-\nu-\frac{1}{2}}_{-\frac{1}{2}}[(1-a^{-2})^{\frac{1}{2}}]\}^2, \quad a>1, \quad \nu>-1$	$\mathfrak{Q}_\nu(2a^2\cosh^2x-1)$
330	$\frac{1}{4}\pi^2a^{-1}\operatorname{sech}(\frac{1}{2}\pi y)$ $\cdot P_{-\frac{1}{2}+\frac{1}{2}iy}[(1-a^{-2})^{\frac{1}{2}}]P_{-\frac{1}{2}+\frac{1}{2}iy}[-(1-a^{-2})^{\frac{1}{2}}]$ $N=a^{-1}K\{[\frac{1}{2}-\frac{1}{2}(1-a^{-2})^{\frac{1}{2}}]^{\frac{1}{2}}\}$ $\cdot K\{[\frac{1}{2}+\frac{1}{2}(1-a^{-2})^{\frac{1}{2}}]^{\frac{1}{2}}\}, \quad a>1$	$(1+a^2\sinh^2x)^{-\frac{1}{2}}$ $\cdot K[(1+a^2\sinh^2x)^{-\frac{1}{2}}]$
327	$\frac{1}{4}\pi^2(\operatorname{sech}\frac{1}{2}\pi y)^2P_{-\frac{1}{2}+\frac{1}{2}iy}[(1-a^2)^{\frac{1}{2}}]$ $\cdot P_{-\frac{1}{2}+\frac{1}{2}iy}[-(1-a^2)^{\frac{1}{2}}]$ $N=K\{2^{-\frac{1}{2}}[1-(1-a^2)^{\frac{1}{2}}]^{\frac{1}{2}}\}$ $\cdot K\{2^{-\frac{1}{2}}[1+(1-a^2)^{\frac{1}{2}}]^{\frac{1}{2}}\}, \quad a\leq 1$	$\operatorname{sech}x\, K[(1-a^2\operatorname{sech}^2x)^{\frac{1}{2}}]$
355	$\frac{1}{4}a^{-1}\pi\,\vert\Gamma(1+\nu+\frac{1}{2}iy)\vert^2\{\mathfrak{P}^{-\nu-\frac{1}{2}}_{-\frac{1}{2}+\frac{1}{2}iy}[(1+a^{-2})^{\frac{1}{2}}]\}^2$ $N=\frac{1}{2}\{\mathfrak{Q}_\nu[a(1+a^{-2})^{\frac{1}{2}}]\}^2, \quad \nu>-1$	$\mathfrak{Q}_\nu(1+2a^2\cosh^2x)$
346	$\frac{1}{4}(p^2-1)^{-\frac{1}{2}}\tan(\pi\nu)$ $\cdot\{[\vert\Gamma(1+\nu+\frac{1}{2}iy)\vert\,\mathfrak{P}^{-\nu-\frac{1}{2}}_{-\frac{1}{2}+\frac{1}{2}iy}(p)]^2$ $-[\vert\Gamma(-\nu+\frac{1}{2}iy)\vert\,\mathfrak{P}^{\nu+\frac{1}{2}}_{-\frac{1}{2}+\frac{1}{2}iy}(p)]^2\}$ $N=\frac{1}{2}\pi^{-1}\tan(\pi\nu)(p^2-1)^{\frac{1}{2}}$ $\cdot\{[\mathfrak{Q}_\nu(p(p^2-1)^{-\frac{1}{2}})]^2$ $-[\mathfrak{Q}_{-\nu-1}(p(p^2-1)^{-\frac{1}{2}})]^2\}, \quad -1<\nu<0$ $p>1$	$\mathfrak{P}_\nu\{1+[2/(p^2-1)]\cosh^2x\}$
331	$\frac{1}{4}a^{-1}\pi\operatorname{sech}(\frac{1}{2}\pi y)\mathfrak{P}_{-\frac{1}{2}+\frac{1}{2}iy}(p)$ $\cdot[\mathfrak{Q}_{-\frac{1}{2}+\frac{1}{2}iy}(p)+\mathfrak{Q}_{-\frac{1}{2}-\frac{1}{2}iy}(p)]$ $N=2a^{-1}(p+1)^{-1}K[(p-1)^{\frac{1}{2}}(p+1)^{-\frac{1}{2}}]$ $\cdot K[2^{\frac{1}{2}}(p+1)^{-\frac{1}{2}}], \quad p=(1+a^{-2})^{\frac{1}{2}}$	$(1+a^2\cosh^2x)^{-\frac{1}{2}}$ $\cdot K[a\cosh x(1+a^2\cosh^2x)^{-\frac{1}{2}}]$
354	$-\frac{1}{2}\pi[a\sin(\pi\nu)]^{-1}P^{-\nu-\frac{1}{2}}_{-\frac{1}{2}+\frac{1}{2}iy}(s)$ $\cdot\operatorname{Re}[Q^{\nu+\frac{1}{2}}_{-\frac{1}{2}+\frac{1}{2}iy}(s)]$ $N=-\frac{1}{2}\pi[a\sin(\pi\nu)]^{-1}P^{-\nu-\frac{1}{2}}_{-\frac{1}{2}}(s)Q^{\nu+\frac{1}{2}}_{-\frac{1}{2}}(s)$ $s=(1-a^{-2})^{\frac{1}{2}}, \quad a>1, \quad \nu>-1$	$\mathfrak{Q}_\nu(1+2a^2\sinh^2x)$
345	$\frac{1}{2}[a\cos(\pi\nu)]^{-1}\{P^{\nu+\frac{1}{2}}_{-\frac{1}{2}+\frac{1}{2}iy}(s)\operatorname{Re}[Q^{-\nu-\frac{1}{2}}_{-\frac{1}{2}+\frac{1}{2}iy}(s)]$ $-P^{-\nu-\frac{1}{2}}_{-\frac{1}{2}+\frac{1}{2}iy}(s)\operatorname{Re}[Q^{\nu+\frac{1}{2}}_{-\frac{1}{2}+\frac{1}{2}iy}(s)]\}$ $N=\frac{1}{2}[a\cos(\pi\nu)]^{-1}\{P^{\nu+\frac{1}{2}}_{-\frac{1}{2}}(s)Q^{-\nu-\frac{1}{2}}_{-\frac{1}{2}}(s)$ $-P^{-\nu-\frac{1}{2}}_{-\frac{1}{2}}(s)Q^{\nu+\frac{1}{2}}_{-\frac{1}{2}}(s)\}$ $s=(1-a^{-2})^{\frac{1}{2}}, \quad a>1, \quad -1<\nu<0$	$\mathfrak{P}_\nu(1+2a^2\sinh^2x)$

	$Ng(y)$	$2Nf(x)$
328	$\frac{1}{4}\{[\mathfrak{Q}_{-\frac{1}{2}+\frac{1}{2}iy}(z)]^2+[\mathfrak{Q}_{-\frac{1}{2}-\frac{1}{2}iy}(z)]^2\}$ $N=(1+z)^{-1}\{K[(\frac{1}{2}+\frac{1}{2}z)^{-\frac{1}{2}}]\}$ $z=(1+a^{-2})^{\frac{1}{2}}$	$0, \quad \sinh x<a^{-1}$ $\operatorname{csch} x\, K[(1-a^{-2}\operatorname{csch}^2 x)^{\frac{1}{2}}], \quad \sinh x>a^{-1}$
347	$\frac{1}{2}[a\cosh(\frac{1}{2}\pi y)]^{-1}$ $\cdot\{P^{-\nu-\frac{1}{2}}_{-\frac{1}{2}+\frac{1}{2}iy}(s)\ \mathrm{Re}[Q^{\nu+\frac{1}{2}}_{-\frac{1}{2}+\frac{1}{2}iy}(s)]$ $+P^{\nu+\frac{1}{2}}_{-\frac{1}{2}+\frac{1}{2}iy}(s)\ \mathrm{Re}[Q^{-\nu-\frac{1}{2}}_{-\frac{1}{2}+\frac{1}{2}iy}(s)]\}$ $N=\frac{1}{2}a^{-1}[P^{-\nu-\frac{1}{2}}_{-\frac{1}{2}}(s)Q^{\nu+\frac{1}{2}}_{-\frac{1}{2}}(s)$ $+P^{\nu+\frac{1}{2}}_{-\frac{1}{2}}(s)Q^{-\nu-\frac{1}{2}}_{-\frac{1}{2}}(s)],$ $s=(1-a^{-2})^{\frac{1}{2}}, \quad a>1, \quad -1<\nu<0$	$\mathfrak{P}_\nu(2a^2\cosh^2 x-1)$
344	$\frac{1}{2}\operatorname{sech}(\frac{1}{2}\pi y)$ $\cdot\mathrm{Re}\{P_\nu^{-\frac{1}{2}iy}(r)[Q_\nu^{\frac{1}{2}iy}(r)+Q^{\frac{1}{2}iy}_{-\nu-1}(r)]\}$ $r=(1-a^2)^{\frac{1}{2}}, \quad a<1, \quad -1<\nu<0$ $N=\frac{1}{2}P_\nu(r)[Q_\nu(r)+Q_{-\nu-1}(r)]$	$\mathfrak{P}_\nu(1+2a^2\sinh^2 x)$
353	$\frac{1}{2}\pi\operatorname{csch}(\frac{1}{2}\pi y)$ $\cdot[P_\nu^{\frac{1}{2}iy}(r)Q_\nu^{-\frac{1}{2}iy}(r)-P_\nu^{-\frac{1}{2}iy}(r)Q_\nu^{\frac{1}{2}iy}(r)]$ $N=\frac{1}{8}\pi^2[P_\nu(r)]^2+\frac{1}{2}[Q_\nu(r)]^2$ $r=(1-a^2)^{\frac{1}{2}}, \quad a<1, \quad \nu>-1$	$\mathfrak{Q}_\nu(1+2a^2\sinh^2 x)$

10. Bessel Functions

	$Ng(y)$	$2Nf(x)$
24	$\frac{1}{2}\pi J_0(ay)$ $N=\frac{1}{2}\pi$	$(a^2-x^2)^{-\frac{1}{2}}, \quad x<a$ $0, \quad x>a$
124	$\frac{1}{2}y^{-1}[\frac{1}{2}\pi J_0(ay)+\mathrm{si}(ay)]$ $N=\frac{1}{2}a$	$0, \quad x<a$ $-\log\left[\frac{(x+a)^{\frac{1}{2}}+(x-a)^{\frac{1}{2}}}{2x^{\frac{1}{2}}}\right], \quad x>a$
50	$2^{\nu-1}a^\nu\pi^{\frac{1}{2}}\Gamma(\frac{1}{2}+\nu)y^{-\nu}J_\nu(ay)$ $N=\frac{1}{2}a^{2\nu}\pi^{\frac{1}{2}}\Gamma(\frac{1}{2}+\nu)[\Gamma(1+\nu)]^{-1}, \quad \nu>-\frac{1}{2}$	$(a^2-x^2)^{\nu-\frac{1}{2}}, \quad x<a$ $0, \quad x>a$
52	$(2a)^\nu\pi^{\frac{1}{2}}\Gamma(\frac{1}{2}+\nu)y^{-\nu}J_\nu(ay)\cos(ay)$ $N=a^{2\nu}\pi^{\frac{1}{2}}\Gamma(\frac{1}{2}+\nu)][\Gamma(1+\nu)]^{-1}, \quad \nu>-\frac{1}{2}$	$(2ax-x^2)^{\nu-\frac{1}{2}}, \quad x<a$ $0, \quad x>a$

	$Ng(y)$	$2Nf(x)$
56	$-\frac{1}{2}(2a)^{-\nu}\pi^{\frac{1}{2}}\Gamma(\frac{1}{2}-\nu)y^{\nu}Y_{\nu}(ay)$ $N=\frac{1}{2}a^{-2\nu}\pi^{\frac{1}{2}}\Gamma(\nu)\Gamma(\frac{1}{2}-\nu), \quad 0<\nu<\frac{1}{2}$	$(x^2-a^2)^{-\nu-\frac{1}{2}}, \quad x>a$ $0, \quad x<a$
43	$\frac{1}{2}(2a)^{-\nu}\pi^{\frac{1}{2}}\Gamma(\frac{1}{2}-\nu)$ $\cdot y^{\nu}[J_{\nu}(ay)\sin(ay)-Y_{\nu}(ay)\cos(ay)]$ $N=\frac{1}{2}a^{-2\nu}\pi^{-\frac{1}{2}}\Gamma(\nu)\Gamma(\frac{1}{2}-\nu), \quad \nu<\frac{1}{2}$	$(x^2+2ax)^{-\nu-\frac{1}{2}}$
57	$-\frac{1}{2}(2a)^{-\nu}\pi^{\frac{1}{2}}\Gamma(\frac{1}{2}-\nu)$ $\cdot y^{\nu}[J_{\nu}(ay)\sin(ay)+Y_{\nu}(ay)\cos(ay)]$ $N=\frac{1}{2}a^{-2\nu}\pi^{\frac{1}{2}}\Gamma(\nu)\Gamma(\frac{1}{2}-\nu), \quad \nu<\frac{1}{2}$	$0, \quad x<2a$ $(x^2-2ax)^{-\nu-\frac{1}{2}}, \quad x>2a$
119	$\frac{1}{2}\pi\{[\gamma+\log(2y/a)]J_0(ay)-\frac{1}{2}\pi Y_0(ay)\}$ $N=\pi\log(2/a), \quad a\leq 1$	$-(a^2-x^2)^{-\frac{1}{2}}\log(a^2-x^2), \quad x<a$ $0, \quad x>a$
121	$-\pi\cos(\frac{1}{2}y)[\frac{1}{2}\pi Y_0(\frac{1}{2}y)$ $-(\gamma+\log 4y)J_0(\frac{1}{2}y)]$ $N=4\pi\log 2$	$-[x(1-x)]^{-\frac{1}{2}}\log[x(1-x)], \quad x<1$ $0, \quad x>1$
319	$\frac{1}{4}a\pi^2[J_0(\frac{1}{2}ay)]^2$ $N=\frac{1}{4}a\pi^2$	$K[(1-a^{-2}x^2)^{\frac{1}{2}}], \quad x<a$ $0, \quad x>a$
26	$\pi(\frac{1}{2}\pi y)^{\frac{1}{2}}[J_{-\frac{1}{4}}(\frac{1}{2}ay)]^2$ $N=\pi(\frac{1}{2}\pi)^{\frac{1}{2}}[\Gamma(\frac{3}{4})]^{-2}$	$x^{-\frac{1}{2}}(a^2-x^2)^{-\frac{1}{2}}, \quad x<a$ $0, \quad x>a$
321	$\frac{1}{2}\pi^2 J_0[\frac{1}{2}y(a+b)]J_0[\frac{1}{2}y(a-b)]$ $N=\frac{1}{2}\pi^2$	$(a^2-x^2)^{-\frac{1}{2}}K\left[\left(\frac{a^2-b^2}{a^2-x^2}\right)^{\frac{1}{2}}\right], \quad x<b$ $(a^2-b^2)^{-\frac{1}{2}}K\left[\left(\frac{a^2-x^2}{a^2-b^2}\right)^{\frac{1}{2}}\right], \quad b<x<a$ $0, \quad x>a$
55	$(2b)^{2\nu}\pi(\frac{1}{2}\pi y)^{\frac{1}{2}}J_{\nu-\frac{1}{4}}(\frac{1}{2}by)J_{-\nu-\frac{1}{4}}(\frac{1}{2}by)$ $N=2^{2\nu}\pi(\pi/b)^{\frac{1}{2}}[\Gamma(\frac{3}{4}+\nu)\Gamma(\frac{3}{4}-\nu)]^{-1}$	$x^{-\frac{1}{2}}(b^2-x^2)^{-\frac{1}{2}}\{[(b+x)^{\frac{1}{2}}+i(b-x)^{\frac{1}{2}}]^{4\nu}$ $+[(b+x)^{\frac{1}{2}}-i(b-x)^{\frac{1}{2}}]^{4\nu}\}, \quad x<b$ $0, \quad x>b$
335	$\frac{1}{2}a\pi J_{\nu+\frac{1}{2}}(\frac{1}{2}ay)J_{-\nu-\frac{1}{2}}(\frac{1}{2}ay)$ $N=a\cos(\pi\nu)(2\nu+1)^{-1}, \quad -\frac{1}{2}<\nu<\frac{1}{2}$	$P_{\nu}(2x^2/a^2-1), \quad x<a$ $0, \quad x>a$

	$Ng(y)$	$2Nf(x)$
30	$-\pi(\frac{1}{2}\pi y)^{\frac{1}{2}}J_{-\frac{1}{4}}(\frac{1}{2}ay)Y_{-\frac{1}{4}}(\frac{1}{2}ay)$ $N=\pi(\frac{1}{2}\pi)^{\frac{1}{2}}[\Gamma(\frac{3}{4})]^{-2}$	$0, \quad x<a$ $x^{-\frac{1}{2}}(x^2-a^2)^{-\frac{1}{2}}, \quad x>a$
336	$\frac{1}{2}\pi(ab)^{\frac{1}{2}}[J_{\nu+\frac{1}{2}}(by)Y_{\nu+\frac{1}{2}}(ay)$ $-J_{\nu+\frac{1}{2}}(ay)Y_{\nu+\frac{1}{2}}(by)]$ $N=(ab)^{\frac{1}{2}}(2\nu+1)^{-1}[(a/b)^{\nu+\frac{1}{2}}-(b/a)^{\nu+\frac{1}{2}}],$ $a>b$	$\mathfrak{P}_\nu[(a^2+b^2-x^2)/2ab], \quad x<a-b$ $0, \quad x>a-b$
59	$-\frac{1}{2}a^\nu\pi(\frac{1}{2}\pi y)^{\frac{1}{2}}$ $\cdot[J_{\frac{1}{2}\nu-\frac{1}{4}}(\frac{1}{2}ay)Y_{-\frac{1}{2}\nu-\frac{1}{4}}(\frac{1}{2}ay)$ $+J_{-\frac{1}{2}\nu-\frac{1}{4}}(\frac{1}{2}ay)Y_{\frac{1}{2}\nu-\frac{1}{4}}(\frac{1}{2}ay)]$ $N=(2\pi)^{-\frac{1}{2}}a^{\nu-\frac{1}{2}}\Gamma(\frac{1}{4}-\frac{1}{2}\nu)\Gamma(\frac{1}{4}+\frac{1}{2}\nu),$ $-\frac{1}{2}<\nu<\frac{1}{2}$	$0, \quad x<a$ $x^{-\frac{1}{2}}(x^2-a^2)^{-\frac{1}{2}}$ $\cdot\{[x+(x^2-a^2)^{\frac{1}{2}}]^\nu$ $+[x-(x^2-a^2)^{\frac{1}{2}}]^\nu\}, \quad x>a$
505	$\frac{1}{8}\pi a^{-1}y\{[J_{\frac{1}{4}}(\frac{1}{8}y^2/a)]^2+[Y_{\frac{1}{4}}(\frac{1}{8}y^2/a)]^2\}$ $N=\frac{1}{2}(2a)^{-\frac{1}{2}}\Gamma(\frac{1}{4})[\Gamma(\frac{3}{4})]^{-1}$	$\mathbf{H}_0(ax^2)-Y_0(ax^2)$
496	$\frac{1}{8}\pi a^{-1}y\{[J_{\frac{1}{4}}(\frac{1}{8}y^2/a)]^2-[Y_{-\frac{1}{4}}(\frac{1}{8}y^2/a)]^2\}$ $N=\frac{1}{2}(2a)^{-\frac{1}{2}}\Gamma(\frac{1}{4})[\Gamma(\frac{3}{4})]^{-1}$	$\mathbf{H}_0(ax^2)$
375	$-\frac{1}{4}a^{-1}(\frac{1}{2}\pi y)^{\frac{1}{2}}J_{-1/8}(\frac{1}{16}y^2/a)Y_{1/8}(\frac{1}{16}y^2/a)$ $N=2^{-5/4}\pi^{-\frac{1}{2}}\Gamma(\frac{1}{8})[\Gamma(\frac{7}{8})]^{-1}$	$x^{\frac{1}{2}}[J_{-1/8}(ax^2)]^2$
92	$\frac{1}{2}\pi(a/y)^{\frac{1}{2}}\{J_{\frac{1}{4}}(\frac{1}{8}a^2/y)\sin(\frac{1}{8}a^2/y+\frac{1}{8}\pi)$ $-Y_{\frac{1}{4}}(\frac{1}{8}a^2/y)\cos(\frac{1}{8}a^2/y+\frac{1}{8}\pi)\}$ $N=2(\pi/a)^{\frac{1}{2}}$	$x^{-\frac{3}{4}}\exp(-ax^{\frac{1}{2}})$
433	$-\frac{1}{4}\pi\sec(\pi\nu)(\pi/y)^{\frac{1}{2}}$ $\cdot[\sin(\frac{1}{2}\pi\nu-\frac{1}{4}\pi-\frac{1}{8}a^2/y)J_\nu(\frac{1}{8}a^2/y)$ $+\cos(\frac{1}{2}\pi\nu-\frac{1}{4}\pi-\frac{1}{8}a^2/y)Y_\nu(\frac{1}{8}a^2/y)]$ $N=\pi a^{-1}\sec(\pi\nu), \quad -\frac{1}{2}<\nu<\frac{1}{2}$	$x^{-\frac{1}{2}}K_{2\nu}(ax^{\frac{1}{2}})$
141	$\frac{1}{2}\pi J_0[a(b^2+y^2)^{\frac{1}{2}}]$ $N=\frac{1}{2}\pi J_0(ab), \quad ab\leq\frac{1}{2}\pi$	$(a^2-x^2)^{-\frac{1}{2}}\cos[b(a^2-x^2)^{\frac{1}{2}}], \quad x<a$ $0, \quad x>a$
143	$\pi\cos(\frac{1}{2}by)J_0[\frac{1}{2}b(a^2+y^2)^{\frac{1}{2}}]$ $N=\pi J_0(\frac{1}{2}ab), \quad ab\leq\pi$	$x^{-\frac{1}{2}}(b-x)^{-\frac{1}{2}}$ $\cdot\cos[ax^{\frac{1}{2}}(b-x)^{\frac{1}{2}}], \quad x<b$ $0, \quad x>b$
139	$\frac{1}{2}ab\pi(b^2+y^2)^{-\frac{1}{2}}J_1[a(b^2+y^2)^{\frac{1}{2}}]$ $N=\frac{1}{2}a\pi J_1(ab), \quad ab\leq\pi$	$\sin[b(a^2-x^2)^{\frac{1}{2}}], \quad x<a$ $0, \quad x>a$

	$Ng(y)$	$2Nf(x)$
378	$(\frac{1}{2}\pi a)^{\frac{1}{2}}(ab)^{\nu}(b^2+y^2)^{-\frac{1}{2}\nu-\frac{1}{4}}J_{\nu+\frac{1}{2}}[a(b^2+y^2)^{\frac{1}{2}}]$ $N=(\frac{1}{2}\pi a/b)^{\frac{1}{2}}J_{\nu+\frac{1}{2}}(ab)a^{\nu}, \quad \nu>-1,$ $ab\leq\tau_{\nu,1}$	$(a^2-x^2)^{\frac{1}{2}\nu}$ $\cdot J_{\nu}[b(a^2-x^2)^{\frac{1}{2}}], \quad x<a$ $0, \quad x>a$
382	$(\pi a)^{\frac{1}{2}}(\frac{1}{2}ab)^{\nu}\cos(\frac{1}{2}ay)$ $\cdot(b^2+y^2)^{-\frac{1}{2}\nu-\frac{1}{4}}J_{\nu+\frac{1}{2}}[\frac{1}{2}a(b^2+y^2)^{\frac{1}{2}}]$ $N=(\frac{1}{2}a)^{\nu}(\pi a/b)^{\frac{1}{2}}J_{\nu+\frac{1}{2}}(\frac{1}{2}ab), \quad \nu>-1,$ $ab\leq 2\tau_{\nu,1}$	$(ax-x^2)^{\frac{1}{2}\nu}$ $\cdot J_{\nu}[b(ax-x^2)^{\frac{1}{2}}], \quad x<a$ $0, \quad x>a$
377	$\frac{1}{2}\pi J_{\frac{1}{2}\nu}(z_1)J_{\frac{1}{2}\nu}(z_2)$ $N=\frac{1}{2}\pi[J_{\frac{1}{2}\nu}(\frac{1}{2}ab)]^2, \quad \nu>-1, \quad ab\leq\tau_{\nu,1}$ $z_{1 \atop 2}=\frac{1}{2}a[(b^2+y^2)^{\frac{1}{2}}\pm y]$	$(a^2-x^2)^{-\frac{1}{2}}J_{\nu}[b(a^2-x^2)^{\frac{1}{2}}], \quad x<a$ $0, \quad x>a$
381	$\pi\cos(\frac{1}{2}ay)J_{\frac{1}{2}\nu}(\frac{1}{2}z_1)J_{\frac{1}{2}\nu}(\frac{1}{2}z_2)$ $N=\pi[J_{\frac{1}{2}\nu}(\frac{1}{4}ab)]^2, \quad \nu>-1, \quad ab\leq 2\tau_{\nu,1}$ $z_{1 \atop 2}=\frac{1}{2}a[(b^2+y^2)^{\frac{1}{2}}\pm y]$	$(ax-x^2)^{-\frac{1}{2}}$ $\cdot J_{\nu}[b(ax-x^2)^{\frac{1}{2}}], \quad x<a$ $0, \quad x>a$
457	$\frac{1}{8}\pi^2\sec(\frac{1}{2}\pi\nu)[J_{\frac{1}{2}\nu}(z_1)J_{\frac{1}{2}\nu}(z_2)$ $+Y_{\frac{1}{2}\nu}(z_1)Y_{\frac{1}{2}\nu}(z_2)]$ $N=\frac{1}{8}\pi^2\sec(\frac{1}{2}\pi\nu)\{[J_{\frac{1}{2}\nu}(\frac{1}{2}ab)]^2$ $+[Y_{\frac{1}{2}\nu}(\frac{1}{2}ab)]^2\}, \quad -1<\nu<1$ $z_{1 \atop 2}=\frac{1}{2}a[(b^2+y^2)^{\frac{1}{2}}\pm y]$	$0, \quad x<a$ $(x^2-a^2)^{-\frac{1}{2}}$ $\cdot K_{\nu}[b(x^2-a^2)^{\frac{1}{2}}], \quad x>a$
436	$\frac{1}{2}\pi J_{\nu}(z_1)J_{\nu}(z_2)$ $N=\frac{1}{2}\pi[I_{\nu}(\frac{1}{2}ab)]^2, \quad \nu>-\frac{1}{2}$ $z_{1 \atop 2}=\frac{1}{2}a[y\pm(y^2-b^2)^{\frac{1}{2}}]$	$(a^2-x^2)^{-\frac{1}{2}}$ $\cdot I_{2\nu}[b(a^2-x^2)^{\frac{1}{2}}], \quad x<a$ $0, \quad x>a$
437	$\pi\cos(\frac{1}{2}ay)J_{\nu}(z_1/2)J_{\nu}(z_2/2)$ $N=\pi[I_{\nu}(\frac{1}{4}ab)]^2, \quad \nu>-\frac{1}{2}$ $z_{1 \atop 2}=\frac{1}{2}a[y\pm(y^2-b^2)^{\frac{1}{2}}]$	$(ax-x^2)^{-\frac{1}{2}}$ $\cdot I_{2\nu}\{[b(ax-x^2)^{\frac{1}{2}}]\}, \quad x>a$ $0, \quad x<a$
389	$\pi J_{\frac{1}{2}\nu-y}(\frac{1}{2}a)J_{\frac{1}{2}\nu+y}(\frac{1}{2}a)$ $N=\pi[J_{\frac{1}{2}\nu}(\frac{1}{2}a)]^2, \quad \nu>-1, \quad a\leq\tau_{\nu,1}$	$J_{\nu}(a\cos\frac{1}{2}x), \quad x<\pi$ $0, \quad x>\pi$
390	$\pi\cos(\frac{1}{2}\pi y)J_{\frac{1}{2}\nu-\frac{1}{2}y}(\frac{1}{2}a)J_{\frac{1}{2}\nu+\frac{1}{2}y}(\frac{1}{2}a)$ $N=\pi[J_{\frac{1}{2}\nu}(\frac{1}{2}a)]^2, \quad \nu>-1, \quad a\leq\tau_{\nu,1}$	$J_{\nu}(a\sin x), \quad x<\pi$ $0, \quad x>\pi$

	$Ng(y)$	$2Nf(x)$
392	$\frac{1}{2}\pi a\nu^{-1}[J_{\frac{1}{2}(\nu-1-y)}(\frac{1}{2}a)J_{\frac{1}{2}(\nu-1+y)}(\frac{1}{2}a)$ $+J_{\frac{1}{2}(\nu+1-y)}(\frac{1}{2}a)J_{\frac{1}{2}(\nu+1+y)}(\frac{1}{2}a)]$ $N=\frac{1}{2}\pi a\nu^{-1}\{[J_{\frac{1}{2}\nu-\frac{1}{2}}(\frac{1}{2}a)]^2+[J_{\frac{1}{2}\nu+\frac{1}{2}}(\frac{1}{2}a)]^2\},$ $\nu>0,\quad a\leq\tau_{\nu,1}$	$\sec(\frac{1}{2}x)$ $\cdot J_\nu(a\cos\frac{1}{2}x),\quad x<\pi$ $0,\quad x>\pi$
503	$\frac{1}{2}\pi\,\mathrm{sech}(\pi y)\{[J_{iy}(a)]^2+[Y_{iy}(a)]^2\}$ $N=\frac{1}{2}\pi\{[J_0(a)]^2+[Y_0(a)]^2\}$	$\mathbf{H}_0(2a\cosh\frac{1}{2}x)$ $-Y_0(2a\cosh\frac{1}{2}x)$
502	$\frac{1}{2}\pi\,\mathrm{sech}(\pi y)\{[J_{iy}(a)]^2+[J_{-iy}(a)]^2\}$ $N=\pi[J_0(a)]^2$	$\mathbf{H}_0(2a\cosh\frac{1}{2}x)$
391	$\frac{1}{2}\pi a\nu^{-1}\cos(\frac{1}{2}\pi y)$ $\cdot[J_{(\frac{1}{2}\nu-1-y)}(\frac{1}{2}a)J_{\frac{1}{2}(\nu-1+y)}(\frac{1}{2}a)$ $+J_{\frac{1}{2}(\nu+1-y)}(\frac{1}{2}a)J_{\frac{1}{2}(\nu+1+y)}(\frac{1}{2}a)]$ $N=\frac{1}{2}\pi a\nu^{-1}\{[J_{\frac{1}{2}\nu-\frac{1}{2}}(\frac{1}{2}a)]^2$ $+[J_{\frac{1}{2}\nu+\frac{1}{2}}(\frac{1}{2}a)]^2\},\quad v>0,\quad a\leq\tau_{\nu,1}$	$\csc x$ $\cdot J_\nu(a\sin x),\quad x<\pi$ $0,\quad x>\pi$
478	$\frac{1}{4}\pi^2\{J_{iy-\nu}(a)J_{iy+\nu}(a)$ $+Y_{iy-\nu}(a)Y_{iy+\nu}(a)+\tan(\pi\nu)$ $\cdot[J_{iy+\nu}(a)Y_{iy-\nu}(a)-J_{iy-\nu}(a)Y_{iy+\nu}(a)]\}$ $N=\frac{1}{4}\pi^2\sec(\pi\nu)$ $\cdot\{[J_\nu(a)]^2+[Y_\nu(a)]\},\quad -\frac{1}{2}<\nu<\frac{1}{2}$	$K_{2\nu}(2a\sinh\frac{1}{2}x)$
396	$\frac{1}{2}\pi[J_{iy}(z_1)Y_{iy}(z_2)-J_{iy}(z_2)Y_{iy}(z_1)]$ $N=\frac{1}{2}\pi[J_0(z_1)Y_0(z_2)-J_0(z_2)Y_0(z_1)],$ $a\leq b\leq\tau_{0,1}$ $z_{1,2}=\frac{1}{2}[(b+a)^{\frac{1}{2}}\pm(b-a)^{\frac{1}{2}}]$	$J_0[(b-a\cosh x)^{\frac{1}{2}}],\quad \cosh x\leq b/a$ $0,\quad \cosh x>b/a$

11. Modified Bessel Functions

	$Ng(y)$	$2Nf(x)$
35	$2^{-\frac{1}{2}}a^{-1}\pi e^{-ay}I_0(\frac{1}{2}ay)$ $N=2^{-\frac{1}{2}}a^{-1}\pi$	$x^{-\frac{1}{2}}(a^2+x^2)^{-\frac{1}{2}}$ $\cdot[x+(a^2+x^2)^{\frac{1}{2}}]^{-\frac{1}{2}}$
42	$\pi^{\frac{1}{2}}(2a)^{-\nu}[\Gamma(\nu+\frac{1}{2})]^{-1}y^\nu K_\nu(ay)$ $N=\frac{1}{2}a^{-2\nu}\pi^{\frac{1}{2}}\Gamma(\nu)[\Gamma(\frac{1}{2}+\nu)]^{-1},\quad \nu>0$	$(a^2+x^2)^{-\nu-\frac{1}{2}}$

	$Ng(y)$	$2Nf(x)$
29	$2^{-\frac{1}{2}}a^{-2}\sinh(\frac{1}{2}ay)K_1(\frac{1}{2}ay)$ $N=2^{-\frac{1}{2}}a^{-2}$	$x^{-\frac{1}{2}}(a^2+x^2)^{-\frac{1}{2}}$ $\cdot[x+(a^2+x^2)^{\frac{1}{2}}]^{-\frac{3}{2}}$
339	$(ab)^{\frac{1}{2}}\pi I_{\nu+\frac{1}{2}}(by)K_{\nu+\frac{1}{2}}(ay)$ $N=a^{-\nu}b^{\nu}\pi(2\nu+1)^{-1},\quad \nu>-\frac{1}{2},\quad a>b$	$\mathfrak{Q}_{\nu}[(x^2+a^2+b^2)/2ab]$
46	$a^{\nu}(\frac{1}{2}\pi y)^{\frac{1}{2}}I_{-\frac{1}{4}-\frac{1}{2}\nu}(\frac{1}{2}ay)K_{\frac{1}{4}-\frac{1}{2}\nu}(\frac{1}{2}ay)$ $N=a^{\nu}(2a/\pi)^{-\frac{1}{2}}\Gamma(\frac{1}{4}-\frac{1}{2}\nu)[\Gamma(\frac{3}{4}-\frac{1}{2}\nu)]^{-1},\quad \nu<\frac{1}{2}$	$x^{-\frac{1}{2}}(a^2+x^2)^{-\frac{1}{2}}$ $\cdot[x+(a^2+x^2)^{\frac{1}{2}}]^{\nu}$
426	$(\frac{1}{2}\pi)^{\frac{3}{2}}\exp(-y^2/16)I_0(y^2/16)$ $N=(\frac{1}{2}\pi)^{\frac{3}{2}}$	$\exp(-x^2)K_0(x^2)$
74	$\frac{1}{2}\pi(2a/y)^{-\frac{1}{2}}\exp(-\frac{1}{8}y^2/a)I_{-\frac{1}{4}}(\frac{1}{8}y^2/a)$ $N=2^{-\frac{1}{2}}a^{-\frac{1}{4}}\pi[\Gamma(\frac{3}{4})]^{-1}$	$x^{-\frac{1}{2}}\exp(-ax^2)$
422	$\frac{1}{8}\pi a^{-1}yK_{\frac{1}{4}}(\frac{1}{8}y^2/a)[I_{\frac{1}{4}}(\frac{1}{8}y^2/a)+I_{-\frac{1}{4}}(\frac{1}{8}y^2/a)]$ $N=\frac{1}{4}(2a)^{-\frac{1}{2}}\Gamma^2(\frac{1}{4})$	$K_0(ax^2)$
497	$2^{-\frac{1}{2}}(2\pi a)^{-1}y[K_{\frac{1}{4}}(\frac{1}{8}y^2/a)]^2$ $N=\frac{1}{2}\pi^{-1}(2a)^{-\frac{1}{2}}[\Gamma(\frac{1}{4})]^2$	$I_0(ax^2)-\mathbf{L}_0(ax^2)$
75	$\frac{1}{4}(2a/y)^{-\frac{3}{2}}\exp(-\frac{1}{8}y^2/a)[I_{-\frac{3}{4}}(\frac{1}{8}y^2/a)$ $-I_{\frac{3}{4}}(\frac{1}{8}y^2/a)]$ $N=2^{-\frac{1}{2}}a^{-\frac{3}{4}}[\Gamma(\frac{1}{4})]^{-1}$	$x^{\frac{1}{2}}\exp(-ax^2)$
431	$-\pi a^{-1}yY_1[(2ay)^{\frac{1}{2}}]K_1[(2ay)^{\frac{1}{2}}]$ $N=a^{-2}$	$x^{-3}K_0(ax^{-1})$
438	$(\frac{1}{2}\pi a)^{\frac{1}{2}}(ab)^{\nu}(b^2-y^2)^{-\frac{1}{2}\nu-\frac{1}{4}}I_{\nu+\frac{1}{2}}[a(b^2-y^2)^{\frac{1}{2}}],\quad y<b$ $(\frac{1}{2}\pi a)^{\frac{1}{2}}(ab)^{\nu}(y^2-b^2)^{-\frac{1}{2}\nu-\frac{1}{4}}J_{\nu+\frac{1}{2}}[a(y^2-b^2)^{\frac{1}{2}}],\quad y>b$ $N=(\frac{1}{2}\pi a/b)^{\frac{1}{2}}a^{\nu}I_{\nu+\frac{1}{2}}(ab),\quad \nu>-1$	$(a^2-x^2)^{\frac{1}{2}\nu}$ $\cdot I_{\nu}[b(a^2-x^2)^{\frac{1}{2}}],\quad x<a$ $0,\quad x>a$
443	$(\pi a)^{\frac{1}{2}}(\frac{1}{2}ab)^{\nu}\cos(\frac{1}{2}ay)$ $\cdot\begin{cases}(b^2-y^2)^{-\frac{1}{2}\nu-\frac{1}{4}}I_{\nu+\frac{1}{2}}[\frac{1}{2}a(b^2-y^2)^{\frac{1}{2}}], & y<b\\(y^2-b^2)^{-\frac{1}{2}\nu-\frac{1}{4}}J_{\nu+\frac{1}{2}}[\frac{1}{2}a(y^2-b^2)^{\frac{1}{2}}], & y>b\end{cases}$ $N=(\pi a/b)^{\frac{1}{2}}(\frac{1}{2}a)^{\nu}I_{\nu+\frac{1}{2}}(\frac{1}{2}ab),\quad \nu>-1$	$(ax-x^2)^{\frac{1}{2}\nu}$ $\cdot I_{\nu}[b(ax-x^2)^{\frac{1}{2}}],\quad x<a$ $0,\quad x>a$

	$Ng(y)$	$2Nf(x)$
88	$b^{\nu}\{(a+iy)^{-\frac{1}{2}\nu}K_{\nu}[2b(a+iy)^{\frac{1}{2}}]$ $+(a-iy)^{-\frac{1}{2}\nu}K_{\nu}[2b(a-iy)^{\frac{1}{2}}]\}$ $N=2b^{\nu}a^{-\frac{1}{2}\nu}K_{\nu}(2a^{\frac{1}{2}}b)$	$x^{\nu-1}\exp(-ax-b^2/x)$
298	$2a(a^2+y^2)^{-1}K_0[b(a^2+y^2)^{\frac{1}{2}}]$ $N=2a^{-1}K_0(ab)$	$-e^{ax}\operatorname{Ei}\{-a[(b^2+x^2)^{\frac{1}{2}}+x]\}$ $-e^{-ax}\operatorname{Ei}\{-a[(b^2+x^2)^{\frac{1}{2}}-x]\}$
464	$\pi(4a^2+y^2)^{-\frac{1}{2}}K_{2\nu}[b(4a^2+y^2)^{\frac{1}{2}}]$ $N=\frac{1}{2}\pi a^{-1}K_{2\nu}(2ab)$	$K_{\nu}\{a[(b^2+x^2)^{\frac{1}{2}}+x]\}$ $\cdot K_{\nu}\{a[(b^2+x^2)^{\frac{1}{2}}-x]\}$
449	$(\frac{1}{2}\pi a)^{\frac{1}{2}}(ab)^{-\nu}(b^2+y^2)^{\frac{1}{2}\nu-\frac{1}{4}}K_{\nu-\frac{1}{2}}[a(b^2+y^2)^{\frac{1}{2}}]$ $N=\frac{1}{2}(2\pi a/b)^{\frac{1}{2}}a^{-\nu}K_{\nu-\frac{1}{2}}(ab)$	$(a^2+x^2)^{-\frac{1}{2}\nu}$ $\cdot K_{\nu}[b(a^2+x^2)^{\frac{1}{2}}]$
100	$2a^{\nu}\cos[\nu\arctan(y/b)]K_{\nu}[a(b^2+y^2)^{\frac{1}{2}}]$ $N=2a^{\nu}K_{\nu}(ab)$	$(a^2+x^2)^{-\frac{1}{2}}\{[(a^2+x^2)^{\frac{1}{2}}+x]^{\nu}$ $+[(a^2+x^2)^{\frac{1}{2}}-x]^{\nu}\}\exp[-b(a^2+x^2)^{\frac{1}{2}}]$
448	$\frac{1}{2}K_{\nu}(z_1)K_{\nu}(z_2)$ $N=\frac{1}{2}[K_{\nu}(\frac{1}{2}ab)]^2$ $z_{\substack{1\\2}}=\frac{1}{2}a[(b^2+y^2)^{\frac{1}{2}}\pm y]$	$(a^2+x^2)^{-\frac{1}{2}}$ $\cdot K_{2\nu}[b(a^2+x^2)^{\frac{1}{2}}]$
501	$I_0(z_2)K_0(z_1)$ $N=I_0(\frac{1}{2}ab)-K_0(\frac{1}{2}ab)$ $z_{\substack{1\\2}}=\frac{1}{2}a[(b^2+y^2)^{\frac{1}{2}}\pm y]$	$(a^2+x^2)^{-\frac{1}{2}}$ $\cdot\{I_0[b(a^2+x^2)^{\frac{1}{2}}]$ $-\mathbf{L}_0[b(a^2+x^2)^{\frac{1}{2}}]\}$
514	$\frac{1}{2}\pi I_{\nu}(z_2)K_{\nu}(z_1)$ $N=\frac{1}{2}\pi I_{\nu}(\frac{1}{2}ab)K_{\nu}(\frac{1}{2}ab)$ $z_{\substack{1\\2}}=\frac{1}{2}a[(b^2+y^2)^{\frac{1}{2}}\pm y]$	$(a^2+x^2)^{-\frac{1}{2}}\{\frac{1}{2}\pi\sec(\pi\nu)$ $\cdot I_{2\nu}[b(a^2+x^2)^{\frac{1}{2}}]$ $+is_{0,2\nu}[ib(a^2+x^2)^{\frac{1}{2}}]\}$
455	$\frac{1}{4}b\nu^{-1}[K_{\frac{1}{2}+\frac{1}{2}\nu}(z_1)K_{\frac{1}{2}+\frac{1}{2}\nu}(z_2)$ $-K_{\frac{1}{2}-\frac{1}{2}\nu}(z_1)K_{\frac{1}{2}-\frac{1}{2}\nu}(z_2)]$ $N=\frac{1}{4}b\nu^{-1}\{[K_{\frac{1}{2}+\frac{1}{2}\nu}(\frac{1}{2}ab)]^2-[K_{\frac{1}{2}-\frac{1}{2}\nu}(\frac{1}{2}ab)]^2\}$ $z_{\substack{1\\2}}=\frac{1}{2}a[(b^2+y^2)^{\frac{1}{2}}\pm y]$	$(a^2+x^2)^{-1}$ $\cdot K_{\nu}[b(a^2+x^2)^{\frac{1}{2}}]$
500	$\pi^{-1}K_0\{\frac{1}{2}a[y+(y^2-b^2)^{\frac{1}{2}}]\}$ $\cdot K_0\{\frac{1}{2}a[y-(y^2-b^2)^{\frac{1}{2}}]\}$ $N=\frac{1}{4}\pi\{[J_0(\frac{1}{2}ab)]^2+[Y_0(\frac{1}{2}ab)]^2\}$	$(a^2+x^2)^{-\frac{1}{2}}$ $\cdot\{\mathbf{H}_0[b(a^2+x^2)^{\frac{1}{2}}]$ $-Y_0[b(a^2+x^2)^{\frac{1}{2}}]\}$

	$Ng(y)$	$2Nf(x)$
490	$\frac{1}{2}\pi K_0[a(2\cosh y)^{\frac{1}{2}}]$ $N=\frac{1}{2}\pi K_0(2^{\frac{1}{2}}a)$	$K_{ix}(ae^{i\pi/4})$ $\cdot K_{ix}(ae^{-i\pi/4})$
491	$\frac{1}{2}\pi K_{2\nu}(2a\cosh\frac{1}{2}y)$ $N=\frac{1}{2}\pi K_{2\nu}(2a)$	$K_{\nu+ix}(a)K_{\nu-ix}(a)$
468	$\pi I_{\nu-y}(a)I_{\nu+y}(a)$ $N=\pi[I_\nu(a)]^2, \quad \nu>-\frac{1}{2}$	$I_{2\nu}(2a\cos\frac{1}{2}x), \quad x<\pi$ $0, \quad x>\pi$
467	$\pi\cos(\frac{1}{2}\pi y)I_{\nu-\frac{1}{2}y}(a)I_{\nu+\frac{1}{2}y}(a)$ $N=\pi[I_\nu(a)]^2, \quad \nu>-\frac{1}{2}$	$I_{2\nu}(2a\sin x), \quad x<\pi$ $0, \quad x>\pi$
472	$\frac{1}{2}\pi a\nu^{-1}[I_{\nu-\frac{1}{2}-y}(a)I_{\nu-\frac{1}{2}+y}(a)$ $-I_{\nu+\frac{1}{2}-y}(a)I_{\nu+\frac{1}{2}+y}(a)]$ $N=\frac{1}{2}\pi a\nu^{-1}\{[I_{\nu-\frac{1}{2}}(a)]^2-[I_{\nu+\frac{1}{2}}(a)]^2\}, \quad \nu>0$	$\sec(\frac{1}{2}x)$ $\cdot I_{2\nu}(2a\cos\frac{1}{2}x), \quad x<\pi$ $0, \quad x>\pi$
471	$\frac{1}{2}\pi a\nu^{-1}\cos(\frac{1}{2}\pi y)[I_{\nu-\frac{1}{2}-\frac{1}{2}y}(a)I_{\nu-\frac{1}{2}+\frac{1}{2}y}(a)$ $-I_{\nu+\frac{1}{2}-\frac{1}{2}y}(a)I_{\nu+\frac{1}{2}+\frac{1}{2}y}(a)]$ $N=\frac{1}{2}\pi a\nu^{-1}\{[I_{\nu-\frac{1}{2}}(a)]^2+[I_{\nu+\frac{1}{2}}(a)]^2\}, \quad \nu>0$	$\csc x$ $\cdot I_{2\nu}(2a\sin x), \quad x<\pi$ $0, \quad x>\pi$
475	$\frac{1}{2}\pi^2\csc(2\pi\nu)[I_{-\nu-y}(a)I_{-\nu+y}(a)$ $-I_{\nu-y}(a)I_{\nu+y}(a)]$ $N=\frac{1}{2}\pi^2\csc(2\pi\nu)\{[I_{-\nu}(a)]^2-[I_\nu(a)]^2\},$ $-\frac{1}{2}<\nu<\frac{1}{2}$	$K_{2\nu}(2a\cos\frac{1}{2}x), \quad x<\pi$ $0, \quad x>\pi$
474	$\frac{1}{2}\pi^2\csc(2\pi\nu)\cos(\frac{1}{2}\pi y)[I_{-\nu-\frac{1}{2}y}(a)I_{-\nu+\frac{1}{2}y}(a)$ $-I_{\nu-\frac{1}{2}y}(a)I_{\nu+\frac{1}{2}y}(a)], \quad -\frac{1}{2}<\nu<\frac{1}{2}$ $N=\frac{1}{2}\pi^2\csc(2\pi\nu)\{[I_{-\nu}(a)]^2-[I_\nu(a)]^2\}$	$K_{2\nu}(2a\sin x), \quad x<\pi$ $0, \quad x>\pi$
287	$\frac{1}{4}\pi\exp(-\frac{1}{2}a^2)\operatorname{sech}(\frac{1}{2}\pi y)$ $\cdot[I_{iy}(\frac{1}{2}a^2)+I_{-iy}(\frac{1}{2}a^2)]$ $N=\frac{1}{2}\pi\exp(-\frac{1}{2}a^2)I_0(\frac{1}{2}a^2)$	$-i\exp[-(a\cosh x)^2]$ $\cdot\operatorname{Erf}(ia\cosh x)$
251	$K_{iy}(a)$ $N=K_0(a)$	$\exp(-a\cosh x)$
222	$K_{iy}[(a^2-b^2)^{\frac{1}{2}}]\cos[y\arctan(b/a)]$ $N=K_0[(a^2-b^2)^{\frac{1}{2}}], \quad a>b$	$\exp(-a\cosh x)\cosh(b\sinh x)$

	$Ng(y)$	$2Nf(x)$
286	$\frac{1}{2}\exp(\frac{1}{2}a^2)\ \mathrm{sech}(\frac{1}{2}\pi y)K_{i\frac{1}{2}y}(\frac{1}{2}a^2)$ $N=\frac{1}{2}\exp(\frac{1}{2}a^2)K_0(\frac{1}{2}a^2)$	$\exp[(a\cosh x)^2]$ $\cdot\mathrm{Erfc}(a\cosh x)$
504	$\mathrm{sech}(\pi y)K_{iy}(a)[I_{iy}(a)+I_{-iy}(a)]$ $N=2I_0(a)K_0(a)$	$I_0(2a\cosh\frac{1}{2}x)$ $-\mathbf{L}_0(2a\cosh\frac{1}{2}x)$
484	$K_{iy}(ae^{i\pi/4})K_{iy}(ae^{-i\pi/4})$ $N=K_0(ae^{i\pi/4})K_0(ae^{-i\pi/4})$	$K_0[a(2\cosh x)^{\frac{1}{2}}]$
487	$K_{iy}(a)K_{iy}(b)$ $N=K_0(a)K_0(b)$	$K_0[(a^2+b^2+2ab\cosh x)^{\frac{1}{2}}]$
479	$K_{\nu+iy}(a)K_{\nu-iy}(a)$ $N=[K_\nu(a)]^2$	$K_{2\nu}(2a\cosh\frac{1}{2}x)$
480	$\frac{1}{2}a\nu^{-1}[K_{\frac{1}{2}+\nu+iy}(a)K_{\frac{1}{2}+\nu-iy}(a)$ $-K_{\frac{1}{2}-\nu+iy}(a)K_{\frac{1}{2}-\nu-iy}(a)]$ $N=\frac{1}{2}a\nu^{-1}\{[K_{\frac{1}{2}+\nu}(a)]^2-[K_{\frac{1}{2}-\nu}(a)]^2\}$	$\mathrm{sech}(\frac{1}{2}x)$ $\cdot K_{2\nu}(2a\cosh\frac{1}{2}x)$
488	$K_{\frac{1}{2}\nu+iy}(a)K_{\frac{1}{2}\nu-iy}(b)$ $+K_{\frac{1}{2}\nu+iy}(b)K_{\frac{1}{2}\nu-iy}(a)$ $N=2K_{\frac{1}{2}\nu}(a)K_{\frac{1}{2}\nu}(b)$	$\{[(a+be^x)(b+ae^x)^{-1}]^{\frac{1}{2}\nu}$ $+[(b+ae^x)(a+be^x)^{-1}]^{\frac{1}{2}\nu}\}$ $\cdot K_\nu[(a^2+b^2+2ab\cosh x)^{\frac{1}{2}}]$

12. Functions Related to Bessel Functions

	$Ng(y)$	$2Nf(x)$
174	$\frac{1}{2}\pi y^{-1}\mathbf{H}_0(y)$ $N=1$	$\arccos x,\quad x<1$ $0,\quad x>1$
173	$\frac{1}{2}\pi y^{-1}[\sin y-\mathbf{H}_0(y)]$ $N=\frac{1}{2}\pi-1$	$\arcsin x,\quad x<1$ $0,\quad x>1$
25	$a[1-\frac{1}{2}\pi\mathbf{H}_1(ay)]$ $N=a$	$x(a^2-x^2)^{-\frac{1}{2}},\quad x<a$ $0,\quad x>a$
58	$\frac{1}{2}a^{-2\nu-1}\pi\sec(\pi\nu)\{1-\frac{1}{2}a\pi y[J_\nu(ay)\mathbf{H}_{\nu-1}(ay)$ $-\mathbf{H}_\nu(ay)J_{\nu-1}(ay)]\}$ $N=\frac{1}{2}a^{-2\nu-1}\pi\sec(\pi\nu),\quad -\frac{1}{2}<\nu<\frac{1}{2}$	$0,\quad x<a$ $x^{-1}(x^2-a^2)^{-\nu-\frac{1}{2}},\quad x>a$

	$Ng(y)$	$2Nf(x)$
51	$a^{2\nu+1}(2\nu+1)^{-1}[1-(\frac{1}{2}a)^{-\nu}\pi^{\frac{1}{2}}\Gamma(\frac{3}{2}+\nu)y^{-\nu}$ $\cdot\mathbf{H}_{\nu+1}(ay)]$ $N=a^{2\nu+1}(2\nu+1)^{-1}, \quad \nu>-\frac{1}{2}$	$x(a^2-x^2)^{\nu-\frac{1}{2}}, \quad x<a$ $0, \quad x>a$
112	$\frac{1}{4}\pi^2[I_0(ay)-\mathbf{L}_0(ay)]$ $N=\frac{1}{4}\pi^2$	$(a^2+x^2)^{-\frac{1}{2}}$ $\cdot\log[a/x+(1+a^2/x^2)^{\frac{1}{2}}]$
113	$\frac{1}{2}\pi y^{-1}[1+\mathbf{L}_0(ay)-I_0(ay)]$ $N=a$	$\log[\frac{1}{2}+(\frac{1}{4}+\frac{1}{4}a^2/x^2)^{\frac{1}{2}}]$
430	$2^{-\frac{5}{2}}\pi^2a^{-1}[I_0(\frac{1}{8}y^2/a)-\mathbf{L}_0(\frac{1}{8}y^2/a)]$ $N=2^{-\frac{5}{2}}\pi^2a^{-1}$	$x[K_{\frac{1}{4}}(ax^2)]^2$
423	$\frac{1}{4}\pi a^{-1}(\frac{1}{2}\pi y)^{\frac{1}{2}}[I_{-\frac{1}{4}}(\frac{1}{4}y^2/a)-\mathbf{L}_{-\frac{1}{4}}(\frac{1}{4}y^2/a)]$ $N=2^{-\frac{1}{4}}(a/\pi)^{-\frac{3}{4}}[\Gamma(\frac{3}{4})]^{-1}$	$x^{\frac{1}{2}}K_{\frac{1}{4}}(ax^2)$
424	$\frac{1}{4}a^{-2}(\frac{1}{2}\pi y)^{\frac{3}{2}}[I_{-\frac{3}{4}}(\frac{1}{4}y^2/a)-\mathbf{L}_{-\frac{3}{4}}(\frac{1}{4}y^2/a)]$ $N=2^{-5/4}\pi^{\frac{3}{2}}a^{-5/4}[\Gamma(\frac{1}{4})]^{-1}$	$x^{\frac{3}{2}}K_{\frac{1}{4}}(ax^2)$
465	$\frac{1}{8}\pi^2y^{-1}[\mathbf{H}_0(\frac{1}{2}a^2/y)-Y_0(\frac{1}{2}a^2/y)]$ $N=\frac{1}{2}\pi a^{-2}$	$K_0[a(ix)^{\frac{1}{2}}]$ $\cdot K_0[a(-ix)^{\frac{1}{2}}]$
461	$\frac{1}{2}\pi^2(4a^2+y^2)^{-\frac{1}{2}}$ $\cdot\{\mathbf{H}_0[b(4a^2+y^2)^{\frac{1}{2}}]-Y_0[b(4a^2+y^2)^{\frac{1}{2}}]\}$ $N=\frac{1}{4}\pi^2a^{-1}[\mathbf{H}_0(2ab)-Y_0(2ab)]$	$K_0\{a[x+(x^2-b^2)^{\frac{1}{2}}]\}$ $\cdot K_0\{a[x-(x^2-b^2)^{\frac{1}{2}}]\}$
459	$\frac{1}{4}\pi(4a^2+y^2)^{-\frac{1}{2}}$ $\cdot\{I_0[b(4a^2+y^2)^{\frac{1}{2}}]-\mathbf{L}_0[b(4a^2+y^2)^{\frac{1}{2}}]\}$ $N=\frac{1}{4}\pi a^{-1}[I_0(2ab)-\mathbf{L}_0(2ab)]$	$I_0\{a[(b^2+x^2)^{\frac{1}{2}}-x]\}$ $\cdot K_0\{a[(b^2+x^2)^{\frac{1}{2}}+x]\}$
398	$\mathbf{H}_0(2a\cosh\frac{1}{2}y)-Y_0(2a\cosh\frac{1}{2}y)$ $N=\mathbf{H}_0(2a)-Y_0(2a)$	$\operatorname{sech}(\pi x)$ $\cdot\{[J_{ix}(a)]^2+[Y_{ix}(a)]^2\}$
164	$\frac{1}{2}\pi[\mathbf{J}_y(a)+\mathbf{J}_{-y}(a)]$ $N=\pi J_0(a), \quad a\leq\frac{1}{2}\pi$	$\cos(a\sin x), \quad x<\pi$ $0, \quad x>\pi$
163	$\frac{1}{4}\pi\sec(\frac{1}{2}\pi y)[\mathbf{J}_y(a)+\mathbf{J}_{-y}(a)]$ $N=\frac{1}{2}\pi J_0(a), \quad a\leq\frac{1}{2}\pi$	$\cos(a\cos x), \quad x<\frac{1}{2}\pi$ $0, \quad x>\frac{1}{2}\pi$

	$Ng(y)$	$2Nf(x)$
161	$\frac{1}{2}\pi \operatorname{ctn}(\frac{1}{2}\pi y)[\mathbf{J}_y(a)-\mathbf{J}_{-y}(a)]$ $N=\pi\mathbf{H}_0(a), \quad a\le\pi$	$\sin(a \sin x), \quad x<\pi$ $0, \quad x>\pi$
162	$\frac{1}{4}\pi \csc(\frac{1}{2}\pi y)[\mathbf{J}_y(a)-\mathbf{J}_{-y}(a)]$ $N=\frac{1}{2}\pi\mathbf{H}_0(a), \quad a\le\pi$	$\sin(a \cos x), \quad x<\frac{1}{2}\pi$ $0, \quad x>\frac{1}{2}\pi$
259	$\frac{1}{2}\pi[\mathbf{J}_y(ia)+\mathbf{J}_{-y}(ia)]$ $N=\pi I_0(a)$	$\cosh(a \sin x), \quad x<\pi$ $0, \quad x>\pi$
260	$\frac{1}{4}\pi \operatorname{sech}(\frac{1}{2}\pi y)[\mathbf{J}_y(ia)+\mathbf{J}_{-y}(ia)]$ $N=\frac{1}{2}\pi I_0(a)$	$\cosh(a \cos x), \quad x<\frac{1}{2}\pi$ $0, \quad x>\frac{1}{2}\pi$
258	$-\frac{1}{2}i\pi \operatorname{ctn}(\frac{1}{2}\pi y)[\mathbf{J}_y(ia)-\mathbf{J}_{-y}(ia)]$ $N=\pi\mathbf{L}_0(a)$	$\sinh(a \sin x), \quad x<\pi$ $0, \quad x>\pi$
257	$-\frac{1}{4}i\pi \csc(\frac{1}{2}\pi y)[\mathbf{J}_y(ia)-\mathbf{J}_{-y}(ia)]$ $N=\frac{1}{2}\pi\mathbf{L}_0(a)$	$\sinh(a \cos x), \quad x<\frac{1}{2}\pi$ $0, \quad x>\frac{1}{2}\pi$
526	$\frac{1}{2}\pi^{\frac{3}{2}}[a\Gamma(-\nu)]^{-1}\sec(\pi\nu)$ $\cdot[J_{-\nu-\frac{1}{2}}(\frac{1}{2}y^2/a^2)-\mathbf{J}_{-\nu-\frac{1}{2}}(\frac{1}{2}y^2/a^2)]$ $N=-\frac{1}{2}\pi^{\frac{1}{2}}a^{-1}[(\nu+\frac{1}{2})\Gamma(-\nu)]^{-1}, \quad \nu<-\frac{1}{2}$	$D_\nu[ax(i)^{\frac{1}{2}}]D_\nu[ax(-i)^{\frac{1}{2}}]$
333	$2^{\mu-1}\pi^{\frac{1}{2}}\{\Gamma[\frac{1}{2}(3-\mu+\nu)]\Gamma[\frac{1}{2}(2-\mu-\nu)]\}^{-1}$ $\cdot(\mu+\nu)(\mu-\nu-1)y^{\mu-\frac{1}{2}}s_{-\mu-\frac{1}{2},\nu+\frac{1}{2}}(y)$ $N=2^{\mu-1}\pi^{\frac{1}{2}}\{\Gamma[\frac{1}{2}(3-\mu+\nu)]\Gamma[\frac{1}{2}(2-\mu-\nu)]\}^{-1}$ $-\frac{1}{2}<\nu<\frac{1}{2}, \quad \mu<\frac{1}{2}$	$(1-x^2)^{-\frac{1}{2}\mu}P_\nu^\mu(x), \quad x<1$ $0, \quad x>1$
519	$\frac{1}{4}\pi^{\frac{1}{2}}a^{-\frac{3}{2}}\Gamma(2+\mu)[\Gamma(\frac{1}{2}-\mu)]^{-1}$ $\cdot yS_{-\mu-\frac{3}{2},\frac{1}{2}}(\frac{1}{4}y^2/a)$ $N=-\frac{1}{2}\pi^{\frac{1}{2}}a^{-1}\cos(\frac{1}{2}\pi\mu)\Gamma(-1-\mu)$ $\cdot\Gamma(2+\mu)[\Gamma(\frac{1}{2}-\mu)]^{-1}, \quad -2<\mu<0$	$xS_{\mu,\frac{1}{2}}(ax^2)$
466	$\frac{1}{4}\pi^2y^{-1}\sec(\frac{1}{2}\pi\nu)S_{0,\nu}(\frac{1}{2}a^2/y)$ $N=\frac{1}{2}\pi a^{-2}\sec(\frac{1}{2}\pi\nu), \quad -1<\nu<1$	$K_\nu[a(ix)^{\frac{1}{2}}]K_\nu[a(-ix)^{\frac{1}{2}}]$
462	$(4a^2+y^2)^{-\frac{1}{2}}\{\frac{1}{2}\pi\sec(\pi\nu)I_{2\nu}[b(4a^2+y^2)^{\frac{1}{2}}]$ $+is_{0,2\nu}[ib(4a^2+y^2)^{\frac{1}{2}}]\}$ $N=a^{-1}[\frac{1}{4}\pi\sec(\pi\nu)I_{2\nu}(2ab)$ $+\frac{1}{2}is_{0,2\nu}(i2ab)]$	$I_\nu\{a[(b^2+x^2)^{\frac{1}{2}}-x]\}$ $\cdot K_\nu\{a[(b^2+x^2)^{\frac{1}{2}}+x]\}$

	$Ng(y)$	$2Nf(x)$
463	$\pi(4a^2+y^2)^{-\frac{1}{2}}S_{0,2\nu}[b(4a^2+y^2)^{\frac{1}{2}}]$ $N=\frac{1}{2}\pi a^{-1}S_{0,2\nu}(2ab)$	$K_\nu\{a[x+(x^2-b^2)^{\frac{1}{2}}]\}$ $\cdot K_\nu\{a[x-(x^2-b^2)^{\frac{1}{2}}]\}$
250	$S_{0,iy}(a)=-i\frac{1}{2}\pi\,\mathrm{csch}(\pi y)$ $\cdot[J_{iy}(a)-J_{-iy}(a)-\mathbf{J}_{iy}(a)+\mathbf{J}_{-iy}(a)]$ $N=\frac{1}{2}\pi[\mathbf{H}_0(a)-Y_0(a)]$	$\exp(-a\sinh x)$
513	$(2a)^{-\frac{1}{2}}2^{-\mu-1}[\Gamma(\frac{1}{2}-\mu)]^{-1}$ $\cdot\Gamma(\frac{1}{4}-\frac{1}{2}\mu-\frac{1}{2}iy)\Gamma(\frac{1}{4}-\frac{1}{2}\mu+\frac{1}{2}iy)S_{\mu+\frac{1}{2},iy}(a)$ $N=(2a)^{-\frac{1}{2}}2^{-\mu-1}[\Gamma(\frac{1}{2}-\mu)]^{-1}[\Gamma(\frac{1}{4}-\frac{1}{2}\mu)]^{-1}$ $\cdot S_{\mu+\frac{1}{2},0}(a),\quad \mu<\frac{1}{2}$	$(\cosh x)^{\frac{1}{2}}S_{\mu,\frac{1}{2}}(a\cosh x)$

13. Parabolic Cylindrical Functions and Whittaker Functions

	$Ng(y)$	$2Nf(x)$
506	$(2\pi a)^{-\frac{1}{2}}\Gamma(\frac{1}{2}+\nu)$ $\cdot D_{-\nu-\frac{1}{2}}[y(2ai)^{\frac{1}{2}}]D_{-\nu-\frac{1}{2}}[y(-2ai)^{\frac{1}{2}}]$ $N=\frac{1}{2}(2a)^{-\frac{1}{2}}\Gamma(\frac{1}{4}+\frac{1}{2}\nu)[\Gamma(\frac{3}{4}+\frac{1}{2}\nu)]^{-1},\quad \nu>-\frac{1}{2}$	$\csc(\pi\nu)$ $\cdot[\mathbf{J}_\nu(ax^2)-J_\nu(ax^2)]$
507	$(2\pi a)^{-\frac{1}{2}}\{\Gamma(\frac{1}{2}-\nu)D_{\nu-\frac{1}{2}}[y(2ai)^{\frac{1}{2}}]$ $\cdot D_{\nu-\frac{1}{2}}[y(-2ai)^{\frac{1}{2}}]+\cos(\pi\nu)\Gamma(\frac{1}{2}+\nu)$ $\cdot D_{-\nu-\frac{1}{2}}[y(2ai)^{\frac{1}{2}}]D_{-\nu-\frac{1}{2}}[y(-2ai)^{\frac{1}{2}}]\}$ $N=\frac{1}{2}(2a)^{-\frac{1}{2}}\{\Gamma(\frac{1}{4}-\frac{1}{2}\nu)[\Gamma(\frac{3}{4}-\frac{1}{2}\nu)]^{-1}$ $+\cos(\pi\nu)\Gamma(\frac{1}{4}+\frac{1}{2}\nu)[\Gamma(\frac{3}{4}+\frac{1}{2}\nu)]^{-1},$ $-\frac{1}{2}<\nu<\frac{1}{2}$	$-Y_\nu(ax^2)-E_\nu(ax^2)$
80	$\frac{1}{2}(2b)^{-\frac{1}{2}\nu}\exp[\frac{1}{8}(a^2-y)/b]\Gamma(\nu)$ $\cdot\{\exp(-\frac{1}{4}iay/b)D_{-\nu}[(2b)^{-\frac{1}{2}}(a-iy)]$ $+\exp(\frac{1}{4}iay/b)D_{-\nu}[(2b)^{-\frac{1}{2}}(a+iy)]\}$ $N=(2b)^{-\frac{1}{2}\nu}\exp(\frac{1}{8}a^2/b)\Gamma(\nu)D_{-\nu}[a(2b)^{\frac{1}{2}}],$ $\nu>0$	$x^{\nu-1}\exp(-ax-bx^2)$
525	$\pi^{\frac{1}{2}}2^{\frac{1}{2}\nu}\exp(-a^2)D_{\frac{1}{2}\nu-\frac{1}{2}+y}(2a)D_{\frac{1}{2}\nu-\frac{1}{2}-y}(2a)$ $N=\pi^{\frac{1}{2}}2^{\frac{1}{2}\nu}\exp(-a^2)[D_{\frac{1}{2}\nu-\frac{1}{2}}(2a)]^2,\quad \nu\le 1$	$(\sec x)^{\frac{1}{2}\nu+\frac{1}{2}}\exp(-a^2\sec x)$ $\cdot D_\nu[2a(1+\sec x)^{\frac{1}{2}}],\quad x<\frac{1}{2}\pi$ $0,\quad x>\frac{1}{2}\pi$
256	$2^{\frac{1}{2}}\,\mathrm{Re}\{\Gamma(\frac{1}{2}+iy)D_{-\frac{1}{2}-iy}[(2ia)^{\frac{1}{2}}]$ $\cdot D_{-\frac{1}{2}-iy}[(-2ia)^{\frac{1}{2}}]\}$ $N=a(\frac{1}{2}\pi)^{\frac{3}{2}}\{[J_{\frac{1}{4}}(\frac{1}{2}a)]^2+[Y_{\frac{1}{4}}(\frac{1}{2}a)]^2\}$	$(\mathrm{csch}\,x)^{\frac{1}{2}}\exp(-a\,\mathrm{csch}\,x)$

	$Ng(y)$	$2Nf(x)$
289	$\pi^{\frac{1}{2}} \operatorname{sech}(\pi y) D_{-\frac{1}{2}+iy}(2^{\frac{1}{2}}a) D_{-\frac{1}{2}-iy}(2^{\frac{1}{2}}a)$ $N=(2\pi)^{-\frac{1}{2}}a[K_{\frac{1}{4}}(\frac{1}{2}a^2)]^2$	$(\operatorname{sech}x)^{\frac{1}{2}} \exp(a^2 \operatorname{sech}x)$ $\cdot \operatorname{Erfc}[a(1+\operatorname{sech}x)^{\frac{1}{2}}]$
44	$2^{-\frac{1}{2}}a^{-1}\Gamma(\frac{1}{4}-\frac{1}{2}\nu) y^{-\frac{1}{2}} W_{\frac{1}{2}\nu,\frac{1}{4}}(ay) M_{-\frac{1}{2}\nu,-\frac{1}{4}}(ay)$ $N=(2a/\pi)^{-\frac{1}{2}}\Gamma(\frac{1}{4}-\frac{1}{2}\nu)[\Gamma(\frac{3}{4}-\frac{1}{2}\nu)]^{-1}, \quad \nu<\frac{1}{2}$	$x^{-\nu-\frac{1}{2}}(a^2+x^2)^{-\frac{1}{2}}$ $\cdot[a+(a^2+x^2)^{\frac{1}{2}}]^{\nu}$
428	$\pi 2^{-\frac{1}{2}\nu}\Gamma(\frac{1}{2}-2\nu)[\Gamma(\frac{1}{2}+\nu)]^{-1}$ $\cdot y^{\nu-1} \exp(y^2/16) W_{\frac{3}{2}\nu,-\frac{1}{2}\nu}(y^2/8)$ $N=2^{\nu-\frac{3}{2}} \cos(\pi\nu)\Gamma(\nu)\Gamma(\frac{1}{2}-2\nu), \quad 0<\nu<\frac{1}{4}$	$x^{-2\nu} \exp(x^2) K_{\nu}(x^2)$
421	$2^{-\frac{1}{2}\nu}y^{\nu-1} \exp(-y^2/16) W_{-\frac{3}{2}\nu,\frac{1}{2}\nu}(y^2/8)$ $N=2^{\nu-\frac{3}{2}}\Gamma(\nu)[\Gamma(\frac{1}{2}+2\nu)]^{-1}, \quad \nu>0$	$x^{-2\nu} \exp(-x^2) I_{\nu}(x^2)$
527	$2^{-\frac{1}{2}\nu-\frac{5}{2}}a^{-1}[\Gamma(-\nu)]^{-1}\Gamma(-\frac{1}{2}\nu+\frac{1}{2}iy)$ $\cdot\Gamma(-\frac{1}{2}\nu-\frac{1}{2}iy) W_{\frac{1}{2}\nu+\frac{1}{2},i\frac{1}{2}y}(2a^2)$ $N=2^{\frac{1}{2}\nu-\frac{3}{2}}a^{-1}\pi^{\frac{1}{2}}\Gamma(-\frac{1}{2}\nu)$ $\cdot[\Gamma(\frac{1}{2}-\frac{1}{2}\nu)]^{-1}W_{\frac{1}{2}\nu+\frac{1}{2},0}(2a^2), \quad \nu<0$	$\exp[(a \sinh x)^2]$ $\cdot D_{\nu}(2a \cosh x)$
528	$2^{\frac{1}{2}\nu-\frac{3}{2}}\pi^{\frac{1}{2}}a^{-1}W_{\frac{1}{2}\nu,i\frac{1}{2}y}(2a^2)$ $N=2^{\frac{1}{2}\nu-\frac{3}{2}}\pi^{\frac{1}{2}}a^{-1}W_{\frac{1}{2}\nu,0}(2a^2), \quad \nu\leq 1$	$\exp[-(a \sinh x)^2]$ $\cdot D_{\nu}(2a \cosh x)$

TABLE IIA

FUNCTIONS VANISHING IDENTICALLY FOR NEGATIVE VALUES OF THE ARGUMENT

Definition

Take $g(y)$ from the inverse (Table I) before (pp. 105–144) and obtain $h(y)$ under the same number from Table II on pp. 74–96.

TABLE IIIA

FUNCTIONS NOT BELONGING TO EITHER OF THESE CLASSES

Definition

This table contains the inverse transforms of Table III from pp. 97–102. Corresponding pairs of formulas have the same number.

	$NG(y)$	$Nf(x)$
2	$n!(iy)^{-n-1}-e^{-iby}\sum_{m=0}^{n} n!b^m(iy)^{m-n-1}/m!$ $n=0, 1, 2, \ldots, \quad N=b^{n+1}(n+1)^{-1}$	$x^n, \quad 0<x<b$ $0, \quad$ otherwise
1	$iy^{-1}(e^{-iby}-e^{-iay})$ $N=b-a$	$1, \quad 0<x<b$ $0, \quad$ otherwise
19	$(c-iy)^{-1}[e^{-a(c-iy)}-e^{-b(c-iy)}]$ $N=c^{-1}(e^{-ac}-e^{-bc})$	$e^{-cx}, \quad a<x<b$ $0, \quad$ otherwise
39	$3^{-\frac{1}{2}}a^{-1}2\pi \exp(-i\frac{4}{27}y^3/a^2)$ $N=2\pi 3^{-\frac{1}{2}}a^{-1}$	$x^{\frac{1}{2}}K_{\frac{1}{3}}(ax^{\frac{3}{2}})$
24	$\pi(\lambda+iy)^{-1}\csc(\pi\lambda+i\pi y)$ $N=\pi\lambda^{-1}\csc(\pi\lambda), \quad -1<\lambda<0$	$e^{-\lambda x}\log(1+e^{-x})$
22	$\pi(b-c)^{-1}\csc(\pi\lambda+i\pi y)(a^{\lambda-1+iy}-b^{\lambda-1+iy})$ $N=\pi(b-a)^{-1}\csc(\pi\lambda)(a^{\lambda-1}-b^{\lambda-1}),$ $0<\lambda<2$	$(a+e^{-x})^{-1}(b+e^{-x})^{-1}e^{-\lambda x}$
20	$\pi a^{\lambda-1+iy}\csc(\pi\lambda+i\pi y)$ $N=\pi a^{\lambda-1}\csc(\pi\lambda), \quad 0<\lambda<1$	$(a+e^{-x})^{-1}e^{-\lambda x}$
21	$\pi a^{\lambda-1+iy}\csc(\pi\lambda+i\pi y)$ $\cdot[\log a-\pi \operatorname{ctn}(\pi\lambda+i\pi y)]$ $N=\pi a^{\lambda-1}\csc(\pi\lambda)(\log a-\pi \operatorname{ctn}\pi\lambda), \quad 0<\lambda<1$	$x(a+e^{-x})^{-1}e^{-\lambda x}$
37	$2b^{-1}\{\cos[a(y^2-b^2)^{\frac{1}{2}}]-\cos(ay)$ $+iy(y^2-b^2)^{-\frac{1}{2}}\sin[a(y^2-b^2)^{\frac{1}{2}}]-i\sin(ay)\}$ $N=4b^{-1}\sinh^2(\frac{1}{2}ab)$	$[(a+x)(a-x)^{-1}]^{\frac{1}{2}}$ $\cdot I_1[b(a^2-x^2)^{\frac{1}{2}}], \quad \lvert x \rvert<a$ $0, \quad \lvert x \rvert>a$
33	$\pi a^{-1}e^{iby/a}\operatorname{sech}(\frac{1}{2}\pi y/a)$ $N=\pi a^{-1}$	$\operatorname{sech}(ax+b)$
28	$a^{-\nu-iy}\Gamma(\nu+iy)$ $N=a^{-\nu}\Gamma(\nu), \quad \nu>0$	$e^{\nu x}\exp(-ae^x)$
34	$2^{\nu-1}a^{-1}[\Gamma(\nu)]^{-1}e^{iby/a}$ $\cdot\Gamma[\frac{1}{2}\nu-\frac{1}{2}iy/a)\Gamma(\frac{1}{2}\nu+\frac{1}{2}iy/a)$ $N=2^{\nu-1}a^{-1}[\Gamma(\frac{1}{2}\nu)]^2[\Gamma(\nu)]^{-1}, \quad \nu<0$	$[\operatorname{sech}(ax+b)]^{\nu}$

	$NG(y)$	$Nf(x)$
23	$ce^{b(a-\nu/c)}e^{iby}B[c(a+iy), \nu-c(a+iy)]$ $N=ce^{b(a-\nu/c)}B(ac, \nu-ac), \quad 0<a<\nu/c$	$(e^{b/c}+e^{-x/c})^{-\nu}e^{-ax}$
25	$i\pi \operatorname{csch}(\pi y)[\gamma+\psi(1-iy)]$ $N=\pi^2/6$	$(1+e^x)^{-1}\log(1+e^x)$
26	$-i\pi \operatorname{csch}(\pi y)[\psi(\nu)-\psi(\nu-iy)]$ $N=\psi'(\nu), \quad \nu>0$	$(1+e^x)^{-\nu}\log(1+e^x)$
29	$a^{-\nu-iy}\Gamma(\nu+iy)[\psi(\nu+iy)-\log a]$ $N=a^{-\nu}\Gamma(\nu)[\psi(\nu)-\log a], \quad \nu>0$	$\nu e^{\nu x}\exp(-ae^x)$
27	$a^{\lambda-\nu+iy}B(\lambda+iy, \nu-\lambda-iy)$ $\cdot[\psi(\nu)-\psi(\nu-\lambda-iy)+\log a]$ $N=a^{\lambda-\nu}B(\lambda, \nu-\lambda)[\psi(\nu)$ $-\psi(\nu-\lambda)+\log a], \quad a\geq 1, \quad \nu>\lambda>0$	$e^{-\lambda x}(a+e^{-x})^{-\nu}$ $\cdot\log(a+e^{-x})$
3	$(iy)^{-\nu-1}\gamma(\nu+1, iby)$ $N=b^{\nu+1}(\nu+1)^{-1}, \quad \nu>-1$	$x^\nu, \quad 0<x<b$ 0, otherwise
6	$(iy)^{-\nu-1}e^{-iby}\gamma(\nu+1, -iby)$ $N=(\nu+1)^{-1}b^{\nu+1}, \quad \nu>-1$	$(b-x)^\nu, \quad 0<x<b$ 0, otherwise
7	$\pi a^{-\nu}\csc(\pi\nu)[\Gamma(\nu)]^{-1}e^{iay}\Gamma(\nu, iay)$ $N=\pi a^{-\nu}\csc(\pi\nu), \quad 0<\nu<1$	$x^{-\nu}(a+x)^{-1}, \quad 0<x<\infty$ $0, \quad -\infty<x<0$
8	$\pi b^{-\nu}\csc(\pi\nu)[\Gamma(\nu)]^{-1}\Gamma(\nu, iby)$ $N=\pi b^{-\nu}\csc(\pi\nu), \quad 0<\nu<1$	$x^{-1}(x-b)^\nu, \quad b<x<\infty$ 0, otherwise
4	$(iy)^{\nu-1}\Gamma(1-\nu, iby)$ $N=(\nu-1)^{-1}b^{1-\nu}, \quad \nu>1$	$x^{-\nu}, \quad b<x<\infty$ $0, \quad -\infty<x<b$
5	$(iy)^{\nu-1}e^{iay}\Gamma(1-\nu, iay)$ $N=(\nu-1)^{-1}a^{1-\nu}, \quad \nu>1$	$(a+x)^{-\nu}, \quad 0<x<\infty$ $0, \quad -\infty<x<0$
30	$\Gamma(\lambda+iy)\zeta(\lambda+iy)$ $N=\Gamma(\lambda)\zeta(\lambda), \quad \lambda>1$	$[\exp(e^{-x})-1]^{-1}e^{-\lambda x}$

	$NG(y)$	$Nf(x)$
31	$(1-2^{1-\lambda-iy})\Gamma(\lambda+iy)\zeta(\lambda+iy)$ $N=(1-2^{1-\lambda})\Gamma(\lambda)\zeta(\lambda),\quad \lambda>0$	$[\exp(e^{-x})+1]^{-1}e^{-\lambda x}$
38	$\Gamma(\nu+iy)(a^2-b^2)^{-\frac{1}{2}iy}$ $\cdot\mathfrak{P}_{-iy}^{-\nu}[a(a^2-b^2)^{-\frac{1}{2}}]$ $N=\nu^{-1}b^{\nu}[a+(a^2-b^2)^{\frac{1}{2}}]^{-\nu},\quad a>b,\quad \nu>0$	$\exp(-ae^x)I_\nu(be^x)$
16	$\pi^{\frac{1}{2}}2^{\mp iby}[\Gamma(\nu)]^{-1}(\frac{1}{2}\mid y\mid/a)^{\nu-\frac{1}{2}}K_{\nu-\frac{1}{2}}(a\mid y\mid)$ $N=\frac{1}{2}\pi^{\frac{1}{2}}a^{1-2\nu}\Gamma(\nu-\frac{1}{2})[\Gamma(\nu)]^{-1},\quad \nu>\frac{1}{2}$	$[a^2+(x\pm b)^2]^{-\nu}$
32	$2(b/a)^{iy}K_{iy}(2ab)$ $N=2K_0(2ab)$	$\exp(-a^2e^x-b^2e^{-x})$
36	$2(b/a)^{iy}\operatorname{sech}(\pi y)K_{iy}(2ab)$ $N=2K_0(2ab)$	$\exp(a^2e^x+b^2e^{-x})\frac{1}{2}x$ $\cdot\operatorname{Erfc}(ae^{\frac{1}{2}x}+be^{-\frac{1}{2}x})$
40	$2K_{\frac{1}{2}\nu-iy}(a)K_{\frac{1}{2}\nu+iy}(b)$ $N=2K_{\frac{1}{2}\nu}(a)K_{\frac{1}{2}\nu}(b)$	$[(a+be^x)(ae^x+b)^{-1}]^{\frac{1}{2}\nu}$ $\cdot K_\nu[(a^2+b^2+2ab\cosh x)^{\frac{1}{2}}]$
17	$ia^{-\nu}y^{-1}\{\pi\nu\csc(\pi\nu)[\mathbf{J}_\nu(iay)-J_\nu(iay)]-1\}$ $N=a^{\nu+1}\nu(\nu^2-1)^{-1},\quad \nu>1$	$[(a^2+x^2)^{\frac{1}{2}}+x]^{-\nu},\quad 0<x<\infty$ $0,\quad -\infty<x<0$
18	$\pi a^{-\nu}\csc(\pi\nu)[\mathbf{J}_\nu(iay)-J_\nu(iay)]$ $N=\nu^{-1}a^{-\nu},\quad \nu>0$	$(a^2+x^2)^{-\frac{1}{2}}[(a^2+x^2)^{\frac{1}{2}}+x]^{-\nu},\quad 0<x<\infty$ $0,\quad -\infty<x<0$
10	$2^{\nu-\frac{1}{2}}b^{-\frac{1}{2}}\Gamma(\nu)D_{-2\nu}[2(iby)^{\frac{1}{2}}]$ $N=(2b/\pi)^{-\frac{1}{2}}\Gamma(\nu)[\Gamma(\frac{1}{2}+\nu)]^{-1},\quad \nu>0$	$(x-b)^{\nu-1}(x+b)^{-\nu-\frac{1}{2}},\quad b<x$ $0,\quad$ otherwise
11	$2^{\nu}a^{-\frac{1}{2}}\Gamma(\nu)e^{i\frac{1}{2}ay}D_{-2\nu}[(2iay)^{\frac{1}{2}}]$ $N=(\pi/a)^{\frac{1}{2}}\Gamma(\nu)[\Gamma(\frac{1}{2}+\nu)]^{-1},\quad \nu>0$	$x^{\nu-1}(a+x)^{-\nu-\frac{1}{2}},\quad x>0$ $0,\quad$ otherwise
15	$2^{\nu+\mu-1}B(\mu,\nu)e^{iy}{}_1F_1(\mu;\nu+\mu;-2iy)$ $N=2^{\nu+\mu-1}B(\mu,\nu),\quad \nu,\mu>0$	$(1-x)^{\nu-1}(1+x)^{\mu-1},\quad -1<x<1$ $0,\quad$ otherwise
14	$(b-a)^{\nu+\mu-1}(iy)^{-\mu-\nu}e^{-\frac{1}{2}iy(a+b)}$ $\cdot M_{\mu-\nu,\mu+\nu-\frac{1}{2}}[iy(b-a)]$ $N=(b-a)^{2\mu+2\nu-1}\Gamma(2\mu)$ $\cdot\Gamma(2\nu)[\Gamma(2\mu+2\nu)]^{-1},\quad \mu,\nu>0$	$(x-a)^{2\mu-1}(b-x)^{2\nu-1},\quad a<x<b$ $0,\quad$ otherwise

	$NG(y)$	$Nf(x)$
13	$(a+b)^{\nu+\mu+1}\Gamma(2\nu)(iy)^{-\mu-\nu}$ $\cdot\exp[i\frac{1}{2}y(a-b)]W_{\mu-\nu,\mu+\nu-\frac{1}{2}}[iy(a+b)]$ $N=(a+b)^{2\mu+2\nu-1}\Gamma(2\nu)$ $\cdot\Gamma(1-2\mu-2\nu)[\Gamma(1-2\mu)]^{-1},$ $0<\nu<\frac{1}{2}-\mu$	$(x+a)^{2\mu-1}(x-b)^{2\nu-1}, \quad b<x<\infty$ $0, \quad$ otherwise
35	$a^{-1}2^{\frac{1}{2}\nu-1}b^{-\frac{1}{2}\nu}[\Gamma(\nu)]^{-1}$ $\cdot\Gamma(\frac{1}{2}\nu+\frac{1}{2}iy/a)\Gamma(\frac{1}{2}\nu-\frac{1}{2}iy/a)M_{\frac{1}{2}iy/a,\frac{1}{2}\nu-\frac{1}{2}}(2b)$ $N=\pi^{\frac{1}{2}}a^{-1}(\frac{1}{2}b)^{\frac{1}{2}-\frac{1}{2}\nu}\Gamma(\frac{1}{2}\nu)I_{\frac{1}{2}\nu-\frac{1}{2}}(b), \quad \nu>0$	$\exp[-b\tanh(ax)]$ $\cdot(\mathrm{sech}\,ax)^{\nu}, \quad x>0$ $0, \quad x<0$
41	$2^{\frac{1}{2}\nu}(2\pi a^2)^{-\frac{1}{2}}$ $\cdot\exp(a^2)\Gamma[\frac{1}{2}(1+\nu+iy)]\Gamma[\frac{1}{2}(1+\nu-iy)]$ $\cdot\cos[\frac{1}{2}\pi(\nu-iy)]W_{-\frac{1}{2},i\frac{1}{2}y}(2a^2)$ $N=2^{\frac{1}{2}\nu}(2\pi a^2)^{-\frac{1}{2}}\exp(a^2)\cos(\frac{1}{2}\pi\nu)$ $\cdot[\Gamma(\frac{1}{2}+\frac{1}{2}\nu)]^2W_{-\frac{1}{2}\nu,0}(2a^2), \quad 0<\nu<1$	$\exp[-(a\sinh x)^2]D_{\nu}(2a\sinh x)$

APPENDIX

DISTRIBUTION FUNCTIONS AND THEIR FOURIER TRANSFORMS FOUND IN THE STATISTICAL LITERATURE

TABLE A

UNIVARIATE DENSITY FUNCTIONS

No.	Probability density function $f(x)$	Characteristic function $\varphi(t)=\int_{-\infty}^{\infty} e^{itx}f(x)\,dx$	Notes
1	$(2\pi)^{-\frac{1}{2}}\exp(-\frac{1}{2}x^2)$ $-\infty<x<\infty$	$\exp(-\frac{1}{2}t^2)$	(1)
2	$[a(2\pi)^{\frac{1}{2}}]^{-1}\exp-\frac{1}{2}\left(\frac{x-b}{a}\right)^2$ $-\infty<x<\infty,\quad a>0$	$\exp(ibt-\frac{1}{2}a^2t^2)$	(2)
3	$(b-a)^{-1}$ $a\leq x\leq b,\quad a<b$	$(e^{ibt}-e^{iat})/i(b-a)t$	(3)
4	$(2a)^{-1}$ $-a\leq x\leq a,\quad a>0$	$\sin(at)/at$	(4)
5	$1-\lvert x\rvert$ $-1\leq x\leq +1$	$(2/t^2)(1-\cos t)$	

No.	Probability density function $f(x)$	Characteristic function $\varphi(t)=\int_{-\infty}^{\infty} e^{itx}f(x)\,dx$	Notes
6	$(1-\cos x)/\pi x^2$	$\begin{cases}1-\|t\|, & \|t\|\le 1\\ 0, & \|t\|>1\end{cases}$	(5)
7	ae^{-ax} $x>0,\quad a>0$	$a/(a-it)$	(6)
8	$[\Gamma(p)]^{-1}x^{p-1}e^{-x}$ $x>0,\quad p>0$	$(1-it)^{-p}$	
9	$a^p[\Gamma(p)]^{-1}x^{p-1}e^{-ax}$ $x>0,\quad p>0,\quad a>0$	$a^p(a-it)^{-p}$	(7)
10	$[\Gamma(p+q)/\Gamma(p)\Gamma(q)]e^{px}(1+e^x)^{-p-q}$ $-\infty<x<\infty,\quad p>0,\quad q>0$	$\Gamma(p+it)\Gamma(q-it)/\Gamma(p)\Gamma(q)$	
11	$[2^{-\frac{1}{2}n}/\Gamma(\frac{1}{2}n)]x^{\frac{1}{2}n-1}e^{-\frac{1}{2}x}$ $x>0,\quad n>0$ (integer)	$(1-2it)^{-\frac{1}{2}n}$	(8)
12	$[\pi(1+x^2)]^{-1}$ $-\infty<x<\infty$	$e^{-\|t\|}$	(9)
13	$a[\pi a^2+\pi(x-m)^2]^{-1}$ $-\infty<x<\infty,\quad a>0$	$\exp(imt-a\|t\|)$	(10)
14	$(2/\pi)(1+x^2)^{-2}$ $-\infty<x<\infty$	$(1+\|t\|)e^{-\|t\|}$	(11)
15	$(2/\pi)x^2(1+x^2)^{-1}$ $-\infty<x<\infty$	$(1-\|t\|)e^{-\|t\|}$	(11)
16	$\frac{1}{2}e^{-\|x\|}$ $-\infty<x<\infty$	$(1+t^2)^{-1}$	(12)
17	$(2a)^{-1}e^{-a^{-1}\|x-m\|}$ $-\infty<x<\infty,\quad a>0$	$(1+a^2t^2)^{-1}e^{imt}$	(13)
18	$[\Gamma(p)]^{-1}\exp(px-e^x)$ $-\infty<x<\infty,\quad p>0$	$\Gamma(p+it)/\Gamma(p)$	(11)

No.	Probability density function $f(x)$	Characteristic function $\varphi(t)=\int_{-\infty}^{\infty} e^{itx}f(x)\,dx$	Notes
19	$p^{p+1}e^{-p}/a\Gamma(p+1)[1+(x/a)]^p \cdot\exp[-p(x/a)]$ $-a\leq x<\infty;\ p>0,\quad a>0$	$e^{-iat}[1-ia(t/p)]^{-p-1}$	(14)
20	$\frac{1}{4}(1+\lvert x\rvert)e^{-\lvert x\rvert}$ $-\infty<x<\infty$	$(1+t^2)^{-2}$	(11)
21	$\exp(-x-e^{-x})$ $-\infty<x<\infty$	$\Gamma(1-it)$	(15)
22	$(a/\pi^{\frac{1}{2}})\exp[2a(b^{\frac{1}{2}})]x^{-\frac{3}{2}}\{\exp[-bx-(a^2/x)]\}$ $x>0;\quad a>0,\quad b>0$	$\exp\{2a[b^{\frac{1}{2}}-(b-it)^{\frac{1}{2}}]\}$	(11)
23	$\frac{1}{2}\,\mathrm{sech}(\frac{1}{2}\pi x)$ $-\infty<x<\infty$	$\mathrm{sech}\,t$	(16)
24	$\frac{1}{4}\pi[\mathrm{sech}(\frac{1}{2}\pi x)]^2$ $-\infty<x<\infty$	$t\,\mathrm{csch}\,t$	(16)
25	$\frac{1}{2}x\,\mathrm{csch}(\frac{1}{2}\pi x)$ $-\infty<x<\infty$	$\mathrm{sech}^2 t$	(16)
26	$[\Gamma(p+q)/\Gamma(p)\Gamma(q)]x^{p-1}(1-x)^{q-1}$ $0\leq x\leq 1;\quad p>0,\quad q>0$	${}_1F_1(p;p+q;-it)$	(17)
27	$[\Gamma(m)/\pi^{\frac{1}{2}}\Gamma(m-\frac{1}{2})](1+x^2)^{-m}$ $-\infty<x<\infty;\quad m>0$	$[(\frac{1}{2}\lvert t\rvert)^{m-\frac{1}{2}}/\pi\Gamma(m-\frac{1}{2})]K_{m-\frac{1}{2}}(\lvert t\rvert)$	(18)
28	$[\pi^{\frac{1}{2}}\Gamma(m+\frac{1}{2})]^{-1}(\frac{1}{2}\lvert x\rvert)^m K_m(\lvert x\rvert)$ $-\infty<x<\infty;\quad m>0$	$(1+t^2)^{-m-\frac{1}{2}}$	(11)
29	$[m^p/\Gamma(p)]x^{-p-1}e^{-m/x}$ $x>0;\quad m>0,\quad p>0$	$[2/\Gamma(p)](imt)^{\frac{1}{2}p}K_p[2(-imt)^{\frac{1}{2}}]$	(11)
30	$[\Gamma(p+q)/\Gamma(p)\Gamma(q)]x^{p-1}(1+x)^{-p-q}$ $x>0;\quad p>0,\quad q>0$	${}_1F_1(p;-q;-it)$	
31	$[\Gamma(p+q)/\Gamma(p)\Gamma(q)]e^{-px}(1-e^{-x})^{q-1}$ $x>0,\quad p>0,\quad q>0$	$\Gamma(p+q)\Gamma(p-it)/\Gamma(p)\Gamma(p+q-it)$	

No.	Probability density function $f(x)$	Characteristic function $\varphi(t)=\int_{-\infty}^{\infty} e^{itx}f(x)\,dx$	Notes
32	$2\Gamma(2p)[\Gamma(p)]^{-2}(e^{x}+e^{-x})^{-2p}$ $-\infty<x<\infty,\quad p>0$	$\lvert\Gamma(p+\frac{1}{2}it)\rvert^{2}[\Gamma(p)]^{-2}$	(11)
33	$[2^{n}\Gamma(m+n)]^{-1}x^{n-1}W_{m,n-\frac{1}{2}}(2x)$ $-\infty<x<\infty,\quad m>0,\quad n>0$	$(1+it)^{m-n}(1-it)^{-m-n}$	(11)
34	$(\frac{1}{2}b)^{1-p}a^{p}\exp(-\frac{1}{4}b^{2}/a)x^{\frac{1}{2}p-\frac{1}{2}}$ $\cdot\exp(-ax)I_{p-1}[b(x)^{\frac{1}{2}}]$ $x>0,\quad a>0,\quad p>0$	$[1-i(t/a)]^{-p}$ $\cdot\exp\{it(b/2a)^{2}[1-i(t/a)]^{-1}\}$	(19)
35	$b^{\frac{1}{2}-\frac{1}{2}p}e^{-b}x^{\frac{1}{2}p-\frac{1}{2}}e^{-x}I_{p-1}[2(bx)^{\frac{1}{2}}]$ $x>0,\quad b>0,\quad p>0$	$(1-ib)^{-p}\exp[ibt/(1-it)]$	(20)
36	$\lambda^{\frac{1}{2}-\frac{1}{4}n}e^{-\frac{1}{2}\lambda}x^{\frac{1}{4}n-\frac{1}{2}}e^{-\frac{1}{2}x}I_{\frac{1}{2}n-1}(\lambda x)^{\frac{1}{2}}$ $x>0,\quad \lambda>0,\quad n>0$ (integer)	$(1-2it)^{-\frac{1}{2}n}\exp[it\lambda/(1-2it)]$	(21)
37	$[2^{-\frac{1}{2}p-\frac{1}{2}q}/\Gamma(p)]x^{\frac{1}{2}p+\frac{1}{2}-1}W_{\frac{1}{2}p-\frac{1}{2}q,\frac{1}{2}p+\frac{1}{2}q-\frac{1}{2}}(2x)$ $-\infty<x<\infty,\quad p>0,\quad q>0$	$(1-it)^{-p}(1+it)^{-q}$	(22)
38	$[2^{\frac{1}{2}-p}/\Gamma(p)\pi^{\frac{1}{2}}]x^{p-\frac{1}{2}}K_{p-\frac{1}{2}}(x)$ $-\infty<x<\infty;\quad p>0$	$(1+t^{2})^{-p}$	(23)
39	$[\Gamma(p+q)/\Gamma(p)\Gamma(q)](1+e^{x})^{-p-q}e^{px}$ $0<x<\infty;\quad p>0,\quad q>0$	$\Gamma(p+it)\Gamma(q-it)/\Gamma(p)\Gamma(q)$	(22)
40	$\exp(-\frac{1}{2}b^{2})2^{-\frac{1}{2}p-\frac{1}{2}q}x^{\frac{1}{2}p+\frac{1}{2}q-1}$ $\cdot\sum_{k=0}^{\infty}(\frac{1}{2}b)^{2k}\frac{x^{k}}{k!\Gamma(p+k)}W_{\frac{1}{2}p-\frac{1}{2}q,\frac{1}{2}p+\frac{1}{2}q+k-\frac{1}{2}}(2x)$ $-\infty<x<\infty;\quad p>0,\quad q>0$	$(1-it)^{-p}(1+it)^{-q}$ $\cdot\exp[-\frac{1}{2}b^{2}t^{2}/(1+t^{2})]$	(21)
41	$\exp(-\frac{1}{2}b^{2})\pi^{-\frac{1}{2}}(\frac{1}{2}x)^{p-\frac{1}{2}}$ $\cdot\sum_{k=0}^{\infty}\frac{(\frac{1}{4}b^{2}x)^{k}}{k!\Gamma(p+k)}K_{p+k-\frac{1}{2}}(x)$ $-\infty<x<\infty,\quad p>0$	$(1+t^{2})^{-p}\exp[-\frac{1}{2}b^{2}t^{2}/(1+t^{2})]$	(21)

No.	Probability density function $f(x)$	Characteristic function $\varphi(t)=\int_{-\infty}^{\infty} e^{itx}f(x)\,dx$	Notes
42	$(3/\pi)^{\frac{1}{2}}x^{-(16/27)x^2}W_{\frac{1}{2},\frac{1}{3}}(\frac{32}{27}x^{-2})$ $x>0$	$\exp\{-\|t\|^{\frac{2}{3}}[1-i\sqrt{3}(t/\|t\|)]\}$	(24)
43	$\frac{1}{6}(3/\pi)^{\frac{1}{2}}x^{-1}\exp(x^3/27)W_{-\frac{1}{2},\frac{1}{6}}(\frac{2}{27}x^3)$ $x>0$	$\exp\{-\|t\|^{\frac{3}{2}}[1+i(t/\|t\|)]\}$	(24)
44	$(3/\pi)^{\frac{1}{2}}\|x\|^{-1}e^{-(2/27)x^2}W_{\frac{1}{2},\frac{1}{6}}(\frac{4}{27}x^{-2})$ $-\infty<x<\infty$	$\exp(-\|t\|^{\frac{2}{3}})$	(24)
45	$\pi^{-1}3^{-\frac{1}{4}}(2/3x)^{\frac{3}{4}}K_{\frac{1}{3}}[\frac{4}{9}3^{\frac{1}{4}}(2/3x)^{\frac{1}{2}}]$ $x>0$	$\exp(-\|t\|^{\frac{1}{2}}\{1-(i/\sqrt{3})[t/(\|t\|)]\})$	(24)
46	$(2\pi)^{-\frac{1}{2}}e^{-\frac{1}{2}x}x^{-\frac{3}{2}}$ $x>0$	$\exp(-\|t\|^{\frac{1}{2}}\{[1+i[t/(\|t\|)]\})$	(25)

Notes:

(1) Standardized normal distribution. (2) Normal distribution.

(3) Rectangular or uniform distribution. (4) Symmetric rectangular distribution.

(5) Khintchine's convex distribution, H. Cramer, "Mathematical Methods of Statistics," p. 94. Princeton Univ. Press, Princeton, New Jersey, 1946.

(6) Exponential distribution. (7) Gamma distribution.

(8) Chi-square distribution with n degrees of freedom.

(9) Standardized Cauchy distribution. (10) Cauchy distribution.

(11) S. Kullback, "Theory and Application of Characteristic Functions," Lecture notes.

(12) Standardized Laplace distribution. (13) Laplace distribution.

(14) Pearsonian type III distribution.

(15) R. A. Fisher and L. H. C. Tippet, *Proc. Cambridge Phil. Soc.* **24,** 180 (1928).

(16) J. Bass and P. Levy, *C. R. Acad. Sci. (Paris)* **230,** 815 (1950).

(17) Beta distribution. (18) Pearsonian type VII distribution.

(19) R. G. Laha, *Bull. Calcutta Math. Soc.* **46,** 60 (1954).

(20) T. A. McKay, *Biometrika* **24,** 39 (1932).

(21) Noncentral chi-square with n degrees of freedom and a noncentral parameter λ. See reference in note (19).

(22) S. Kullback, *Ann. Math. Stat.* **7,** 52 (1936).

(23) K. Pearson, S. A. Stouffer, and F. N. David, *Biometrika* **24** (1932).

(24) V. M. Zolotarev, *Dokl. Akad. Nauk. USSR* **98,** 715 (1954).

(25) B. V. Gnedenko and A. N. Kolmogorov, "Limit Distribution for Sums of Independent Random Variables," p. 171. Addison-Wesley, Reading, Massachusetts, 1954.

TABLE B

UNIVARIATE DISCRETE DISTRIBUTIONS

No.	Probability (discrete) p_x	Characteristic function $\varphi(t)=\sum_x e^{itx}p_x$	Notes
1	$\binom{n}{x}P^x(1-P)^{n-x}$ $x=0, 1, \ldots, n;\quad 0<P<1,$ $n>0$ (integer)	$[1+P(e^{it}-1)]^n$	(1)
2	$e^{-m}\dfrac{m^x}{x!}$ $x=0, 1, \ldots, \infty;\quad m>0$	$\exp[m(e^{it}-1)]$	(2)
3	$\binom{n+x-1}{x}P^x(1-P)^n$ $x=0, 1, \ldots, \infty;\quad 0<P<1,\quad n>0$	$[(1-P)/(1-Pe^{it})]^n$	(3)
4	$\binom{M}{x}\binom{N}{n-x}\Big/\binom{M+N}{n}$ $x=0, 1, 2, \ldots, n;\quad M, N, n$ positive integers	${}_2F_1\,(-M, -n; -M-N; 1-e^{it})$	(4)

No.	Probability (discrete) p_x	Characteristic function $\varphi(t)=\sum_x e^{itx}p_x$	Notes
5	$a_x\theta^x/f(\theta)$ a_x real, $+(\theta)=\sum_x a_x\theta^x$ θ real, $x=0, 1, 2, \ldots, \infty$	$f(\theta e^{it})/f(\theta)$	(5)
6	$-\theta^x/x\log(1-\theta)$ $x=1, 2, \ldots, \infty;\quad 0<\theta<1$	$\log(1-\theta e^{it})/\log(1-\theta)$	(5)
7	$\dfrac{2\theta^{2x+1}}{(2x+1)\log[(1+\theta)/(1-\theta)]}$ $x=1, 2, \ldots, \infty;\quad 0<\theta<1$	$\dfrac{\log(1+\theta e^{it})-\log(1-\theta e^{it})}{\log(1+\theta)-\log(1-\theta)}$	(5)
8	$(\arcsin\theta)^{-1},\quad x=1$ $\dfrac{1\cdot 3\cdots(2x-1)}{2\cdot 4\cdots(2n)}\dfrac{\theta^{2x+1}}{2x+1}(\arcsin\theta)^{-1}$ $x=1, 2, 3, \ldots;\quad 0<\theta<1$	$\dfrac{\arcsin(\theta e^{it})}{\arcsin(\theta)}$	(5)

Notes:
(1) Binomial distribution. (2) Poisson distribution.
(3) Negative binomial distribution. (4) Hypergeometric distribution.
(5) A. Noack, *Ann. Math. Stat.* **21,** 127 (1950).

TABLE C

MULTIVARIATE DENSITY FUNCTIONS

No.	Probability density function $f(x_1, \ldots, x_k)$	Characteristic function $\varphi(t_1, \ldots, t_k) = \int_{-\infty}^{\infty} \exp[i(t_1x_1, \ldots, t_kx_k)] \cdot f(x_1, \ldots, x_k)\, dx_1 \cdots dx_k$	Notes
1	$\lvert A\rvert^{-\frac{1}{2}}(2\pi)^{-\frac{1}{2}k}\exp[-\frac{1}{2}(X-M)A^{-1}(X-M')]$ $X=(x_1, \ldots, x_k)$; $-\infty<x_j<\infty$, $j=1, 2, \ldots, k$ $M=(m_1, m_2, \ldots, m_k)$ $A=(a_{ij})$; $i, j=1, 2, \ldots, n$ symmetric *positive definite*	$\exp[iTM'-\frac{1}{2}TAT']$ $T=(t_1, t_2, \ldots, t_k)$	(1)
2	$\pi^{-\frac{1}{4}k(k-1)} \Big/ \prod_{j=1}^{k}\Gamma(\frac{1}{2}n-\frac{1}{2}j)$ $\cdot \lvert A\rvert^{(n-1)/2}\lvert X\rvert^{(n-k-2)/2}\exp(-\sum_{i,j} a_{ij}x_{ij})$ $x_{ij}=x_{ji}$ and $X=(x_{ij})$, $i, j=1, 2, \ldots, k$ symmetric *positive definite* $A=(a_{i,j})$; $i, j=1, 2, \cdots, k$ symmetric *positive definite*	$\lvert A\rvert^{(n-1)/2}/\lvert A-iT\rvert^{(n-1)/2}$ $T=(\epsilon_{i,j}t_{ij})$, $i, j=1, 2, \ldots, k$ $\epsilon_{ij}=\begin{cases}1, & i=j\\ 0, & i\neq j\end{cases}$	(2)

No.	Probability density function $f(x_1, \ldots, x_k)$	Characteristic function $\varphi(t_1, \ldots, t_k) = \int_{-\infty}^{\infty} \exp[i(t_1x_1, \ldots, t_kx_k)] \cdot f(x_1, \ldots, x_k)\, dx_1 \cdots dx_k$	Notes
3	$\frac{N!}{x_1!x_2!\cdots x_k!} p_1^{x_1}p_2^{x_2}\cdots p_k^{x_k}$ $x_j > 0$ (integer); $j=1, 2, \ldots, k$ $\sum_j x_j = N$; $0 < p_j < 1$ $j = 1, 2, \ldots, k$; $\sum_j p_j = 1$	$(p_1e^{it_1} + p_2e^{it_2} + \cdots + p_ke^{it_k})^N$	(3)

Notes:

(1) Multivariate normal distribution. (2) Wishart distribution.

(3) Multinomial distribution.

LIST OF ABBREVIATIONS, SYMBOLS, AND NOTATIONS

ϵ_n = Neumann's number, $\epsilon_0 = 1, \quad \epsilon_n = 2, \quad n = 1, 2, 3, \ldots$

γ = Euler's constant, $\quad \gamma = 0.57721\cdots$

$\binom{a}{b}$ = binomial coefficient, $\quad \binom{a}{b} = \Gamma(a+1)/\Gamma(b+1)\Gamma(a-b+1)$

$\tau_{\nu,1}$ = first positive root of $J_\nu(x)$, $\quad \zeta_{\nu,1}$ = first positive root of $Y_\nu(x)$

Notations

1. Elementary Functions

Trigonometric and inverse trigonometric functions: $\sin x$, $\cos x$, $\tan x = \sin x/\cos x$, $\operatorname{ctn} x = \cos x/\sin x$, $\sec x = 1/\cos x$, $\csc x = 1/\sin x$; $\arcsin x$, $\arccos x$, $\arctan x$, $\operatorname{arcctn} x$.

Hyperbolic functions: $\sinh x = \frac{1}{2}(e^x - e^{-x})$, $\cosh x = \frac{1}{2}(e^x + e^{-x})$, $\tanh x = \sinh x/\cosh x$, $\operatorname{ctnh} x = \cosh x/\sinh x$, $\operatorname{sech} x = 1/\cosh x$, $\operatorname{csch} x = 1/\sinh x$.

2. Gamma Function and Related Functions

Gamma function: $\Gamma(z) = \int_0^\infty e^{-t} t^{z-1}\, dt, \qquad \operatorname{Re} z > 0$

Beta function: $B(x, y) = \Gamma(x)\Gamma(y)/\Gamma(x+y)$

ψ function: $\psi(z) = d[\log\Gamma(z)]/dz$

3. Riemann and Hurwitz Zeta Function

$$\zeta(s) = \sum_{n=1}^{\infty} n^{-s}, \qquad \text{Re } s > 1$$

$$\zeta(s, \nu) = \sum_{n=1}^{\infty} (n+\nu)^{-s}, \qquad \text{Re } s > 1$$

4. Legendre Functions (Definition after Hobson)

$$\mathfrak{P}_\nu^\mu(z) = [\Gamma(1-\mu)]^{-1}[(z+1)/(z-1)]^{\frac{1}{2}\mu}{}_2F_1(-\nu, \nu+1; 1-\mu; \tfrac{1}{2}-\tfrac{1}{2}z)$$

$$\mathfrak{Q}_\nu^\mu(z) = 2^{-\nu-1}[\Gamma(\nu+\tfrac{3}{2})]^{-1}e^{i\pi\mu}\pi^{\frac{1}{2}}\Gamma(\mu+\nu+1) \cdot z^{-\mu-\nu-1}(z^2-1)^{\frac{1}{2}\mu}{}_2F_1[\tfrac{1}{2}(\mu+\nu+1), \tfrac{1}{2}(\mu+\nu+2), \nu+\tfrac{3}{2}; z^{-2}], \qquad 1 < z < \infty$$

$$P_\nu^\mu(x) = [\Gamma(1-\mu)]^{-1}[(1+x)/(1-x)]^{\frac{1}{2}\mu}{}_2F_1(-\nu, \nu+1; 1-\mu; \tfrac{1}{2}-\tfrac{1}{2}x), \qquad -1 < x < 1$$

$$Q_\nu^\mu(x) = \tfrac{1}{2}\exp(-i\pi\mu)[\exp(-i\tfrac{1}{2}\pi\mu)\mathfrak{Q}_\nu^\mu(x+i0) + \exp(i\tfrac{1}{2}\pi\mu)\mathfrak{Q}_\nu^\mu(x-i0)], \qquad -1 < x < 1$$

5. Bessel Functions

$$J_\nu(z) = \sum_{n=1}^{\infty} (-1)^n[n!\Gamma(\nu+n+1)]^{-1}(\tfrac{1}{2}z)^{\nu+2n}$$

$$Y_\nu(z) = [\sin(\pi\nu)]^{-1}[J_\nu(z)\cos(\nu\pi) - J_{-\nu}(z)]$$

6. Modified Bessel Functions

$$I_\nu(z) = \exp(-i\tfrac{1}{2}\pi\nu)J_\nu(ze^{i\pi}) = \sum_{n=1}^{\infty} [n!\Gamma(\nu+n+1)]^{-1}(\tfrac{1}{2}z)^{\nu+2n}$$

$$K_\nu(z) = \tfrac{1}{2}\pi(\sin\pi\nu)^{-1}[I_{-\nu}(z) - I_\nu(z)]$$

7. Anger–Weber Functions

$$\mathbf{J}_\nu(z) = \pi^{-1}\int_0^\pi \cos(z\sin t - \nu t)\,dt, \qquad \mathbf{E}_\nu(z) = -\pi^{-1}\int_0^\pi \sin(z\sin t - \nu t)\,dt$$

8. Struve Functions

$$\mathbf{H}_\nu(z) = \sum_{n=0}^{\infty} (-1)^n[\Gamma(n+\tfrac{3}{2})\Gamma(\nu+n+\tfrac{3}{2})]^{-1}(\tfrac{1}{2}z)^{\nu+2n+1}$$

$$\mathbf{L}_\nu(z) = -i\exp(-i\tfrac{1}{2}\pi\nu)\mathbf{H}_\nu(ze^{i\frac{1}{2}\pi})$$

9. Lommel Functions

$$s_{\mu,\nu}(z)=[(\mu-\nu+1)(\mu+\nu+1)]^{-1}\,{}_1F_2[1;\tfrac{1}{2}(\mu-\nu+3),\tfrac{1}{2}(\mu+\nu+3);-\tfrac{1}{4}z^2],\quad \mu\pm\nu\neq -1,-2,-3,\ldots$$

$$S_{\mu,\nu}(z)=s_{\mu,\nu}(z)+2^{\mu-1}\Gamma[\tfrac{1}{2}(\mu-\nu+1)]\Gamma[\tfrac{1}{2}(\mu+\nu+1)]\cdot\{\sin[\tfrac{1}{2}\pi(\mu-\nu)]J_\nu(z)-\cos[\tfrac{1}{2}\pi(\mu-\nu)]Y_\nu(z)\}$$

9a. Special Cases of Lommel Functions

$$s_{\nu,\nu}(z)=\pi^{\frac{1}{2}}2^{\nu-1}\Gamma(\tfrac{1}{2}+\nu)\mathbf{H}_\nu(z)$$

$$S_{\nu,\nu}(z)=\pi^{\frac{1}{2}}2^{\nu-1}\Gamma(\tfrac{1}{2}+\nu)[\mathbf{H}_\nu(z)-Y_\nu(z)]$$

$$s_{0,\nu}(z)=\tfrac{1}{2}\pi\csc(\pi\nu)[\mathbf{J}_\nu(z)-\mathbf{J}_{-\nu}(z)]$$

$$S_{0,\nu}(z)=\tfrac{1}{2}\pi\csc(\pi\nu)[\mathbf{J}_\nu(z)-\mathbf{J}_{-\nu}(z)-J_\nu(z)+J_{-\nu}(z)]$$

$$s_{-1,\nu}(z)=-\tfrac{1}{2}\pi\nu^{-1}\csc(\pi\nu)[\mathbf{J}_\nu(z)+\mathbf{J}_{-\nu}(z)]$$

$$S_{-1,\nu}(z)=\tfrac{1}{2}\pi\nu^{-1}\csc(\pi\nu)[J_\nu(z)+J_{-\nu}(z)-\mathbf{J}_\nu(z)-\mathbf{J}_{-\nu}(z)]$$

$$s_{1,\nu}(z)=1-\tfrac{1}{2}\pi\nu\csc(\pi\nu)[\mathbf{J}_\nu(z)+\mathbf{J}_{-\nu}(z)]$$

$$S_{1,\nu}(z)=1+\tfrac{1}{2}\pi\nu\csc(\pi\nu)[J_\nu(z)+J_{-\nu}(z)-\mathbf{J}_\nu(z)-\mathbf{J}_{-\nu}(z)]$$

$$S_{\frac{1}{2},\frac{1}{2}}(z)=z^{-\frac{1}{2}};\qquad S_{\frac{3}{2},\frac{1}{2}}(z)=z^{\frac{1}{2}}$$

$$S_{-\frac{1}{2},\pm\frac{1}{2}}(z)=z^{-\frac{1}{2}}[\sin z\,\mathrm{Ci}(z)-\cos z\,\mathrm{si}(z)]$$

$$S_{-\frac{3}{2},\pm\frac{1}{2}}(z)=-z^{-\frac{3}{2}}[\sin z\,\mathrm{si}(z)+\cos z\,\mathrm{Ci}(z)],\ \lim[\Gamma(\nu-\mu)]^{-1}s_{\mu-1,\nu}(z)=-2^{\nu-1}\Gamma(\nu)J_\nu(z)$$

Lommel functions of two variables:

$$U_\nu(w,z)=\sum_{n=0}^{\infty}(-1)^n(w/z)^{\nu+2n}J_{\nu+2n}(z)$$

$$V_\nu(w,z)=\cos(\tfrac{1}{2}w+\tfrac{1}{2}z^2/w+\tfrac{1}{2}\nu\pi)+U_{2-\nu}(w,z)$$

10. Generalized Hypergeometric Series

$$ {}_mF_n(a_1, a_2, \ldots, a_m; b_1, b_2, \ldots, b_n; z) = \frac{\Gamma(b_1)\cdots\Gamma(b_n)}{\Gamma(a_1)\cdots\Gamma(a_m)} \cdot \sum_{k=0}^{\infty} \frac{\Gamma(a_1+k)\cdots\Gamma(a_m+k)}{\Gamma(b_1+k)\cdots\Gamma(b_n+k)} \frac{z^k}{k!} $$

11. Gaussian Hypergeometric Series

$$ {}_2F_1(a, b; c, z) = \frac{\Gamma(c)}{\Gamma(a)\Gamma(b)} \sum_{k=0}^{\infty} \frac{\Gamma(a+k)\Gamma(b+k)}{\Gamma(c+k)} \frac{z^k}{k!} $$

12. Confluent Hypergeometric Functions

$$ {}_1F_1(a; c; z) = \frac{\Gamma(c)}{\Gamma(a)} \sum_{k=0}^{\infty} \frac{\Gamma(a+k)}{\Gamma(c+k)} \frac{z^k}{k!} $$

Whittaker functions:

$$ M_{k,\mu}(z) = z^{\mu+\frac{1}{2}} e^{-\frac{1}{2}z} {}_1F_1(\tfrac{1}{2}+\mu-k; 2\mu+1; z) $$

$$ W_{k,\mu}(z) = \frac{\Gamma(-2\mu)}{\Gamma(\frac{1}{2}-\mu-k)} M_{k,\mu}(z) + \frac{\Gamma(2\mu)}{\Gamma(\frac{1}{2}+\mu-k)} M_{k,-\mu}(z) $$

Parabolic cylindrical functions:

$$ D_\nu(z) = 2^{\frac{1}{2}\nu+\frac{1}{4}} z^{-\frac{1}{2}} W_{\frac{1}{2}\nu+\frac{1}{4},\pm\frac{1}{4}}(\tfrac{1}{2}z^2) $$

$$ \begin{aligned} D_n(z) &= (-1)^n \exp(\tfrac{1}{4}z^2)\,(d^n/dz^n)[\exp(-\tfrac{1}{2}z^2)] \\ &= \exp[-\tfrac{1}{4}z^2] He_n(z), \qquad n = 0, 1, 2, \ldots \end{aligned} $$

$$ D_{-1}(z) = (\tfrac{1}{2}\pi)^{\frac{1}{2}} \exp(\tfrac{1}{4}z^2)\ \mathrm{Erf}(z2^{-\frac{1}{2}}) $$

$$ D_{-\frac{1}{2}}(z) = (\tfrac{1}{2}z/\pi)^{\frac{1}{2}} K_{\frac{1}{4}}(\tfrac{1}{4}z^2) $$

Incomplete gamma functions:

$$ \gamma(\alpha, z) = \int_0^z t^{\alpha-1} e^{-t}\,dt = \alpha^{-1} z\,{}_1F_1(\alpha; \alpha+1; -z) $$

$$ \Gamma(\alpha, z) = \int_z^\infty t^{\alpha-1} e^{-t}\,dt = \Gamma(\alpha) - \gamma(\alpha, z) = z^{\frac{1}{2}\alpha-\frac{1}{2}} e^{-\frac{1}{2}z} W_{\frac{1}{2}\alpha-\frac{1}{2},\pm\frac{1}{2}\alpha}(z) $$

Error functions:

$$\operatorname{Erf}(z) = 2\pi^{-\frac{1}{2}} \int_0^z \exp(-t^2)\, dt = \begin{cases} 2\pi^{-\frac{1}{2}} z\, {}_1F_1(\tfrac{1}{2}; \tfrac{3}{2}; -z^2) \\ 2\pi^{-\frac{1}{2}} z^{-\frac{1}{2}} \exp(-\tfrac{1}{2}z^2)\, M_{-\frac{1}{4},\frac{1}{4}}(z^2) \end{cases}$$

$$\operatorname{Erfc}(z) = 2\pi^{-\frac{1}{2}} \int_z^\infty \exp(-t^2)\, dt = 1 - \operatorname{Erf}(z) = (\pi z)^{-\frac{1}{2}} \exp(-\tfrac{1}{2}z^2)\, W_{-\frac{1}{4},\pm\frac{1}{4}}(z^2)$$

$$\operatorname{Erf}(x^{\frac{1}{2}} e^{i\frac{1}{4}\pi}) = 2^{\frac{1}{2}} e^{i\frac{1}{4}\pi}[C(x) - iS(x)]$$

$$\operatorname{Erfc}(x^{\frac{1}{2}} e^{i\frac{1}{4}\pi}) = 1 - C(x) - S(x) - i[C(x) - S(x)]$$

Fresnel integrals:

$$C(x) = (2\pi)^{-\frac{1}{2}} \int_0^x t^{-\frac{1}{2}} \cos t\, dt$$

$$S(x) = (2\pi)^{-\frac{1}{2}} \int_0^x t^{-\frac{1}{2}} \sin t\, dt$$

Exponential integral:

$$-\operatorname{Ei}(-z) = \int_z^\infty t^{-1} e^{-t}\, dt = \Gamma(0, z) = z^{-\frac{1}{2}} e^{-\frac{1}{2}z} W_{-\frac{1}{2},0}(z) = -\gamma \log z - \sum_{n=1}^{\infty} (n!n)^{-1}(-z)^n$$

$$\overline{\operatorname{Ei}}(x) = \lim_{\epsilon \to 0} \left[\int_{-x}^{-\epsilon} t^{-1} e^{-t}\, dt + \int_\epsilon^\infty t^{-1} e^{-t}\, dt = \gamma + \log x + \sum_{n=1}^{\infty} (n!n)^{-1} x^n \right], \qquad x > 0$$

$$\operatorname{Ei}(-ix) = \operatorname{Ci}(x) - i\, \operatorname{si}(x)$$

$$\overline{\operatorname{Ei}}(ix) = \operatorname{Ci}(x) + i[\pi + \operatorname{si}(x)]$$

Sine integral:

$$\operatorname{Si}(x) = \int_0^x t^{-1} \sin t\, dt$$

$$\operatorname{si}(x) = -\int_x^\infty t^{-1} \sin t\, dt = \operatorname{Si}(x) - \tfrac{1}{2}\pi$$

Cosine integral:

$$\operatorname{Ci}(x) = -\int_x^\infty t^{-1} \cos t\, dt = \gamma + \log x + \sum_{n=1}^{\infty} [2n(2n)!]^{-1}(-1)^n x^{2n}$$

13. Elliptic Integrals

$$K(k) = \int_0^{\frac{1}{2}} (1-k^2 \sin^2 t)^{-\frac{1}{2}}\, dt = \tfrac{1}{2}\pi\ {}_2F_1(\tfrac{1}{2}, \tfrac{1}{2}; 1; k^2)$$

$$E(k) = \int_0^{\frac{1}{2}} (1-k^2 \sin^2 t)^{\frac{1}{2}}\, dt = \tfrac{1}{2}\pi\ {}_2F_1(-\tfrac{1}{2}, \tfrac{1}{2}; 1; k^2)$$

14. Bessel Integral Function

$$\mathrm{Ji}_0(x) = -\int_x^\infty t^{-1} J_0(t)\, dt = \gamma + \log(\tfrac{1}{2}x) + \tfrac{1}{2} \sum_{n=1}^{\infty} [n(n!)^2]^{-1} (-1)^n (\tfrac{1}{2}x)^{2n}$$